HEATING, VENTILATING, AND AIR-CONDITIONING SYSTEMS ESTIMATING MANUAL

HEATING, VENTILATING, AND AIR-CONDITIONING SYSTEMS ESTIMATING MANUAL

A. M. KHASHAB, P.E.
Senior Cost Consultant, Bechtel Corporation

McGRAW-HILL BOOK COMPANY

New York St. Louis San Francisco Auckland Bogotá Düsseldorf
Johannesburg London Madrid Mexico Montreal New Delhi Panama
Paris São Paulo Singapore Sydney Tokyo Toronto

Library of Congress Cataloging in Publication Data

Khashab, A M, date.
Heating, ventilating, and air-conditioning systems estimating manual.

Bibliography: p.
Includes index.
1. Heating—Estimates. 2. Ventilation—Estimates.
3. Air conditioning—Estimates. I. Title.
TH7335.K48   697   77-4208
ISBN 0-07-034535-X

Copyright © 1977 by McGraw-Hill, Inc. All rights reserved. Printed in the United States of America. No part of this publication may be reproduced, stored in a retrieval system, or transmitted, in any form or by any means, electronic, mechanical, photocopying, recording, or otherwise, without the prior written permission of the publisher.

3456789 HDHD 8654321

The editors for this book were Jeremy Robinson and Joseph Williams, the designer was Elliot Epstein, and the production supervisor was Frank Bellantoni. It was set in Electra by University Graphics, Inc.

Printed and bound by Halliday Lithographic Corporation.

To my family, whose patience and understanding
over the period that it took to complete
this manual are greatly appreciated.

CONTENTS

Foreword, xi
Preface, xiii
Acknowledgments, xv

1 HEATING, VENTILATING, AND AIR-CONDITIONING ESTIMATING CRITERIA   1

General, 1
Estimating and Estimators, 1
Types of Estimates, 2
Estimate Development Procedure, 4
Estimating the Cost of Estimating, 16
Construction Contracts, 20
Estimating Forms, 20
HVAC System Parameters, 21

2 HEAT-GENERATION EQUIPMENT   23

General, 23
Boilers, 23
Boiler Feedwater Equipment, 32
Pressure-Reducing Stations, 36
Fuel Storing and Handling Equipment, 39
Heat Exchangers, 43
Solar Energy, 45

3 COOLING-GENERATION EQUIPMENT   49

General, 49
Mechanical Compression Water Chillers, 50
Absorption Water Chillers, 51
Condensers, 54
Cooling Towers, 59

4 HEAT- AND COOLING-GENERATION AUXILIARY EQUIPMENT 63

General, 63
Water-Treatment Equipment, 63
Pumps, 65
Hydronic Specialties, 74

5 HEAT-DISTRIBUTION EQUIPMENT 79

General, 79
Natural-Convection Units, 80
Forced-Convection Units, 85
Radiant Heating, 88

6 COOLING-DISTRIBUTION EQUIPMENT 93

General, 93
Forced-Convection Terminal Units, 93
Terminal Control Boxes for All-Air High-Pressure Systems, 98
Supply-Air Outlets, 102
Return and Exhaust Inlets, 105

7 AIR-HANDLING EQUIPMENT 107

General, 107
Central Air-Handling Units, 107
Heating, Ventilating, and Air-Conditioning Units, 108
Heating-Ventilating Units, 115
Heat-Recovery Systems, 117
Fans, 119
Air-Treatment Equipment, 128
Dampers, 133

8 PIPING AND ACCESSORIES 135

General, 135
Piping Systems, 135
Piping Materials, 135
Application of Pipe and Tube to HVAC, 138
Pipe Fittings, 140
Methods of Joining Pipe, 141
Underground Piping Systems, 145
Hangers and Supports, 146
Valves, 147
Expansion Joints, 151
Strainers, 151
Steam Traps, 151
Air Vents, 152
Thermometers and Gauges, 152
Pipe Sizes, 153
Estimating, 153

9 SHEET-METAL WORK (DUCTWORK) 167

General, 167
Materials, 168
Pressure and Velocity Classifications, 168
Weights and Thicknesses of Metal Sheets, 169
Duct Types, 169
Recommended Construction and Gauges for Various Types of Ducts, 170
Acoustically Lined Ductwork, 173
Apparatus Casings and Plenums, 174
Access Doors, 174
Belt Guards, 174
Hangers for Ducts, 175
Duct Sizes, 175
Surface Areas of Ducts, 175
Ductwork Takeoff Rules, 177
Estimating of Ductwork, 177

10 THERMAL INSULATION 187

General, 187
Materials, 187
Thickness of Insulation, 188
General Requirements for HVAC-System Insulation, 189
Estimating of Thermal Insulation, 193

11 AUTOMATIC CONTROL SYSTEMS 199

General, 199
Feedback Control Systems, 199
Definitions of Control Terms, 200
Types of Control Systems, 201
Automatic-Control-System Components, 201
Control of Heating Systems, 211
Control of Cooling and Dehumidifying Systems, 212
Sequence of Operation of Automatic Control Systems, 213
Central Control Panels, 216
Estimating, 216

12 TESTING AND BALANCING 221

General, 221
Testing, 221
Balancing, 223
Estimating the Cost of Testing and Balancing, 225

13 MOTOR AND MOTOR STARTERS 227

Motors, 227
Motor Starters, 228

14 FOUNDATIONS AND VIBRATION ISOLATION 229

General, 229
Estimating, 229

15 PAINTING   233
General, 233
Estimating, 233

16 RIGGING   237
General, 237
Estimating, 239

17 EXCAVATION AND BACKFILL   241
General, 241
Estimating, 242

18 CONCEPTUAL ESTIMATES   245
General, 245
General Estimating Data, 245
General Design Data, 245

19 CHANGE ORDERS, ALTERATIONS, AND RENOVATIONS   265
Change Orders, 265
Alterations and Renovations, 266

20 VALUE ENGINEERING FOR HEATING, VENTILATING, AND AIR CONDITIONING   267
General, 267
Value-Engineering Job-Plan Phases, 268
Typical High-Cost Areas, 270
Value Engineering and the Estimator, 270

21 SYSTEMS OF WEIGHTS AND MEASURES   273
General, 273
The International System of Units, 274

22 ABBREVIATIONS AND SYMBOLS   279
Abbreviations, 279
Mechanical Symbols, 283

Appendix LOAD CALCULATIONS   287
Climate Control and Comfort, 287
Ambient Temperature and Relative Humidity, 287
Heating-Load Calculations, 287
Degree-Days and Fuel Consumption, 293
Cooling and Dehumidifying Calculations, 295

Bibliography, 303

Index, 305

FOREWORD

Experience in project control over the years has pointed up the need for reassessing our way of communicating with contractors, vendors, clients, and management regarding the preparation of estimates. The worldwide search for profitability in projects and a firm commitment to meeting budgeting and scheduling demands points up the need for updating our skills in estimating and cost engineering.

With the publication of the *HVAC Systems Estimating Manual*, we have available to us the guidelines for articulating our estimates in a more meaningful representation. The manual is replete with charts, graphs, tabulations, and basic illustrations that will prove to be time-savers and accentuate our skills in the preparation of estimates.

A. M. Khashab has made a notable contribution to cost engineering with this manual. The *HVAC Systems Estimating Manual* fills the discipline gap—and provides the basics in clear, concise terms.

There are many ingredients for absolute project control. One thing is certain, however; the *HVAC Systems Estimating Manual* will be a contributing factor in helping us to manage our projects for control and profitability.

Alfred L. Dellon
Management Consultant

PREFACE

The *Heating, Ventilating, and Air-Conditioning Systems Estimating Manual* is intended to be a reference for HVAC professionals, architects, and owners, as well as a textbook for engineering students, to aid them in understanding the techniques of estimating heating, ventilating, and air-conditioning equipment and installation costs. Each chapter of this book takes into account different aspects of HVAC work, presenting a concise review of system design, application, and installation, followed by detailed estimating data. Labor man-hour tables are included, showing the productive labor required for material and equipment installation. It is not the intention of the author to present material prices; however, an attempt was made to include equipment prices in material cost charts. These charts are designed for reference only and should not be used for bid estimating.

The manual includes many estimating data, which can aid and guide the user in producing accurate conceptual, design-stage, and bid estimates. Conversion factors are also included for use in converting to the International System of weights and measures. It is advisable to give careful attention to Chap. 1, which is the introductory chapter and the key to the manual.

It is hoped that this manual will be useful as a new guide for professionals and as a comprehensive textbook for students.

A. M. *Khashab*

ACKNOWLEDGMENTS

Many of the principles involved in the design, application, installation, and estimating of heating, ventilating, and air-conditioning equipment resulted from the work of HVAC professional engineers, societies, and associations, as well as HVAC manufacturers. The author was fortunate in being able to use information from their publications. Assistance in allowing the use of portions of their material is acknowledged from the American Society of Heating, Refrigerating and Air Conditioning Engineers; Bell & Gossett Company; Carrier Air Conditioning Company; Industrial Press; McGraw-Hill Book Company; the Trane Company; and other contributors too numerous to mention.

Thanks are due also to the publisher's team, who helped with the editing and coordinating of this manual; to Thea Vardakis and Esther Chase, who did a marvelous job in typing the original manuscript; as well as to my colleagues at McKee-Berger-Mansueto, Inc., whose assistance is gratefully acknowledged.

HEATING, VENTILATING, AND AIR-CONDITIONING SYSTEMS ESTIMATING MANUAL

1

HEATING, VENTILATING, AND AIR-CONDITIONING ESTIMATING CRITERIA

GENERAL

Estimating is essentially a formulation of the science and tools required to calculate the approximate cost of a project during its delivery period.

Estimating is a complex task under the best of conditions. Variations in construction volume, labor cost, supplies of material and equipment, and delivery schedules have an impact on costs, and thus complicate the estimating process further. Government agencies and private owners require cost information and analyses, for which estimating provides the basic data. The desired result is the ability to control construction costs within a reasonable framework that represents a realistic project cost.

ESTIMATING AND ESTIMATORS

Estimating is the process of developing a project's construction cost. This process should include the following:

1. Reviews of project documents (drawings and specifications)
2. Quantity takeoff of material and equipment, as indicated on project documents
3. Local market analysis for material and labor
4. Pricing of material and equipment based on the quantity takeoff
5. Estimating the labor cost for the bill of materials
6. Calculating general conditions, overhead, and profit
7. Arriving at a total project cost

A professional estimator must be able to produce accurate estimates for a variety of building applications within a particular discipline. The goal usually is *to be a low bidder and realize a profit for the job*.

The estimator must have broad experience in his or her field of specialization, and must keep constantly attuned to the rapid changes within the trade. The estimator should have good judgment, keep categorized records for trade publications, and check present data, price lists, labor rates, and data from previous jobs, since they are the most important materials used by the estimator to develop an estimate.

Definition *Contractor.* A party who has legally agreed to perform the work or furnish and install the equipment detailed in the contract documents. For practical purposes, contractors may be classified as general or prime contractors and subcontractors. The general or prime contractor usually makes an agreement directly with the owner. The general contractor may utilize the services of specialty subcontractors for particular parts of the job. If so, the contractor is fully responsible and liable to the owner for the acts of the subcontractors.

TYPES OF ESTIMATES

Of the many types of estimates, two are commonly used in the building industry. They are contractor bid estimates and design-phase estimates (used for cost management during design).

Contractor's Estimate

1. The *complete* estimate covers the cost of all heating, ventilating, and air conditioning (HVAC) items shown on the project documents and must be very comprehensive. It is used for submitting bids. The difference between the estimated cost and actual cost of a job is a measure of the skill of the estimator.

2. The *progress* estimate is made to determine the amount of progress payment due a job under construction.

Design-Phase Estimate

This estimate is used in the planning and design process to enable the owners and engineers to control a project's costs during its design period.

Four basic types of estimates are used in developing the cost framework:

1. Cost model estimate
2. Schematic estimate
3. Preliminary (design-development) estimate
4. Final (prebid) estimate

Estimates 2, 3, and 4 are developed at the ends of particular stages of design, and their values should be in agreement with the project's budget. If they are higher, remedial action must be taken to reduce the estimate values so they are within the budget.

Cost Model Estimate

A project cost model is prepared from the project program, master plan, or preliminary schematic design documents. The cost model presents the project cost based upon parameters expressed in dollars per gross square foot of building area or other appropriate units. The amounts of the markups, contingencies, and escalation should be delineated.

Schematic Estimate

A project schematic estimate is prepared when the schematic design documents are completed. The schematic estimate presents the project cost by HVAC system parameters, with quantity backup.

The parameters and design documents should be of sufficient scope to enable quantity takeoffs and unit pricing. The less defined conceptual parameters should be expressed in dollars per square foot of gross building area or other appropriate units. The amounts of the markups, contingencies, and escalation should be delineated.

Preliminary (Design-Development) Estimate

A project preliminary estimate is prepared when the preliminary design documents are completed. The preliminary estimate presents the project cost by HVAC system parameters, with quantity survey backup and pricing conmensurate with the level of detail of the design-development documents. The amounts of markups, contingencies, and escalation should be delineated.

Final (Prebid) Estimate

A project final estimate is prepared when the final design documents are completed. The final estimate presents the project cost by HVAC system parameters, with quantity survey backup and pricing. The amounts of markups, contingencies (if applicable), and escalation should be delineated. The final estimate is used by the owner and design team to check the project cost prior to release for bid and to review the bids received.

Estimates are also made for the purposes of claims analysis, cost validation, and feasibility studies.

ESTIMATE DEVELOPMENT PROCEDURE

The following steps should be followed in developing a job estimate:

Specifications and Drawings

Specifications and drawings are part of the contract documents by which the contractor will be bound legally. Therefore, it is absolutely essential that the estimator understand them thoroughly, to avoid any mistakes in the estimate and as an aid in planning the general estimating process.

Specifications
The specifications determine the mechanical contractor's responsibilities and liabilities. According to the Uniform Construction Index's standard construction specifications, job specifications are divided into 16 divisions. Division 15, Mechanical, includes the following sections:

Section 15A, General Requirements—Mechanical

Section 15B, Site Utilities

Section 15C, Plumbing Systems

Section 15D, Heating, Ventilating, Air Conditioning and Refrigeration

Sections 15A, 15B, and 15D are the most important to the HVAC estimator. In general, the specifications for the general construction, plumbing, and electrical trades include many items that affect HVAC construction. The HVAC estimator should search through other trade specifications for the provisions that affect HVAC estimates.

Under the general-construction specifications, for example, the estimator must determine whether the general constructor or the mechanical contractor is responsible for such items as

- Temporary heat
- Outside louvers
- Painting
- Concrete pads
- Excavation and backfill
- Cutting and patching
- Enclosures for radiation and fan coil units

Under the plumbing specifications, the estimator must determine whether the plumbing contractor or the HVAC contractor is responsible for such items as

- Condensate drain system
- Sprinkler system
- Hookups for special equipment

Under the electrical specifications, the estimator must determine whether the electrical contractor or the mechanical contractor is responsible for the following items:

- Control wiring system
- Furnishing and/or installing the electric duct heater, electric baseboard radiation, and electric wall heaters
- Furnishing motors, starters, and motor control centers
- Furnishing air boots for air-handling light fixtures

During a careful reading of the specifications, the estimator should write down the information that will be of use later, in the takeoff and pricing procedures.

Section 15A, General Requirements—Mechanical, will cover the following items:

- Scope
- Intent
- Approvals
- Equipment deviations
- Codes, rules, permits, and fees
- Surveys and measurements
- Drawings
- Cooperation with other trades
- Protection
- Scaffolding, rigging, and hoisting
- Material and workmanship
- Quiet operation
- Accessibility
- Waterproofing
- Guarantee
- Progress charts

- Abbreviations
- References
- Definitions
- Shop drawings
- Samples
- Existing services
- Equipment design and installation
- Equipment supports, foundations, and stands
- Pipe sleeves
- Escutcheons and plates
- Spare parts
- Motors, motor controls, and electrical work
- Coordinating electrically operated equipment and controls
- Concrete work
- Temporary openings
- Temporary heat
- Identification
- Tests
- Cutting and patching
- Painting
- Excavation and backfill
- Operating instructions
- Access doors

Section 15B, Site Utilities, will cover the following heating-utilities items:

- Work of this section
- Work of other sections
- Codes and standards
- Excavation and backfill
- Pipe and/or conduit systems
- Underground fuel-oil storage tanks
- Foundations, supports, and concrete thrust blocks

- Inspection and testing
- Construction of manholes
- Approval or rejection of work
- Final inspection

Section 15D, Heating, Ventilating, Air Conditioning and Refrigeration, will include the following items:

- Work of this section
- Work of other sections
- Design criteria
- Shop drawings
- Pipe and fittings
- Pipe hangers and supports
- Valves, strainers, traps, and specialties
- Thermometers and pressure gauges
- Expansion joints
- Orifice plates and flowmeters
- Boilers and burner units
- Boiler feedwater systems
- Pressure-reducing stations
- Refrigeration equipment
- Cooling towers
- Heat exchangers
- Water-circulating pumps
- Condensate water pumps
- Air separators
- Expansion tanks
- Fuel-oil transfer pumps
- Water-treatment equipment and systems
- Direct radiation
- Air-conditioning units
- Heating and ventilating units

- Heating transfer coils
- Air filters
- Fans
- Unit heaters and cabinet unit heaters
- Fan-coil units and terminal units
- Flexible connections
- V-belt drivers and belt guards
- Acoustical treatment
- Vibration treatment
- Ductwork
- Dampers and access doors
- Diffusers, registers, and grilles
- Thermal insulation
- Temperature control systems
- Motors, motor controllers, and electrical work
- Motor control centers
- Testing, cleaning, and balancing

Drawings

All the work included within a job contract must be indicated graphically on a set of drawings. These drawings should be produced according to current standards of construction drafting and in an understandable manner. HVAC drawings are normally organized as follows:

- Drawing lists, general notes, symbols, and abbreviations
- Floor plans indicating the layout of the various HVAC systems, i.e., ductwork, piping, equipment, etc.
- Mechanical room layouts
- Views and sections
- Riser diagrams
- Flow diagrams
- Control systems and control diagrams
- Installation details for ductwork, piping, and equipment
- Equipment schedules

The HVAC estimator should refer to the drawings for other trades to coordinate HVAC construction with those trades and to obtain additional information affecting the takeoff procedure.

During a careful review of the HVAC drawings, the estimator should check the following items, which affect the quantity takeoff and pricing procedures:

Checklist for review of HVAC drawings

- Job location
- Design stage, or percentage of completion
- Scale
- Systems layout degree of completion
- Indication of sizes and dimensions
- Difficulty in understanding the drawings
- Possible interference with the work of other trades
- Missing items

The HVAC estimator should communicate with the consulting engineer who designed the work to clear up any misunderstandings or to obtain additional information that might affect job costs.

Quantity Takeoff

The quantity takeoff procedure is the act of surveying, measuring, and counting all materials and equipment indicated on a set of drawings. The individual steps in the HVAC takeoff are:

- Takeoff equipment (boilers, chillers, pumps, air-handling units, fans, etc.)
- Takeoff air devices and air terminals
- Takeoff radiation system
- Takeoff piping and accessory systems
- Takeoff ductwork, dampers, and louvers
- Takeoff piping, ductwork, and equipment insulation
- Takeoff thermometers and gauges
- Takeoff motor starters, motor control centers, and any electrical work to be furnished by HVAC contractor

- Takeoff temperature control system
- Takeoff other special systems to be provided (furnished and installed) by HVAC contractor

Takeoff sheets and tools. For an accurate takeoff, well-designed standard takeoff sheets must be used. These sheets provide standardization, continuity, and a permanent record, and they reduce the work load as well as the chance of error. The HVAC estimator should use the following takeoff sheets:

1. Equipment and air devices takeoff sheets. These are 8½ × 11 in standard sheets ruled into four columns, indicating item number, classification, quantity, and remarks.
2. Ductwork takeoff sheets. A typical sheet appears in Chap. 9.
3. Piping and accessories takeoff sheets. These include the following individual takeoff sheets:
    a. Pipe takeoff sheets.
    b. Fitting takeoff sheets.
    c. Valves, strainers, traps, and joints takeoff sheets. Typical sheets appear in Chap. 8.

For measuring, counting, and calculating, the estimator should use the following takeoff items:

1. Colored pencils, for checking off items on drawings as they are taken off. Use the same color for all the material included within a system to enable following up this system.
2. Automatic mechanical counter, for counting typical diffusers, registers, grilles, and similar items.
3. Rotometer, for measuring duct and pipe runs. It should include ⅛, ¼, and ½ scales.
4. Metallic tape, for measuring duct and pipe runs. One side of this tape is ⅛ scale; the other side is ¼ scale.
5. Two rules (architect's rule, with ⅛ scale, ¼ scale, etc., and engineer's rule, with 1/20 scale, 1/40 scale, etc.), for accurate measuring.
6. Adding machine or electronic calculator, for handling the many additions and multiplications.
7. Magnifying glass, for examining drawings in detail.
8. Electronic takeoff machine (if available).

The estimator should have a library, including a large collection of catalogues, technical books, installation manuals, price lists, labor rates, previous job files, and check data information.

Scope and design contingency. After the takeoff is completed, the estimator should be in a position to decide the percentage of completion of the drawings, in order to estimate the design contingency. This design percentage is applied to direct costs, and its value depends on the design stage. That is, the schematic stage requires a high percentage, to allow for the many unpredictable items, but the final design stage, in which the documents are complete, requires no design contingency.

Local Market Analysis

The market analysis is the procedure of gathering information from local sources on conditions which may have an impact on project costs during the construction period. The analysis should include a review of the following items:

1. *Construction volume.* The construction volume in a project's local area may affect the job construction cost. As a rule, the project construction cost will be directly affected by the level of local construction volume; i.e., when the local construction volume is high, the project cost may be above normal, owing to possible material and labor shortages during the job construction period, and vice versa. Information may be obtained from local government agencies and suppliers.

2. *Labor resources and rates.* The labor situation in the project's local area must be researched by the HVAC estimator to determine the current and anticipated labor rates and availability of pipe fitters, sheet-metal workers, and asbestos workers during the project's construction period. This information may be obtained from local unions.

3. *Contractor resources.* Contractor availability in the project's local and surrounding area must be determined from local contractor associations and government agencies. The local contractors are either union or open-shop contractors or both. This information helps the estimator to know the competitive contractor as well as the choices of subcontractors.

4. *Material.* The anticipated availability and prices of materials in the project's local area must be determined from local suppliers. This information affects the strategy of materials pricing.

5. *Escalation percentage.* The estimator must determine the escalation percentage, which will be applied to the current total direct cost to bring

it up to the anticipated bid date or midconstruction period. The estimator should use information obtained from labor unions and material suppliers, which anticipate changes in labor rates and material price trends. The escalation indexes compiled by contractors, government agencies, and trade publications (e.g., *Engineering News-Record*, published by McGraw-Hill) may also be used.

6. *Market contingency.* Sometimes the estimator cannot accurately predict changes in the market, especially during inflationary periods, when the demand is higher than the rate of production, or when the construction volume is low. In such cases the estimator should apply a market contingency percentage to the total direct cost to cover the unpredictable market. The market contingency percentage is essentially a judgment made by the estimator, based upon a feel for the future market resulting from the local market analysis.

Pricing Material and Equipment

After the takeoff is complete, the estimator should write up all material and equipment classifications on the job estimating sheets and transpose to the same estimating sheets the total quantity of each item as it appears in the takeoff sheets. The estimator then should be ready to price the individual items. This can be done by using available price lists and suppliers' and manufacturers' quotations. Normally, material prices should include the freight costs for job-site delivery and the sales tax (if applicable). Most suppliers offer contractors a discount, which is a percentage reduction applied to their price lists. Material prices are generally estimated as costs per unit quantity, as given in Table 1-1.

TABLE 1-1 Units for Material Costs

Classification	Cost per Unit Quantity
Air outlets and air-terminal units	Dollars each
Equipment	Dollars each
Piping accessories (fittings, valves, strainers, hangers, etc.)	Dollars each
Pipes and pipe insulation	Dollars per foot
Fin-tube radiation	Dollars per foot
Flexible duct	Dollars per foot
Ductwork	Dollars per pound
Duct insulation	Dollars per square foot
Temperature controls	Dollars per point

Estimating Labor Costs

A labor cost is the dollar equivalent of an estimated labor time required to install a piece of equipment or a unit of material or to complete a particular activity. The estimator should estimate the time required to install each individual item included on the job estimating sheets. Estimating labor costs is a difficult task requiring both experience and a knowledge of how HVAC work is accomplished in the field.

The estimator may use man-hour labor tables, which are based on the performance of average skilled craftsmen under normal working conditions and normal supervision. These tables can be obtained from the following sources:

- Contractor's past historical data
- *Estimating Labor Manual*, published by the Mechanical Contractors Association of America
- Cost manuals published by technical publishing firms

The labor unit used in this handbook is the man-hours, and estimates are based on an assumption of normal productivity. Man-hour estimates must include the labor for handling, erecting, and joining a unit of material. Many factors affect the rate of production, including weather conditions, height, area, existing conditions, shift work, overtime, and market conditions. The estimator should analyze the job conditions carefully to determine the factors which will affect the rate of production, and use the applicable correction factors to adjust the labor estimates given in the tables.

Team formation is generally determined by the volume of work included in the job. Normally, for a small installation, a team of two workers may be used; for larger installations, the team is formed differently and may be called a *crew*. The crew is comprised of one apprentice and one foreman for each five journeymen.

The labor rate includes wage rate per hour and jurisdictional fringe benefits (see Table 1-2). Fringe benefits may include the following items:

- Health and welfare insurance
- Pension fund
- Vacation fund
- Apprenticeship training fund
- Travel allowance
- Annuity funds

TABLE 1-2 1976 Wage Rate in Principal Cities for HVAC Trades*

Location		Asbestos Workers			Steam Fitters			Sheet-Metal Workers		
City	State	Hourly Wage	Fringe	Total	Hourly Wage	Fringe	Total	Hourly Wage	Fringe	Total
Atlanta	Ga.	$9.05	$0.75	$9.80	$9.80	$1.62	$11.42	$9.27	$0.85	$10.12
Baltimore	Md.	7.90	2.14	10.04	10.65	1.10	11.75	9.91	1.19	11.10
Boston	Mass.	11.11	2.81	13.92	12.56	1.21	13.77	10.08	1.91	11.99
Chicago	Ill.	10.96	1.55	12.51	10.85	1.52	12.37	11.50	1.29	12.79
Cleveland	Ohio	10.03	1.86	11.89	10.12	2.05	12.17	10.58	1.49	12.07
Dallas	Tex.	9.58	0.84	10.42	8.00	2.06	10.06	7.88	1.43	9.27
Denver	Colo.	8.60	2.12	10.72	10.35	1.90	12.25	10.74	1.99	12.73
Detroit	Mich.	10.32	3.36	13.68	9.22	4.00	13.22	9.32	3.16	12.48
Minneapolis	Minn.	10.57	1.18	11.75	8.13	2.13	10.26	10.23	1.14	11.37
New Orleans	La.	7.09	0.30	7.39	10.30	1.06	11.36	10.16	0.87	11.03
New York	N.Y.	10.60	3.47	14.07	10.40	4.47	14.87	12.17	2.66	14.93
St. Louis	Mo.	12.18	1.07	13.25	10.45	2.78	13.23	10.44	2.32	12.76
San Francisco	Calif.	12.15	2.95	15.10	11.05	5.70	16.75	9.81	3.17	12.98
Seattle	Wash.	10.91	1.32	12.23	10.80	2.55	13.35	10.59	2.02	12.61
Washington	D.C.	11.16	1.04	12.20	11.88	1.32	13.20	10.93	1.57	12.50

*These data are intended for general reference only and should not be considered reliable for bid estimating.

The estimator estimates the total hourly rate by adding the following items:

1. Hourly base rate
2. Fringe benefits
3. Workmen's comprehensive and liability insurance (normally 18 to 18.5 % of the total amount for the above items)

The following are the trades usually used by HVAC contractors:

- Asbestos workers for insulation work
- Pipe fitters, for piping work, equipment, radiation, and coils
- Sheet-metal workers for ductwork, air outlets, and air-handling equipment.

Contractor's Markup

Three elements are involved in a contractor's markup:

- Job overhead costs
- General and administrative overhead costs
- Profit

Job Overhead Costs

Job overhead costs vary considerably from job and are dependent upon preconstruction and construction indirect expenses. These costs are most often estimated as a percentage applied to the subtotal of direct costs. Job overhead costs typically include the following (but there may be other items):

- Wages of general superintendent, engineers, and draftsmen
- Shop drawings and instruction manuals
- Field office and related costs
- Truck and equipment expenses
- Small-tool expenses
- Job insurance
- Permits and fees
- Taxes
- Warranty and guarantee, callback
- Delay penalties
- Telephone and telegraph
- Construction and performance bonds
- Miscellaneous costs (tags, cleanup, move-on–move-off expenses, etc.)

General and Administrative Overhead Costs

General and administrative overhead costs are relatively fixed for each contractor; they depend upon the size of the contractor's organization and its volume of work. These costs should be shown as a percentage. They include, but are not limited to, the following:

- Estimating and engineering
- Secretarial and accounting
- Consulting fees
- Officers' salaries
- Office rent
- Travel expenses
- Miscellaneous costs (office supplies, insurance, telephone, etc.)

Profit   Profit should be applied as a separate percentage, after all costs are included. Usually, the profit percentage is determined by the contractor's executives. Profit margin varies considerably and is dependent upon the total project cost; as a rule, a higher profit margin is used for smaller projects and vice versa. The profit margin is the average of the individual percentages applied to material, labor, and subcontractors' costs. Once the estimator has calculated the job profit and has arrived at a total job cost, the assignment has been completed.

ESTIMATING THE COST OF ESTIMATING

The cost of estimating is the direct salary expense of the technical personnel who have been assigned to provide a job estimate, multiplied by the general and administrative overhead factor. This factor normally ranges from 1.40 to 3.00, depending upon the firm size. Thus,

Estimating fees = direct salary expense × overhead factor   (1-1)

Normally, for a multimillion-dollar construction cost, the estimating fee would run from 0.05 to 0.5% of the project value. For a construction cost below a million dollars, this fee would run from 0.5 to 1% of the project value. As a rule, a smaller percentage is used for a higher project value.

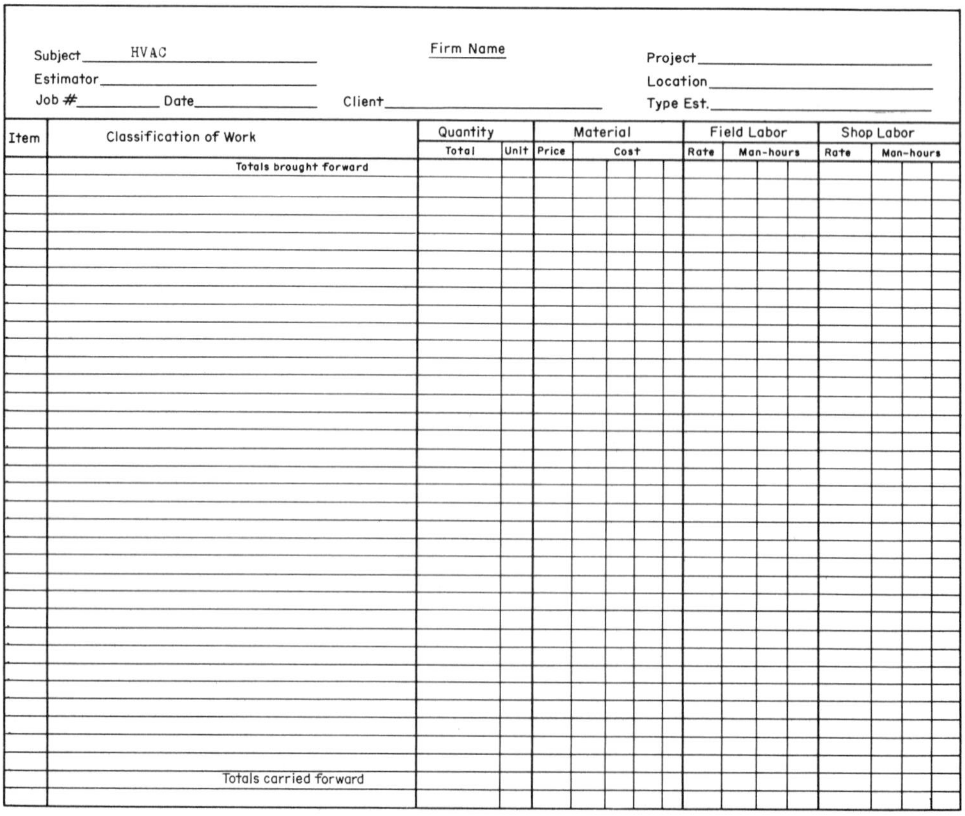

FIGURE 1-1 Typical estimating sheet.

FIGURE 1-2 Typical summary sheet.

Item	Classification of Work	Man-hours	Labor cost Rate	Labor cost Cost	Material cost	Total cost
	Totals brought forward					
1	HEAT-GENERATION EQUIPMENT					
2	COOLING GENERATION EQUIPMENT					
3	HEAT-DISTRIBUTION EQUIPMENT					
4	COOLING DISTRIBUTION EQUIPMENT					
5	AIR HANDLING EQUIPMENT					
6	PIPING AND ACCESSORIES					
7	SHEET METAL WORK					
8	TESTING AND BALANCING					
9	INSULATION					
10	TEMPERATURE CONTROL SYSTEMS					
11	SPECIAL SUBSYSTEMS					
12	SITE UTILITY					
	Subtotal (1)					
13	SALES TAX					
	Subtotal (2)					
14	JOB OVERHEAD					
	Total direct cost					
15	ESTIMATING CONTINGENCY					
	Subtotal (3)					
16	MARKET CONTINGENCY					
	Subtotal (4)					
17	ESCALATION TO BID DATE					
	Subtotal (5)					
18	GENERAL OVERHEAD					
	Subtotal (6)					
19	PROFIT					
	Total cost					

FIGURE 1-3 Typical summary sheet of HVAC system parameters.

Subject	HVAC		Firm Name		Project	
Estimator					Location	
Job #	Date		Client		Type Est.	

Item	Classification of Work	Quantity		Material		Field Labor		Shop Labor	
		Total	Unit	Price	Cost	Rate	Man-hours	Rate	Man-hours
	Totals brought forward								
1	HEAT-GENERATION EQUIPMENT								
01	Boilers		EA						
02	Stacks and breeching		LF						
03	Boiler feed water equipment		EA						
04	Heat exchangers		EA						
05	Condensate and circulating pumps		EA						
06	Expansion tanks and air separators		EA						
07	Pressure reducing stations		EA						
08	Water treatment system		JOB						
09	Fuel storage tanks		EA						
10	Fuel transfer pumps		EA						
11	Solar energy system (if applicable)		JOB						
2	COOLING GENERATION EQUIPMENT								
01	Chillers		EA						
02	Cooling towers		EA						
03	Condensing units		EA						
04	Circulating pumps		EA						
05	Expansion tanks and air separators		EA						
06	Water treatment system		JOB						
3	HEAT-DISTRIBUTION EQUIPMENT								
01	Radiators and convectors		EA						
02	Baseboard and finned tube units		LF						
03	Unit heaters and cabinet unit heaters		EA						
04	Fan coil units and unit ventilators		EA						
	Totals carried forward								

Subject	HVAC		Firm Name		Project	
Estimator					Location	
Job #	Date		Client		Type Est.	

Item	Classification of Work	Quantity		Material		Field Labor		Shop Labor	
		Total	Unit	Price	Cost	Rate	Man-hours	Rate	Man-hours
	Totals brought forward								
4	COOLING DISTRIBUTION EQUIPMENT								
01	Registers, grilles and diffusers		EA						
02	Terminal reheat units		EA						
03	Mixing boxes		EA						
04	Variable volume boxes		EA						
05	Pressure reducing boxes		EA						
06	Fan coil units		EA						
07	Induction units		EA						
5	AIR HANDLING EQUIPMENT								
01	Air conditioning units		EA						
02	Humidifiers and dehumidifiers		EA						
03	Filters		EA						
04	Make-up air units (H-V)		EA						
05	Furnaces		EA						
06	Heat Recovery units		EA						
07	Return fans		EA						
08	Exhaust and ventilation fans		EA						
09	Heat pump units		EA						
10	Louvers, dampers, and access doors		SF						
6	PIPING AND ACCESSORIES								
01	Piping systems		LF						
02	Valves, strainers, traps, and miscell.		EA						
03	Hangers, sleeves, and supports		EA						
04	Thermometers and gauges		EA						
	Totals carried forward								

FIGURE 1-4 Typical estimating sheets of HVAC system parameters.

Item	Classification of Work	Quantity Total	Unit	Material Price	Cost	Field Labor Rate	Man-hours	Shop Labor Rate	Man-hours
	Totals brought forward								
7	SHEET METAL WORK								
01	Ductwork (galvanized steel, stainless steel, aluminum and fiberglass)		LBS						
02	Transite duct		LF						
03	Flexible duct		LF						
04	Plenums and casing		SF						
05	Acoustic lining		SF						
8	TESTING AND BALANCING								
01	For all systems		JOB						
9	INSULATION								
01	Equipment		EA						
02	Piping systems		LF						
03	Ductwork		SF						
10	TEMPERATURE CONTROL SYSTEMS								
01	Terminal or zone controls		PT						
02	Equipment controls		PT						
03	Central control console		JOB						
04	Motors control centers		JOB						
11	SPECIAL SUBSYSTEMS								
01	Motor starters		JOB						
02	Foundation and vibration isolation		JOB						
03	Rigging		TON						
04	Painting		SF						
	Totals carried forward								

Item	Classification of Work	Quantity Total	Unit	Material Price	Cost	Field Labor Rate	Man-hours	Shop Labor Rate	Man-hours
	Totals brought forward								
12	SITE UTILITY								
01	Cooling and heating underground piping or conduits		LF						
02	Manholes and associated piping		JOB						
03	Anchors and supports		EA						
04	Excavation and backfills		CY						
	Totals carried forward								

FIGURE 1-4 Typical estimating sheets of HVAC system parameters *(Continued).*

(The percentages given on page 16 are intended for reference only and should not be considered reliable for an official proposal.)

CONSTRUCTION CONTRACTS

Construction contracts are usually awarded through competitive bidding or through negotiation.

Competitive Bidding

Competitive bidding is the method generally used in the construction industry to award contracts. It is only effective when complete working drawings and specifications are available. For private jobs the list of bidders is selected by the owner, but for public works any qualified bidder can get on a bidding list. A qualified bidder is often a contractor that has a good reputation for quality and a good financial status. Public works are always publicly advertised. Private jobs are sometimes advertised, but the architect or the owner can contact the bidders directly. Bids are normally opened in the presence of all bidders; the bidder who submitted the low (and reasonable) price gets the job and is normally known as the *low bidder*.

Negotiated Contract

This method allows the owner to control the selection of the contractor. The contractor that is selected by the owner submits a proposal, including the total cost of the job as a single lump sum. The owner and the contractor then negotiate the proposed job cost. Once they agree on a reasonable cost, they sign a contract. This type of contract is known as a *negotiated* contract. It is often used for multiphase construction, the remodeling of occupied buildings, and when mechanical equipment is a major job factor and the owner may prefer one manufacturer to another.

ESTIMATING FORMS

For accurate estimating and fast production, well-designed forms must be used. Wherever possible, the forms should provide continuity and reduce the work load (transposing, cross-referencing, etc.). Normally, the estimator needs the following forms:

1. Estimating work sheet—includes columns for quantity, material cost, and labor man-hours for each item (see Fig. 1-1).
2. Summary sheet—includes columns for material costs and labor cost for each system, as well as all contingencies and contractor's markups (see Fig. 1-2).

TABLE 1-3 Check Figures for Contractor's Markup, Contingencies, and Escalation

Item	Normal Range, %	Apply to Subtotal
Job overhead	3–10	(2)
Estimating contingency	0–10	Direct cost
Market contingency	0–5	(3)
Escalation	½–¾ per month	(4)
General overhead	1–3	(5)
Profit	10	(6)

HVAC SYSTEM PARAMETERS

All the systems necessary for heating, ventilating, air conditioning, or otherwise maintaining a controlled environment within the building may be classified by specific parameters for estimating purposes. The summary and estimating sheets in Figs. 1-3 and 1-4 indicate the various parameters as well as the items which may be encountered under each parameter.

Important Notes for Fig. 1-3

1. Normally, the following items of work are performed by subcontractors:

    a. Sheet-metal work

    b. Insulation

    c. Temperature control systems

2. Sales tax is applied to the material costs included in subtotal (1); the tax depends on the locality.

3. Contingency and escalation to bid date should not be estimated in a contractor's type estimate.

4. Table 1-3 indicates percentage *check* figures for job and general overhead, contingencies, escalation, and profit. The data are intended for general reference only and should not be used for bid estimating.

5. Sometimes site utility work is bid as a separate contract. In this case, it should not be included in the summary sheet.

2

HEAT-GENERATION EQUIPMENT

GENERAL Heat-generation equipment is normally used in a heating system to heat or generate the fluid (i.e., hot water or steam) which conveys heat to the various heating units. Heat generation is a process of heat transfer. During this process, the heating medium transfers its heat energy to the heated fluid to increase its temperature or change its state. The types of heat generators that are commonly used in HVAC are boilers, heat exchangers (converters), and solar collectors.

Material Costs and Labor Man-Hours for Heat-Generation Equipment

Material costs for heat-generation equipment may be obtained from the material cost charts that follow. Each chart shows the relationship of the material price per unit of capacity (in thousands of Btu's per hour, kilowatts, gallons per minute, gallons, or feet) to the total capacity of the unit, based on 1976 average costs. These costs are intended for reference only and should not be considered reliable for estimating actual material costs.

Labor man-hours for heat-generation equipment may be obtained from the labor tables that follow. The labor figures include handling, assembling, erecting, and making the unit ready for test. These figures are based on normal conditions; i.e., the unit is received within 100 ft of its final location and is to be mounted on a flat slab foundation. For final hookup, see Chap. 8; for rigging for heavy equipment, see Chap. 16.

BOILERS

Types of Boilers

Steam and hot-water boilers are built of steel or cast iron in a wide variety of types and sizes. Boilers are classified in a number of different ways, such as those shown in Table 2-1.

TABLE 2-1 Types of Boilers

By Material	By Fluid	By Pressure/Temperature	By Fuel	By Design	By Combustion Air	By Application
Steel	Hot water	Low pressure/low temperature	Coal	Fire tube	Natural draft	Space heating
Cast iron	Steam	High pressure/high temperature	Oil	Water tube	Induced draft	Domestic hot water
			Gas	Scotch	Forced draft	
			Electricity	Sectional		
				Round		

Definitions

1. *Boiler horsepower.* The heat required to evaporate 34.5 pounds of water per hour from and to 212°F, or the equivalent of 33,475 Btu per hour.

2. *British thermal unit (Btu).* The heat required to raise the temperature of one pound of water one degree Fahrenheit.

3. *Calorie.* The heat required to raise the temperature of one gram of water one degree Celsius.

Ratings of Boilers

The gross load (MBH) of the boiler is the sum of the following items:

1. Net heating load required for the building, which includes

   a. Estimated heat losses, in accordance with the data given in the appendix.

   b. Estimated maximum heat required to heat water for domestic use.

2. Piping tax, or the estimated heat emission of the piping connecting the heating equipment. There is little need to apply the pipe tax to highly developed system designs, equipment, and materials installations.

3. Warming-up or pickup allowance, or the estimated increase in the normal load required to heat up the cold system, i.e., the boiler and heating equipment.

Heating boilers are usually rated according to test codes developed by the Hydronic Institute (formerly the Institute of Boiler and Radiator Manufacturers). This Institute has adopted the SBI rating for steel boilers and the I=B=R rating for cast-iron boilers. Their ratings may be obtained from manufacturers' catalogues or published tables of boiler ratings.

Boiler Efficiency

Boiler efficiency is the ratio of the boiler gross output to the heat in the fuel fired. The efficiencies that are obtained under conditions of steady-state operation will be higher than the seasonal efficiencies actually obtained in service.

Combustion Air-Handling Systems

There are three systems for providing combustion air through the boiler.

1. *Natural draft.* The stack or chimney creates a negative pressure great enough to overcome the draft loss through the boiler and draw sufficient quantities of combustion air through the burner or fuel bed.

2. *Induced draft.* A power-driven fan overcomes the draft loss of the boiler and draws the combustion air through the burner or fuel bed.

3. *Forced draft.* Firing requires that the burner provide all the required combustion air at a pressure that will overcome the boiler draft loss and any losses in the breeching or stack.

Burners

There are three types of burners:

1. *Oil burner.* Air-atomizing type approved for operation with CS 12-48 commercial standard Nos. 2, 4, 5, and 6 fuel oils (see the fuel-handling section of this chapter).

2. *Gas burner.* The two types of gas burners are the high-radiant annular-entry type for 15- to 200-hp boilers, and the high-radiant multiport type for 250 hp and above. These types must be approved for operation with natural, manufactured, or mixed gas.

3. *Oil-gas burner.* This burner is a combination of oil burner and gas burner. The burner must be approved for operation with either CS 12-48 commercial oils or natural, manufactured, or mixed gas.

Burner Estimating Data

1. For gas-fired burners without pilot and controls, the estimated material cost per MBH of burner input is $0.20.

2. For oil-fired burners with basic control and fuel-oil pump, see Estimating Chart 2-1.

ESTIMATING CHART 2-1 Oil-Fired Burner with Basic Control and Fuel-Oil Pump.

1. Material cost

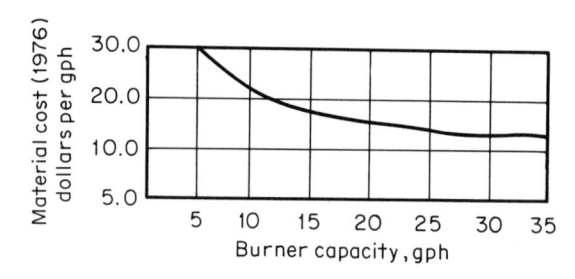

[Graph: Material cost (1976) dollars per gph vs Burner capacity, gph. Curve decreases from ~28 at 5 gph to ~10 at 35 gph.]

2. Labor man-hours for packaged burner with control (all types)

Boiler hp	Weight, lb	Man-hours
25-50	310	32
60	330	36
80	385	36
100	500	40
150	540	40
200	700	44

Boiler hp	Weight, lb	Man-hours
250	800	44
350	1000	48
400	1200	52
500	1400	56
600	1600	60
700	1800	64

Control of Gas- or Oil-Fired Boiler Output

Control of the energy output of a boiler is obtained by regulating the input. Oil- or gas-fired boilers are controlled by burner application as follows:

1. *On-off principle*, for 15- to 20-hp sizes. When the boiler is operating, it is at full capacity.

2. *High-low-off principle*, for 30- to 40-hp sizes.

3. *Full-modulation principle*, for 50-hp units and above. The low setting of either the modulating or high-low type is selected, based on the expected loads as well as the minimum firing limitations of the burner.

Burner operation is controlled by steam pressure or water temperature and/or outdoor ambient temperature.

Cast-Iron Boilers

Cast-iron boilers are generally of the sectional type. Their capacities range from those required for small residences to about 13,000 MBH gross output. Cast-iron boilers are constructed for low-pressure steam at 15 psi and low-temperature water up to 250°F at 30 psi.

For cast-iron boilers, the *I=B=R rating* has been adopted by the Hydronic Institute. The rating code is based upon performance obtained under controlled test conditions and shows the gross I=B=R output and net I=B=R rating. According to the I=B=R rating, the amount of piping tax and pickup allowance varies from 33.3 to 28.8% of the net load for automatically fired steam boilers, and is 15% for automatically fired hot-water boilers. That is, the gross output for steam boilers is the net output times a factor that varies from 1.333 to 1.288. The higher factor is used for smaller boilers, and vice versa. The gross output for hot-water boilers is the net output times 1.15. Cast-iron oil- and gas-fired boilers usually operate in an efficiency range of 70 to 80%. Estimating Charts 2-2*a* and *b* give the estimating data for cast-iron boilers.

Steel Boilers

In *fire-tube* boilers, the gases of combustion pass through the tubes and the water circulates around them. In *water-tube* boilers, the gases of combustion circulate around the tubes and the water passes through them. Either the fire-tube or water-tube type may be shipped in one package, ready for piping connections.

A packaged boiler is a boiler unit including the burner, boiler, controls, and auxiliary equipment. Medium-sized units are usually of a modified scotch marine type. Smaller units are usually of the firebox type (integral water-jacketed furnaces). A packaged boiler may be ordered as a steam or hot-water generator. Steel boilers range in size from small residential units to those with gross heat outputs of 23,500 MBH.

For steel boilers, both residential and commercial, the *SBI rating* has been adopted by the Steel Boiler Industry Division of the Hydronic Institute. The rating code shows the SBI gross outputs and net SBI ratings for steel boilers. The factors used to determine gross I=B=R ratings are also used to determine gross SBI ratings. Steel gas- and oil-fired boilers normally operate in an efficiency range of 70 to 80%. For steel-boiler estimating data, see Estimating Charts 2-3*a* to *c*.

Steam Boilers

Low-pressure steam boilers normally operate at 15 psig pressure and temperatures below 250°F. High-pressure steam boilers operate at pressures of 125 psig and temperatures above 350°F.

Fire-tube and scotch marine boilers generally range in size from 40 to 700 bhp (boiler horsepower). Water-tube boilers, which are popular for oil and gas firing, range from 100 bhp to 1,000,000 lb/h of steam. High-pressure steam boilers are always constructed of steel; they are usually of the package type, with burner and controls, ready for operation.

ESTIMATING CHART 2-2a   Packaged Cast-Iron Hot-Water Residential Boilers with Burners, Controls, Circulator, Extended Jacket, and Tankless Heaters.

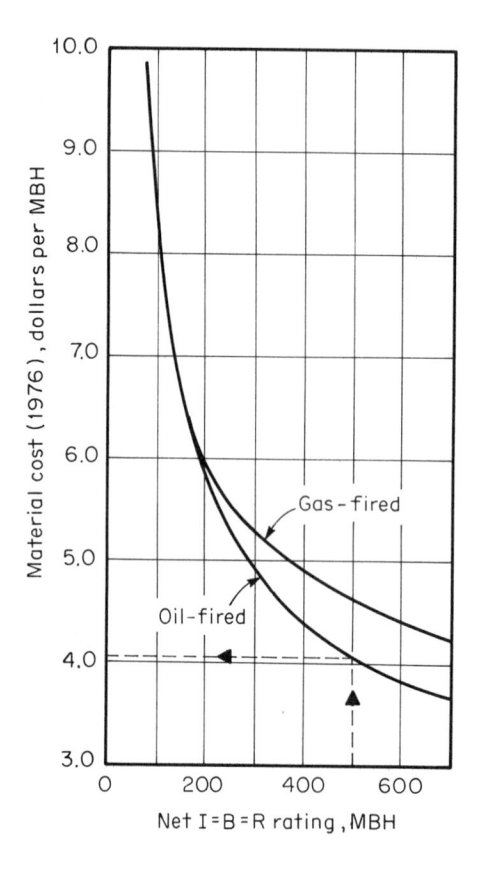

1. Material cost

2. Labor man-hours

Net MBH	Weight, lb	Man-hours
73.0	600	12
97.4	650	12
121.7	700	12
146.1	775	12
194.8	935	12
250.4	980	12
375.7	1470	12
500.0	1725	18
709.6	2500	18

Example: For oil-fired, 500 net MBH cast-iron boiler:

Cost per MBH is $4.04.

Material cost = 500 MBH @ $4.04 = $2,020.

Hot-Water Boilers

Direct-Fired Hot-Water Generators

Direct-fired hot-water generators are steam boilers operating at similar pressures. The basic differences are in the accessories and the fact that the hot-water generators are filled with water. Boilers operating at or below 250°F and 30 psig are called *low-temperature* hot-water boilers. Boilers that operate above 250°F and 160 psig are called *high-temperature* hot-water boilers.

Low-temperature hot-water boilers may be either steel or cast iron and of the fire-tube or water-tube type. High-temperature hot-water boilers should be steel water-tube boilers.

ESTIMATING CHART 2-2b Cast-Iron Sectional Boilers with Burners and Normal Basic Controls. (Units are field-assembled.)

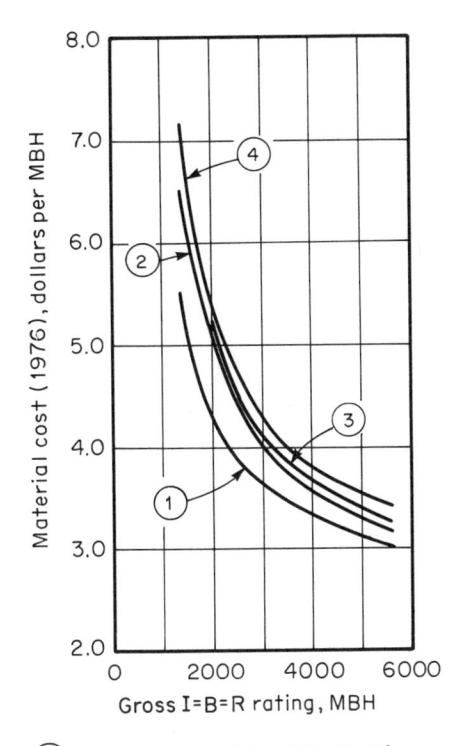

1. Material costs
2. Labor man-hours

Boiler hp	Gross MBH	No. of sections	Man-hours
40	1339	12	72
50	1674	14	84
60	2009	16	96
70	2343	18	108
80	2678	21	126
90	3013	23	138
100	3348	26	156
110	3683	28	168
120	4018	31	186
130	4392	33	198
140	4687	36	216
150	5022	38	228
160	5357	40	240
170	5692	42	250

① For oil-fired boilers (No. 2 oil)
② For gas-fired boilers
③ For oil-fired boilers (No. 6 oil)
④ For oil or gas-fired boilers (No. 2 oil)

Indirect Hot-Water Heaters

Indirect hot-water heaters generally consist of steam boilers connected to heat exchangers, of the coil or tube type, which transmit the heat from the steam to the water. This type of installation has the following advantages:

1. The boiler can operate at low pressure.
2. The boiler is protected from scale and corrosion.
3. The scale is formed in the heat exchanger, so that the parts to which the scale clings can be cleaned or replaced.

Electric Boilers

Electric boilers generally can be classified as steam boilers or hot-water boilers. Their capacities range from those required for small residences up

ESTIMATING CHART 2-3a   Packaged Scotch Marine Boilers (Steam or Hot Water; Low Pressure) with Burners and Normal Basic Controls.

1. Material costs

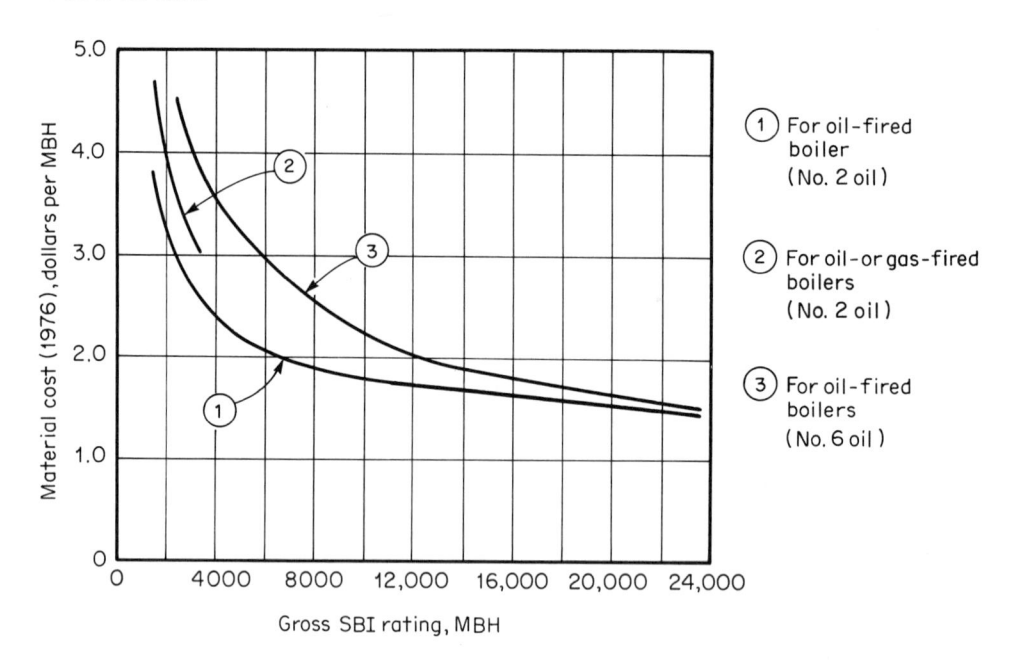

① For oil-fired boiler (No. 2 oil)

② For oil- or gas-fired boilers (No. 2 oil)

③ For oil-fired boilers (No. 6 oil)

Y-axis: Material cost (1976), dollars per MBH
X-axis: Gross SBI rating, MBH

2. Labor man-hours

Boiler hp	Gross MBH	Weight, lb	Man-hours
40	1,339	6,000	24
50	1,674	7,000	24
60	2,009	8,000	24
70	2,343	10,000	36
80	2,678	11,000	36
100	3,348	13,000	36
125	4,184	15,000	48
150	5,021	17,000	48
200	6,695	21,000	48

Boiler hp	Gross MBH	Weight, lb	Man-hours
250	8,369	26,000	60
300	10,043	31,000	60
350	11,716	32,000	60
400	13,390	36,000	72
500	16,738	43,000	72
600	20,085	48,000	80
700	23,433	53,000	80

to 3000 kW for steam boilers and 2000 kW for hot-water boilers. Electric steam boilers may be of the immersion-heater type or electrode type. Immersion electric resistance heaters are normally used for hot-water boilers. Electric boilers normally operate in an efficiency range of 90 to 99%.

Control of Electric Boilers

1. Steam and hot-water boilers with resistance-type heaters are controlled by sequencing the unit on and off the line to modulate the output.

ESTIMATING CHART 2-3b   Steel Firebox Boilers (Steam or Hot Water; Low Pressure) without Windbox, Burner, and Draft Inducer.

1. Material costs

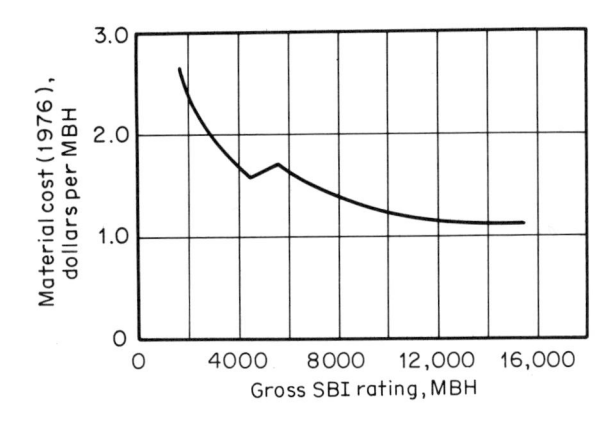

2. Labor man-hours

BHP	Gross MBH	Weight, lb	Man-hours
53	1,774	6,600	20
60	2,009	7,500	20
79	2,658	9,000	30
93	3,100	10,300	30
113	3,769	12,800	30
132	4,432	14,000	40
165	5,537	16,500	40

BHP	Gross MBH	Weight, lb	Man-hours
199	6,648	19,000	40
232	7,746	21,000	54
265	8,857	25,000	54
331	11,074	28,000	68
397	13,290	33,000	68
463	15,499	38,000	68

2. Electrode-type steam boilers are controlled by regulating the amount of water in contact with the electrodes.

For electric-boiler estimating data, see Estimating Chart 2-4.

Boiler Breeching and Stack

Figure 2-1 indicates a typical layout for the boiler breeching and stack. The *boiler breeching* should be gastight and constructed of minimum 12-gauge steel plate. The breeching should be insulated to reduce heat-loss and condensation problems. In most cases, the breeching is field-fabricated (see Chaps. 9 and 10). Prefabricated breeching may be used.

The *boiler stack* can be terminated several feet above the top of the roof, but local codes may govern the stack height above the roof. When steel stacks are used, minimum 12-gauge steel with an acid-resistant refractory lining is recommended for stack sections. A rain cap or hood should be used at the top of the stack to keep out rain and snow. Prefabricated stack is

ESTIMATING CHART 2-3c Packaged Steel Firebox Boilers (Steam and Hot Water; Low Pressure) with Burners and Normal Basic Controls.

1. Material costs

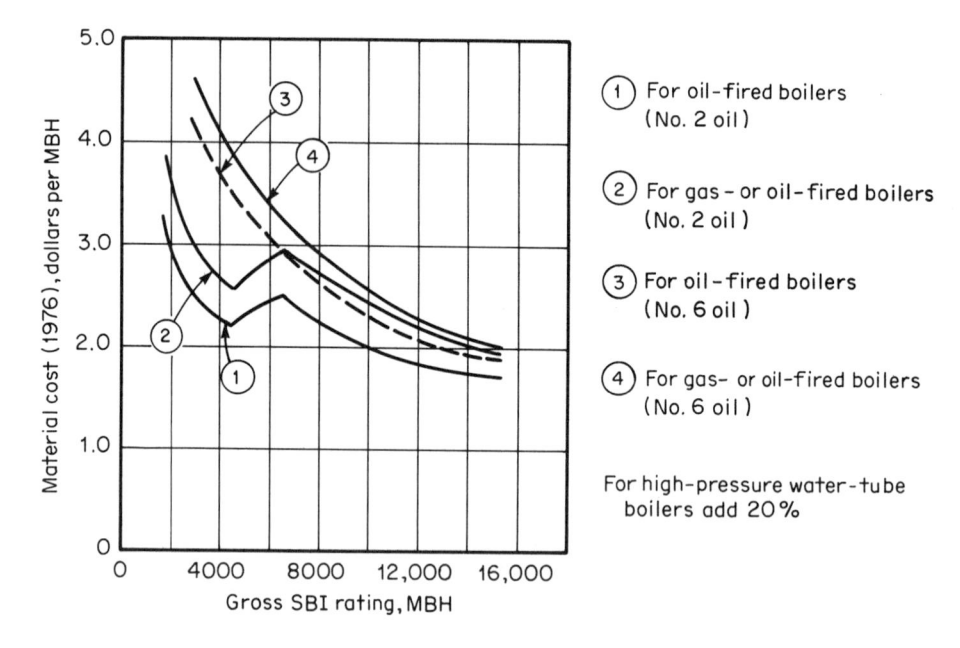

① For oil-fired boilers (No. 2 oil)

② For gas- or oil-fired boilers (No. 2 oil)

③ For oil-fired boilers (No. 6 oil)

④ For gas- or oil-fired boilers (No. 6 oil)

For high-pressure water-tube boilers add 20%

2. Labor man-hours

BHP	Gross MBH	Weight, lb	Man-hours
50	1,674	6,650	24
60	2,009	6,750	24
80	2,678	8,000	30
100	3,348	8,800	30
125	4,184	11,750	42
150	5,021	12,650	42
200	6,695	14,850	42

BHP	Gross MBH	Weight, lb	Man-hours
250	8,369	21,300	54
300	10,043	23,600	54
350	11,716	26,800	54
400	13,390	33,000	66
500	16,738	38,000	66
600	20,085	43,800	72
700	23,433	50,600	72

normally furnished by manufacturers in 3- and 4-ft lengths; all the stack components have to be field-assembled.

Estimating Chart 2-5 gives estimating data for prefabricated stack.

BOILER FEEDWATER EQUIPMENT

The boiler feedwater system is used to return water to the boiler at the highest temperature possible, to maximize efficiency. The use of feedwater equipment assures the recovery of all condensate from low- and high-pressure returns without back pressure, as the receiver is vented to the atmosphere.

Packaged Boiler Feedwater Units (Condensate Return Units)

All units are factory-assembled and include pump and receiver combinations suitable for feedwater temperatures up to 210°F at sea level.

Simplex unit. This is a complete condensate return (boiler feed) set consisting of steel (or cast-iron) receiver, one pump, motor and motor starter (see Fig. 2-2a and c). Units are available with float switch in receiver, or make-up water device and boiler level control. Units comprising float switches are normally used in heating systems only, whereas those comprising make-up water valves

ESTIMATING CHART 2-4 Packaged Electric Boilers (Steam and Hot Water) with Jacket, Circulator, and Controls.

1. Material costs

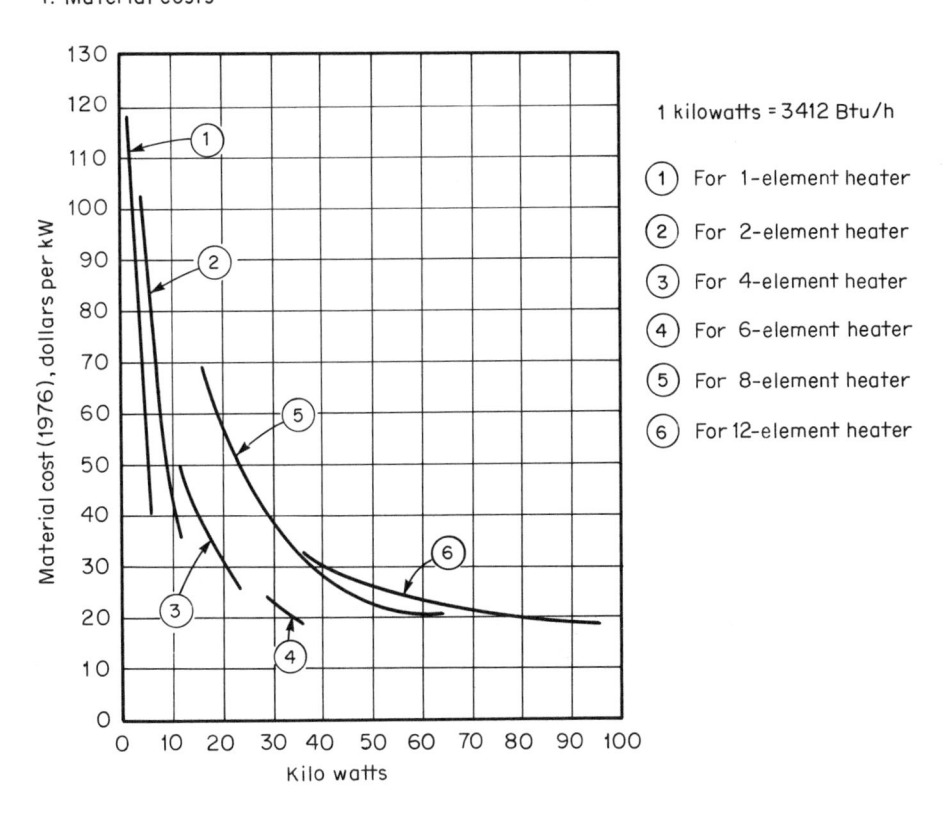

1 kilowatts = 3412 Btu/h

① For 1-element heater
② For 2-element heater
③ For 4-element heater
④ For 6-element heater
⑤ For 8-element heater
⑥ For 12-element heater

2. Labor man-hours

kW	MBH	Weight, lb	Man-hours
2-6	7-20	25	6
4-12	14-40	85	6
12-24	40-82	150	6
28-36	96-123	175	8
36-64	123-218	325	8

kW	MBH	Weight, lb	Man-hours
72-96	245-327	350	8
120	410	650	12
360	1228	1800	16
960	3270	5200	24
2400	8200	7000	30

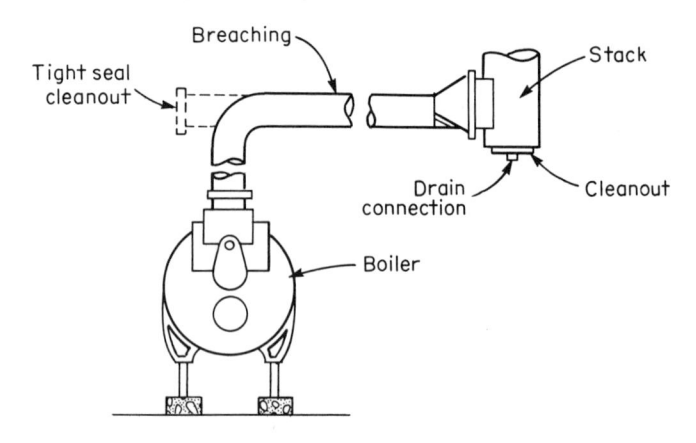

FIGURE 2-1 Boiler breeching and stack.

are used in heating/processing systems to replenish process losses, and maintain uniform boiler water level.

Duplex unit. This is similar to the above, but it has two pumps and one receiver (see Fig. 2-2b and d).

For estimating data on condensate return units, see Estimating Chart 2-6.

ESTIMATING CHART 2-5 Prefabricated Refractory-Lined-Stack Galvanized-Steel Jacket with Joint Cement and Drawbands.

1. Material costs 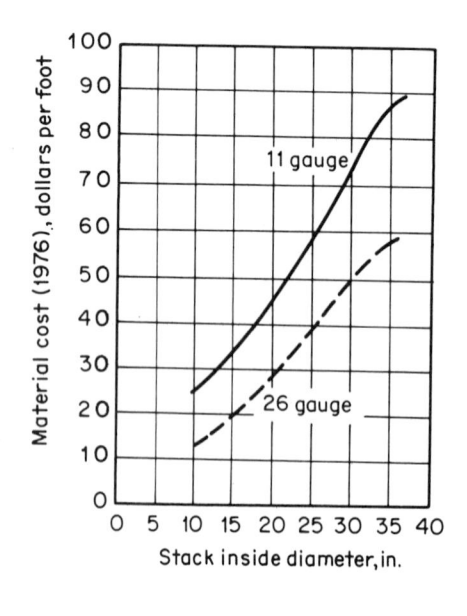
2. Labor man-hours

Inside diameter, in.	Weight per 4ft length, lb*	Man-hours per foot
10	281	1.80
12	357	2.15
15	469	2.70
18	591	3.25
21	765	4.60
24	870	5.25
27	1055	6.00
30	1315	6.50
33	1485	7.25
36	1650	9.00

*Weight for section with 11 gauge steel jackets.

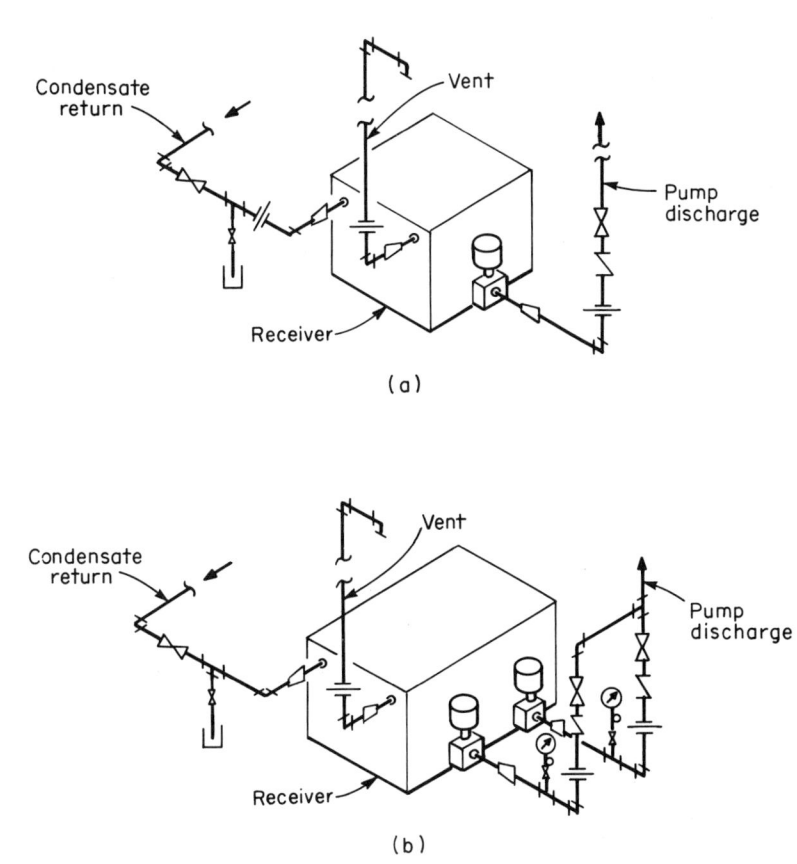

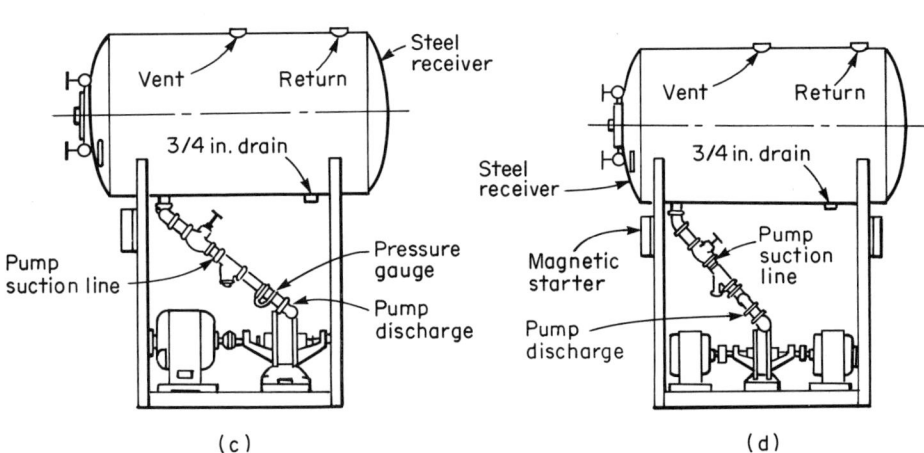

FIGURE 2-2 (*a*) Single condensate pump with cast-iron receiver. (*b*) Duplex condensate pump with cast-iron receiver. (*c*) Feedwater packaged unit (single pump). (*d*) Feedwater packaged unit (duplex pump).

Deaerators

Deaerators are used to remove and vent oxygen and other dissolved gases contained in the feedwater, in order to reduce corrosion in the boiler, steam and condensate lines, and the steam system components. Dissolved gases are removed from water either by counterflow of steam and water or by strike water with high-velocity steam. Packaged units normally include a

ESTIMATING CHART 2-6 Condensate Return Units. (Units include cast-iron receiver, pumps, and standard controls.)

1. Material costs at 20 psi and 1750 – single phase
2. Labor man-hours

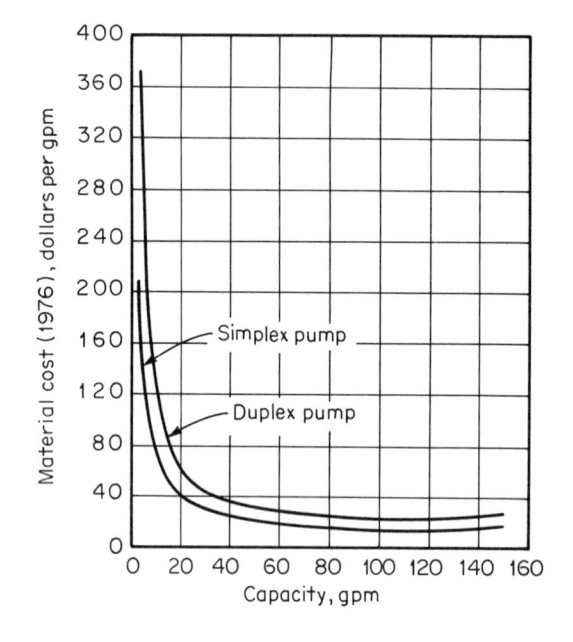

ft² EDR	Capacity, gpm	Weight, lb	Man-hours
Simplex pump			
Up to 10,000	Up to 15	Up to 275	12
15,000	22.5	530	16
30,000	45	660	16
50,000	75	780	18
75,000	112.5	1200	24
100,000	150	1550	24
Duplex pump			
Up to 10,000	Up to 18	Up to 350	12
15,000	22.5	630	16
30,000	45	830	18
50,000	75	1020	18
75,000	112.5	1440	24
100,000	150	1770	24

For 30 psi add 10%.
For 40 psi add 20%.
For 50 psi add 55%.
For 60 psi add 85%.

deaerating heater, feed-water pumps, suction piping, and normal basic controls (see Fig. 2-3).

Estimating Chart 2-7 gives the estimating data for packaged deaerators.

PRESSURE-REDUCING STATIONS[1]

Where steam is supplied by a boiler delivering steam at a higher pressure than is required for the heating system, one or more pressure-reducing valves are required (see Figs. 2-4 and 2-5).

[1] The *ASHRAE Handbook—1973 Systems Volume* has been used as a reference, with the permission of ASHRAE.

Heat-Generation Equipment | 37

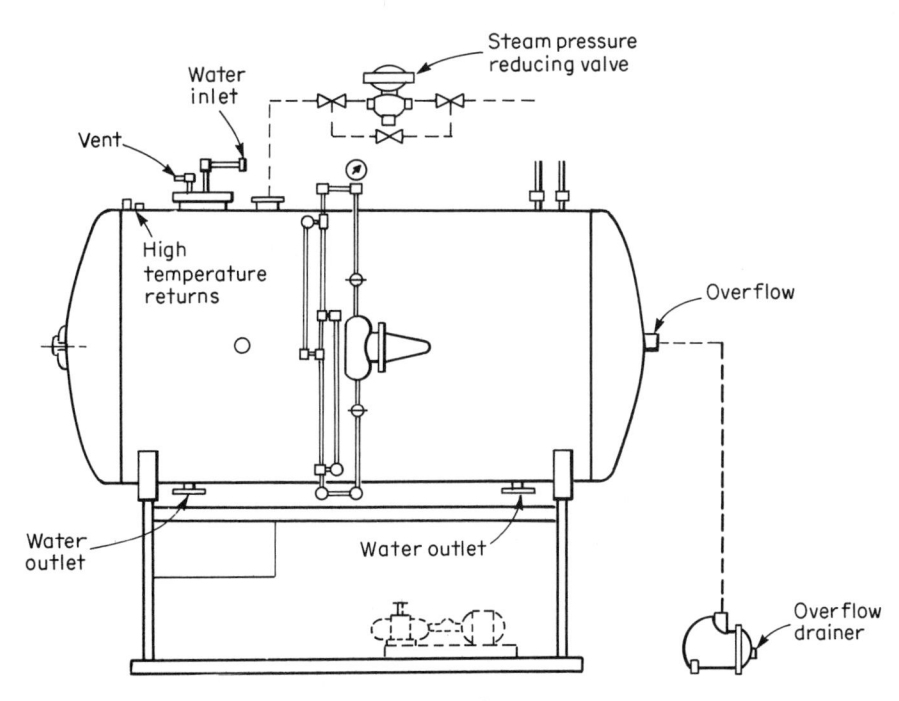

FIGURE 2-3 Typical deaerating tank.

ESTIMATING CHART 2-7 Packaged Deaerators with Feedwater Pumps and Normal Controls.

1. Material costs

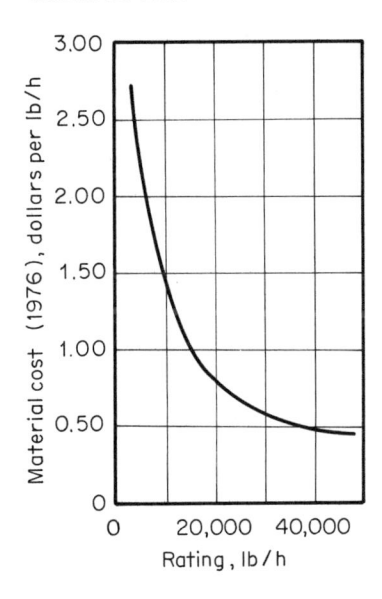

2. Labor man-hours

Rating, lb/h	Weight, lb	Man-hours
15,000	2600	24
30,000	3300	24
45,000	3800	30
70,000	4800	30
100,000	6800	36
140,000	7700	36
200,000	9500	48

Note: For feedwater package unit use the same number of man-hours and deduct 10% from the material cost of the deaerators.

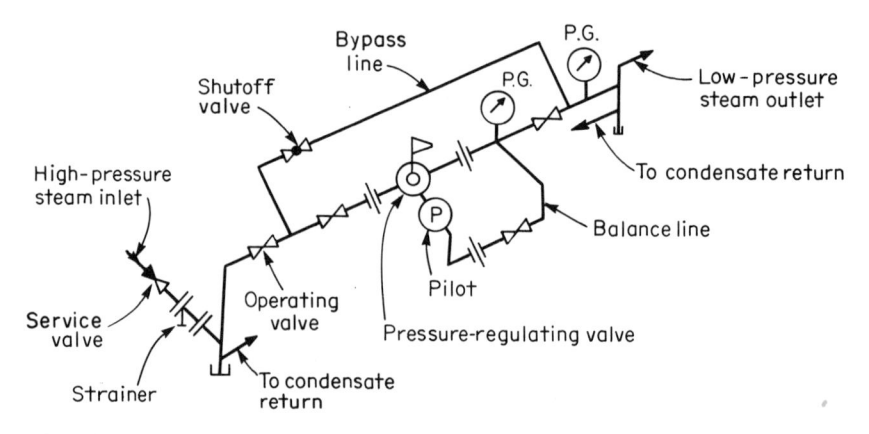

FIGURE 2-4 One-stage pressure-reducing equipment for low-pressure service.

The installation of pressure-reducing valves in pipelines requires detailed planning. They should be installed so that they are accessible for inspection and repair. There should be a bypass around each reducing valve. The globe valve in a bypass line should be of plug-disk construction and must shut off tight. A steam pressure gauge, graduated up to the initial pressure, should be installed in the low-pressure side. The gauge should be located ahead of the shutoff valve because the reducing valve can be more accurately and conveniently adjusted with the shutoff valve closed. A similar gauge should be installed downstream of the shutoff valve, for use during manual operation. Strainers should be installed on the inlet of the primary

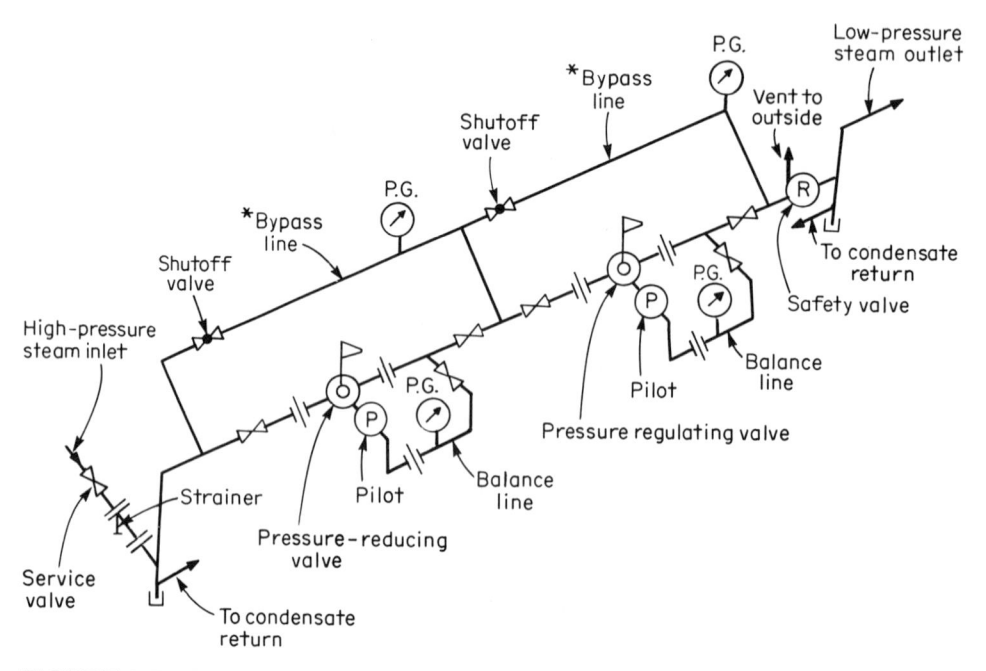

FIGURE 2-5 Two-stage pressure-reducing and pressure-regulating equipment. (*Bypass valve is one size smaller than pressure-reducing valve).

pressure-reducing valve; they are also desirable ahead of the second-stage reduction if there is considerable piping between the two stages.

If a two-stage reduction is made, it is well to install a pressure gauge immediately ahead of the reducing valve of the second reduction stage, for use in setting and checking the operation of the first valve. In all cases, it is advisable to install a drip trap between the two reducing valves. A safety valve installed on the low-pressure side of a reducing valve should be of sufficient capacity and should be set to relieve the entire output of the reducing valves in the event the pressure exceeds the low-pressure setting (see Fig. 2-5).

For estimating data for pressure-reducing-station components, see Chap. 8.

FUEL STORING AND HANDLING EQUIPMENT

Fuels

A fuel may be classified, according to its physical state, as solid, liquid, or gaseous. These three states are represented, respectively, by coal, oil, and gas. Each type of fuel has its advantages, and, in specific applications, each may show advantages over the others. In comparing fuels, compensation must be made for firing efficiencies, storage facilities, and life-cycle costs. The average comparable firing efficiencies are 75% for oil, 72% for gas, 70% for stoker-fired coal, and 65% or less for hand-fired coal.

Heat values for coal. Table 2-2 gives the heat values for coal.

Heat values for common fuel gases. Table 2-3 gives the heat values for fuel gases.

Fuel Oils

The National Bureau of Standards and the American Society of Testing Materials have set up five standards (CS 12-48) for grades of fuel oils, designated as Nos. 1, 2, 4, 5, and 6. The No. 1 and 2 oils are called distillate oils (lighter oil). The No. 4 and 5 oils are blended oils, and No. 6 is a heavy, black residual.

TABLE 2-2 Heat Values for Coal

Type of Coal	Heat Value, Btu/lb
Anthracite	12,910
Semi-anthracite	13,770
Low-volatile bituminous	14,340
Medium-volatile bituminous	13,840
High-volatile bituminous	10,750–13,090
Sub-bituminous	8,940–9,150

TABLE 2-3 Heat Values for Fuel Gases

Type of Gas	Specific Gravity	Heat Value, Btu/ft^3
Natural	0.61	1000
Manufactured	0.55	550
Propane	1.52	2570
Butane	1.95	3225

The heavy oils are cheaper and have greater heating values, but other factors must be considered. Heavy oils require preheating, and only the use of oil in large quantities can justify this expense. The No. 6 oil requires heating equipment to raise its temperature to a point where it is pumpable in all seasons. A fuel-oil heater located in the tank is required, and a maintenance allowance must be made for this. As a rule, a heavy oil is not recommended for a boiler of less than 40 hp or one that is operated for short seasonal periods.

Heat values for fuel oil. Fuel oils consist of carbon, hydrogen, water, and sediment. As the hydrogen content of a fuel increases, more heat is available. Table 2-4 gives the heat values of the various fuel-oil grades at an approximate (typical) analysis of 85% carbon, 14% hydrogen, and 1% water and sediment.

Fuel Storing and Handling

Figure 2-6 shows a typical boiler fuel-oil diagram. It indicates the fuel-storing and fuel-handling equipment, i.e., underground fuel storage tanks for No. 2 oil, fuel-oil transfer pumps, fuel-oil piping, and accessories (see Chap. 8).

TABLE 2-4 Heat Values for Fuel Oils

Grade	API Gravity*	Weight, lb/gal	Heat Value, Btu/gal
1	38–45	6.95–6.67	132,900–137,000
2	30–40	7.29–6.87	135,000–141,800
4	12–32	8.21–7.20	140,600–153,300
5	8–20	8.44–7.77	148,100–155,900
6	6–18	8.57–7.88	149,400–157,300

*API gravity is a parameter widely used in place of the specific gravity of fuel oil.

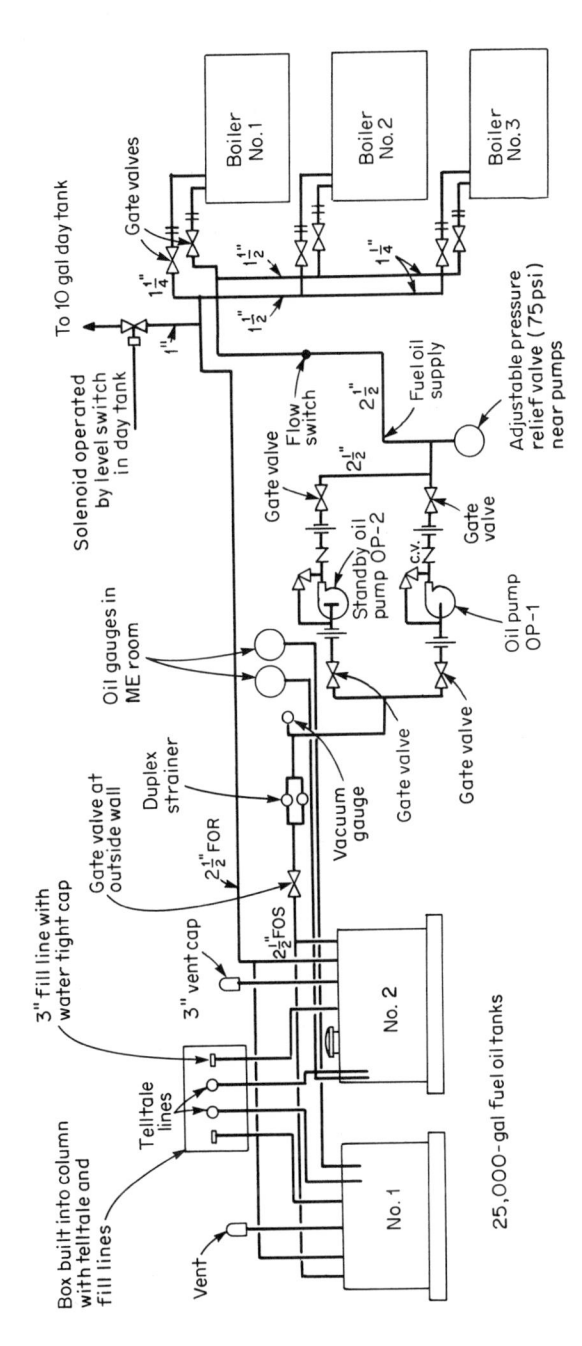

FIGURE 2-6 Boiler fuel-oil diagram.

Fuel-Oil Storage Tanks

Fuel-oil storage tanks are generally of steel construction; recently, fiberglass tanks have become available. There are definite rules and regulations covering the construction of fuel-oil tanks; it is customary to install a tank constructed according to the National Board of Fire Underwriters requirements as well as local codes.

Sizing the tank. In sizing the fuel-oil storage tank, it is first necessary to determine the annual fuel-oil consumption (see the discussion of degree-days in the Appendix). The greatest fuel consumption will occur in January and will amount to approximately 20% of the annual consumption. One-fourth of the January figure gives a week's supply, which is the minimum tank size. Approximately 10% should be added to the minimum tank size to allow for the suction-stub clearance. Additional capacity must be allowed to compensate for oil availability and the method of delivery. If the fuel oil is delivered by rail or truck, the fuel-oil storage tank should be sized to take a full tank car or truck plus one week's fuel consumption. Tank cars or trucks carry from 8000 to 12,000 gal.

Underground tanks. Underground tanks should be located outside the building. An underground tank is always cylindrical; it should be buried so that the top of the tank is at least 2 ft below grade. Underground tanks are generally constructed of steel (fiberglass tanks are also available); the steel tanks should be painted with heavy asphaltum or rust-resistant paint.

In many localities, considerable quantities of groundwater may be encountered. Where such conditions are found, it is necessary to provide an anchor (see Fig. 2-7) to secure the tank in place. The general procedure is to set the tank on a concrete pad and attach the tank to the pad with steel straps or rods.

Estimating Chart 2-8 gives the estimating data for underground steel tanks.

Fuel-Oil Transfer Pumps

Fuel-oil pumps normally transfer fuel oil from storage tanks to the oil burners at the boilers and to the emergency generator day tanks. Fuel-oil pumps are of gear-type design, direct driven, and suitable for pumping all grades of oil. Duplex fuel-oil pump sets are factory piped, wired, and mounted on a steel frame and base. The set should be complete with valves, oil strainers, and pressure gauges. See Estimating Chart 2-9 for fuel-pump estimating data.

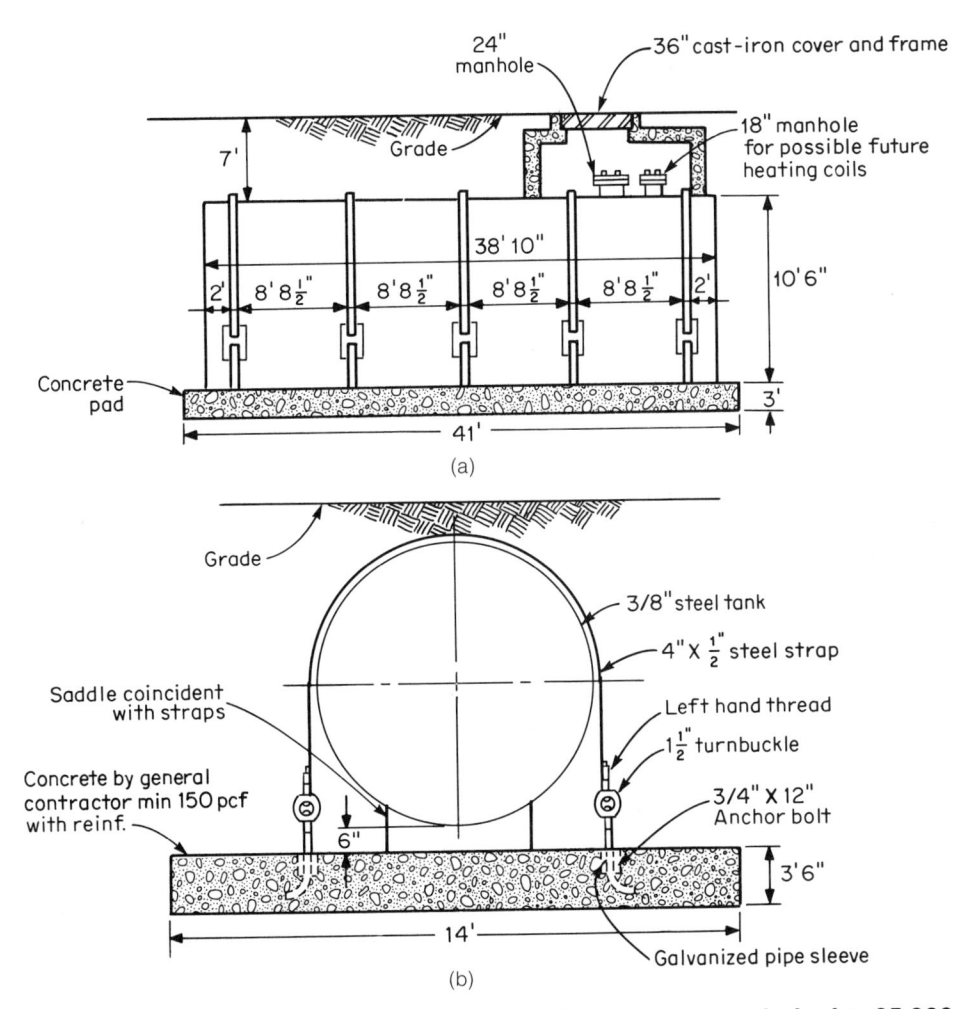

FIGURE 2-7 Method of anchoring fuel-oil tank and manhole for 25,000-gal tank. (*a*) Elevation; (*b*) side view. (Tanks to be located so that they are not subject to bearing pressure of building foundations.)

HEAT EXCHANGERS

General Heat exchangers are of the shell-and-tube type or the coil type. They are used for heating many kinds of liquids, but the most common application is for heating water. When they are used as water heaters or converters, the heated water circulates through the tube (or coil); the heating medium, either steam or water at a higher temperature, is contained in the shell of the heat exchanger. These heat exchangers are often referred to as *instantaneous water heaters* or *converters* when used to provide domestic or process hot water. Since they heat water as needed, a storage tank is not required in most cases.

Types of Heat Exchangers

Water-to-water heat exchangers. These heat exchangers are designed for boiler water to be pumped through the shell and to heat water flowing through the tubes.

ESTIMATING CHART 2-8 Underground Steel Oil-Storage Tanks (including manholes).

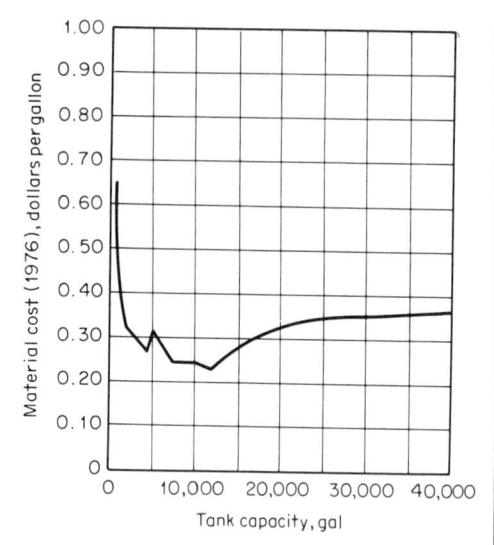

1. Material costs

2. Labor man-hours

Capacity, gal	Thickness, in	Diameter x length	Weight, lb	Man-hours
300	12 GA	35"x 6'	320	8
550	12 GA	48"x 6'	480	8
1,000	3/16	64"x 6'	1,180	16
1,500	3/16	64"x 9'	1,575	16
2,000	3/16	64"x 12'	1,980	20
2,500	3/16	64"x 15'	2,375	20
3,000	3/16	64"x 18'	2,765	24
4,000	3/16	6'x 19'	3,310	24
5,000	1/4	6'x 24'	5,410	32
6,000	1/4	6'x 29'	6,380	32
7,500	1/4	8'x 16'	6,425	36
8,200	1/4	8'x 22'	6,960	36
10,000	1/4	8'x 27'	8,260	40
12,000	1/4	8'x 32'	9,650	48
15,000	3/8	10'x 26'	12,150	54
20,000	3/8	10'x 35'	15,500	60
25,000	3/8	10'x 43'	22,300	64
30,000	3/8	10'x 51'	28,000	64
40,000	1/2	12'x 50'	43,000	72

Prices vary considerably depending on the steel thickness. See thickness in the man-hours table.

ESTIMATING CHART 2-9 Labor Man-Hours for Fuel-Oil Transfer Pumps

Boiler Horsepower	Pump Capacity (based on No. 2 oil and 1750 motor rpm), gph	Man-hours
15–40	30	6
50–100	45	6
125–250	140	8
300–600	270	10
700	400*	12

*In this case, the motor speed is 1150 rpm.

Steam-to-water heat exchangers. In these heat exchangers, steam enters through an opening in the top of the shell and heats water flowing through the tubes. As the steam gives up its latent heat to the heated water, it condenses, collects at the bottom of the shell, and drains off.

See Estimating Chart 2-10 for steam-to-water heat-exchanger estimating data.

Material-Cost-Calculation Procedure

Each curve in Estimating Chart 2-10 indicates the material cost per foot of tube length for a particular shell diameter. Thus, to determine the estimated material cost of a steam-to-water converter with an 18-in shell diameter and a 6-ft tube length, the 18-in-diameter curve is used. The material cost per foot is $630 (see the dotted lines in Estimating Chart 2-10). Therefore, the material cost for this converter is 6 ft × $630/ft = $3780.

For water-to-water heat exchangers, add 5% to the material costs of similar-size steam-to-water heat exchangers. For high-temperature heat exchangers which have ¾-in OD 90-10 copper-nickel tubes, modular cast-iron heads, and steel shells, add 75% to the material costs of similar-size steam-to-water heat exchangers.

SOLAR ENERGY

General

The primary purpose for the collection of solar energy is to use it as a fuel supplement. The direct use of the sun's energy as it penetrates the building in the form of light and heat is the most efficient way to use solar energy. The orientation, amount, and type of fenestration and solar control devices must optimize the use of the sun.

The use of solar energy for heating, cooling, humidification, and domestic hot-water heating via solar collectors and storage systems is technically feasible. Hardware is now available in sufficient quantities to provide such systems in building facilities. The use of solar-energy systems can eliminate the need for significant quantities of fossil fuels. Continued increases in the cost of fossil fuels and continued development of more efficient and lower-cost solar collectors and storage systems will make solar-energy systems economically competitive with many fossil-fuel systems in the near future. The functional and economic feasibility of solar-energy systems depends upon the selection and design of those heating and cooling systems which are most compatible with solar-energy systems.

Solar Collectors

Flat-plate collectors are less costly and are adequate for building use. Solar collectors may be separate from the building; they can be integrated with

ESTIMATING CHART 2-10 Steam-to-Water Heat Exchangers (¾" O.D. Copper Tubes, Cast-Iron Heads, and Steel Shell).

1. Material costs

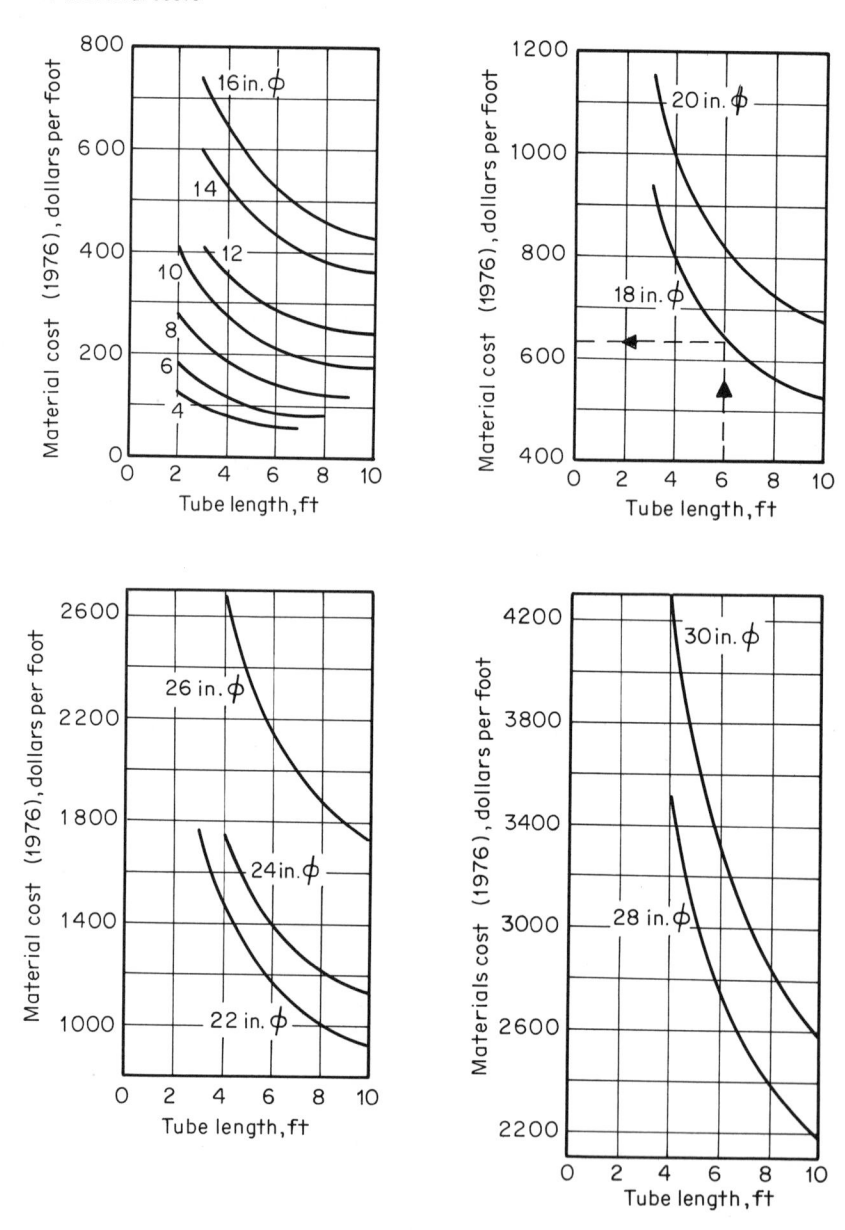

building skins, to function as solar shades to control the direct solar heat gain on walls and windows in summer; or they can be an integral part of the roof.

Figure 2-8 shows a typical flat-plate collector panel. The standard panel is 4 × 8 ft. The metallic absorber plate is coated with nonreflecting black paint to absorb as much heat as possible. Tubes are bonded to the absorber plate so that fluid circulating through the tubes can conduct the absorbed heat. The absorber surface facing the sun is covered with a glass (or polycarbonate plastic) sheet that allows solar radiation to pass through but traps the

Heat-Generation Equipment | 47

**ESTIMATING CHART 2-10 Steam-to-Water Heat Exchangers
(¾" O.D. Copper Tubes, Cast-Iron Heads, and Steel Shell)** *(Continued).*

2. Labor man-hours

Dia.(in.) X Length (ft)	Weight, lb	Man-hours
4 x 2 → 7	56 → 126	4
6 x 2 → 7	68 → 193	4
8 x 2 → 4	112 → 184	4
10 x 2	184	4
6 x 8	218	6
8 x 5 → 9	220 → 364	6
10 x 3 → 10	230 → 552	6
12 x 3 → 6	294 → 501	6
14 x 3 → 4	449 → 534	6
12 x 7 → 10	570 → 777	8
14 x 5 → 10	619 → 1044	8
16 x 3 → 8	570 → 1095	8
18 x 3 → 6	712 → 1084	8
20 x 3 → 4	1001 → 1158	8

Dia.(in.) X Length (ft)	Weight, lb	Man-hours
16 x 9 → 10	1200 → 1305	10
18 x 7 → 10	1208 → 1540	10
20 x 5 → 10	1315 → 2100	10
22 x 3 → 8	1212 → 2132	10
24 x 4 → 6	1710 → 2146	10
26 x 4	2036	10
22 x 9 → 10	2316 → 2500	12
24 x 7 → 10	2364 → 3018	12
26 x 5 → 8	2280 → 3012	12
28 x 4 → 6	2502 → 3068	12
30 x 4	2886	12
26 x 9 → 10	3256 → 3500	14
28 x 7 → 10	3351 → 4200	14
30 x 5 → 8	3205 → 4162	14
30 x 9 → 10	4481 → 4800	16

long-wave radiation inside the collector. The back of the absorber is insulated with fiberglass blanket insulation to minimize the heat loss.

The area required for a plate collector will vary from 25 to 50% of the total floor area of the building it serves. The size is dependent upon climate, size of storage facilities, type of heating and cooling systems, and optimum economic feasibility.

Solar collectors can be used with absorption refrigeration systems for summer cooling. Current absorption refrigeration technology requires a hot-water generating temperature of at least 200°F. This temperature range requires high-efficiency collectors with double glazing and selective surface coating.

Storage Systems

Thermal storage systems are required to store energy, either hot or cold, for use in the building when there is insufficient solar radiation (e.g., nights and rainy days). Storage systems are usually designed for one to three days output from the collector. Storage tanks may be constructed of concrete, steel, or fiberglass.

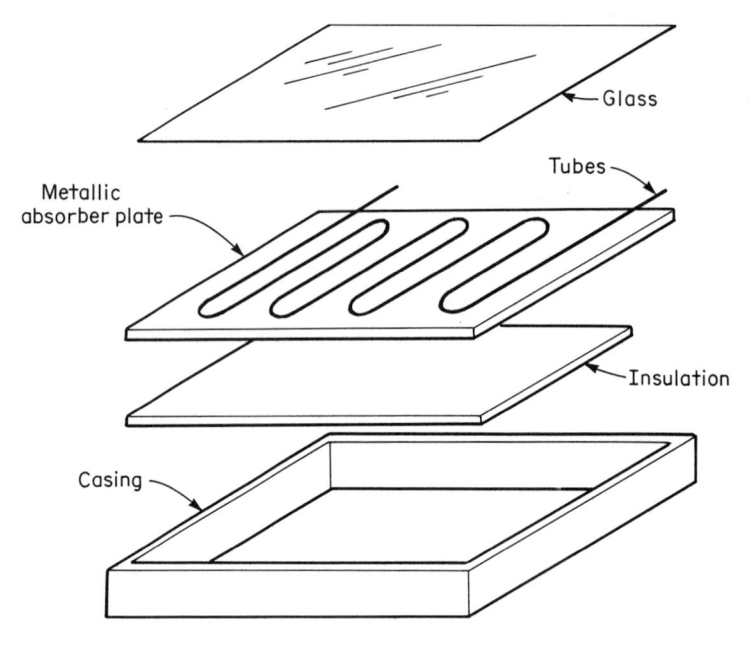

FIGURE 2-8 Typical flat-plate collector panel.

Estimating   With the continued development of more efficient and lower-cost solar collectors, it is hard to predict the initial cost of a solar-energy system. However, the current initial cost of solar collectors is quite high, owing to the limited use of solar energy in the construction industry. The data that follow indicate the cost of solar collectors. They are based on prices obtained in 1976 from solar-collector manufacturers for 8 × 4 ft panels with double-glazed cover. The costs per square foot of panel area, based on quantity, are as follows:

An order of 10 to 30 panels would be priced at $192 to $320 each, f.o.b., or $6 to $10 per square foot of panel. Quantities of 1000 and over would be priced at $160 to $256 each, f.o.b., or $5 to $8 per square foot. (The wide price ranges indicate the differences in manufacturers' quotations.) The installed cost per square foot of panel area ranges from $16 to $20 and includes installation, auxiliary equipment, pumps, and storage tanks.

3

COOLING-GENERATION EQUIPMENT

GENERAL The cooling-generation equipment in a refrigeration system normally supplies the cooling fluid, in the liquid state, to the place where cooling is desired.

Refrigerants are the vital working fluids in refrigeration systems. They absorb the heat gain of the conditioned space and dispose of it elsewhere. Heat is removed from the system by evaporating the liquid refrigerant, and is disposed of by condensing the refrigerant vapor. These processes occur in absorption systems as well as in mechanical compression systems. In applications such as the heat pump, the refrigerant can be used to add heat to the system.

In many refrigeration applications, heat is transferred to a secondary coolant, which may be any liquid cooled by the refrigerant and used to transmit heat without a change of state. These secondary coolants are known as *water*, *brines*, or *secondary refrigerants*.

The components of the refrigeration system are normally chillers, condensers, cooling towers, and auxiliary equipment.

Definition *Ton of refrigeration*. A refrigerating effect equal to 12 MBH, or 12,000 Btu/h.

Material Costs and Labor Man-Hours for Cooling-Generation Equipment

Material costs. Material costs for cooling-generation equipment may be obtained from the material cost charts that follow. Each chart shows the relationship of material price per unit of capacity (i.e., tonnage of cooling load) to the total capacity of the unit, based on 1976 average costs. These costs are intended for reference only and should not be considered reliable for estimating actual material costs.

Labor man-hours. Labor man-hours for cooling-generation equipment may be obtained from the labor tables that follow. The labor man-hours include handling, assembling, erecting, and making the unit ready for test. They are based on normal working conditions; i.e., the unit is received within 100 ft of its final location and is mounted on a flat slab foundation. For final hookup, see Chap. 8. For rigging for heavy equipment, see Chap. 16.

MECHANICAL COMPRESSION WATER CHILLERS

The most common application for the mechanical compression water chiller is water chilling for air conditioning. The basic components are the compressor, evaporator, condenser, and expansion device.

Figure 3-1 shows a simple mechanical water chiller employing a water-cooled condenser. As shown, water enters the evaporator, where it is chilled by liquid refrigerant evaporating at a lower temperature. The refrigerant gas produced by this evaporation is drawn into the compressor, which increases the pressure of the gas so that it may be condensed at a higher temperature in the condenser. The condenser cooling medium is warmed in the process. The condensed liquid then flows to the evaporator through a metering device. Since the pressure of the liquid refrigerant is reduced from the condenser pressure to the evaporator pressure, a fraction changes to vapor in the process.

An air-cooled or evaporative condenser may be used in place of the water-cooled condenser. However, centrifugal chillers most often use water-cooled condensers.

Mechanical compression water chillers are classified in a number of different ways, as indicated in Table 3-1. As a rough guide, Table 3-2 gives the types of mechanical water chillers generally used for air conditioning. For estimating data for packaged mechanical water chillers, see Estimating Charts 3-1 and 3-2.

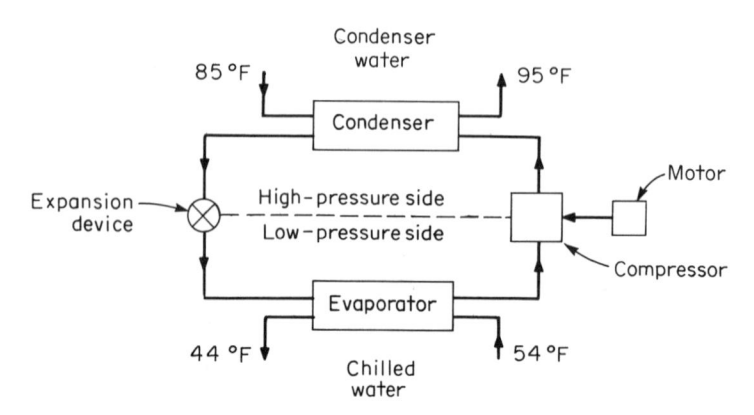

FIGURE 3-1 Diagram for simple mechanical water chiller, employing a water-cooled condenser.

TABLE 3-1 Types of Mechanical Water Chillers

By Compressor		By Condenser
Type	Drive	
Positive Displacement	External drive	Water cooled
Reciprocating	Turbine	Air cooled
Rotary	Engine	
Helical rotary (screw)	External electric motor	Evaporative cooled
Centrifugal	Hermetic motors	

ABSORPTION WATER CHILLERS

Absorption water chillers are heat-operated refrigeration machines. The refrigeration cycle utilizes an absorbent as a secondary fluid to absorb the primary fluid—gaseous refrigerant—which has been vaporized in the evaporator. In both the absorption cycle and the mechanical compression cycle, a refrigerant liquid evaporates and condenses, and these processes occur at two pressure levels within the unit. The two cycles differ in that the absorption cycle uses a heat-operated generator (concentrator) to produce the pressure differential, whereas the mechanical compression cycle uses a compressor.

In most air-conditioning applications, especially in large sizes, water is used as the refrigerant, with lithium bromide (salt solution) as the absorbent. The basic components of the absorption water chiller include the evaporator, absorber, concentrator, condenser, and heat exchanger.

Figure 3-2 shows the flow diagram of a one-shell absorption chiller. As shown, water enters the evaporator, where it is chilled by spraying and evaporating refrigerant water. Liquid refrigerant in the bottom of the evaporator flows by gravity into a sump. The evaporator pump takes refrigerant from this sump and delivers it to spray trees in the evaporator. The refrigerant vapor produced in the evaporator flows to the absorber

TABLE 3-2 Mechanical Water Chillers for Air Conditioning

Type of Chiller	Capacity Range, tons
Reciprocating	Up to 80
Reciprocating or centrifugal*	80–120
Reciprocating, centrifugal, or screw*	120–200
Centrifugal or screw	200–800
Centrifugal	Above 200†

*For air-cooled condenser duty from 80 to 200 tons, reciprocating and screw water chillers are more frequently employed than centrifugal.
†Factory packaged water chillers are available to about 1300 tons, and field-assembled machines to about 10,000 tons.

because of the lower pressure in this area. The refrigerant vapor condenses into liquid as it comes in contact with the absorbent solution.

In the absorber, three quantities of heat are released: the heat of condensation from vapor condensing into the absorbent; the heat of dilution as the vapor goes into solution with absorbent; and sensible heat. In order to remove this heat and maintain a constant temperature in the absorber, the absorbent solution falls over a cooling coil after being sprayed into the absorber. Cooling water is supplied to this coil to remove heat from the absorber. After falling over the cooling coil, the solution of refrigerant and absorbent drops into the bottom of the absorbers shell.

Noncondensable gases may be present in the refrigeration system. These gases must be removed (purged) for proper operation of the absorption chiller. Without proper purging, the pressure in the absorber can increase sufficiently to stop the flow of refrigerant vapor from the evaporator.

A concentrator pump continuously removes part of the solution from the absorber and delivers it, through a heat-exchanger–flash-chamber combination, to the concentrator. There, steam or hot-water coils supply heat to boil the refrigerant out of the solution, leaving the concentrated absorbent in the bottom of the concentrator (the refrigerant has a lower boiling temperature than the absorbent.

ESTIMATING CHART 3-1 Packaged Reciprocating Water Chillers. (All Units Include Insulated Chiller, Condenser, Compressor, Motor, Heat interchanger, and Normal Basic Controls, and Are Completely Piped).

1. Material costs
2. Labor man-hours

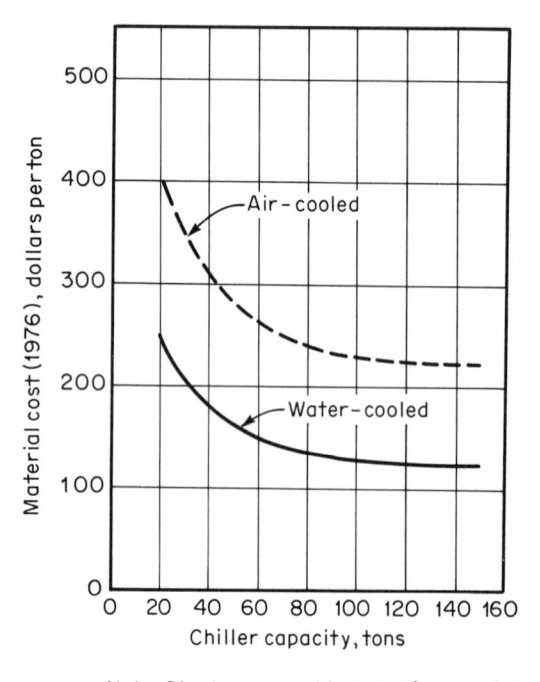

Capacity, tons	Weight, lb	Man-hours
15	1400	16
20	1700	16
25	2200	16
30	2500	16
45	3000	16
55	4500	20
65	4700	20
80	5700	20
100	6400	20
115	6900	24
125	8400	28

Note: Starters are not included in material cost.

ESTIMATING CHART 3-2   Packaged Centrifugal Water Chillers. (All Units Include Chiller, Condenser, Hermetic Compressor, Lubricating System, and Control Panel, and Are Completely Piped.)

1. Material costs

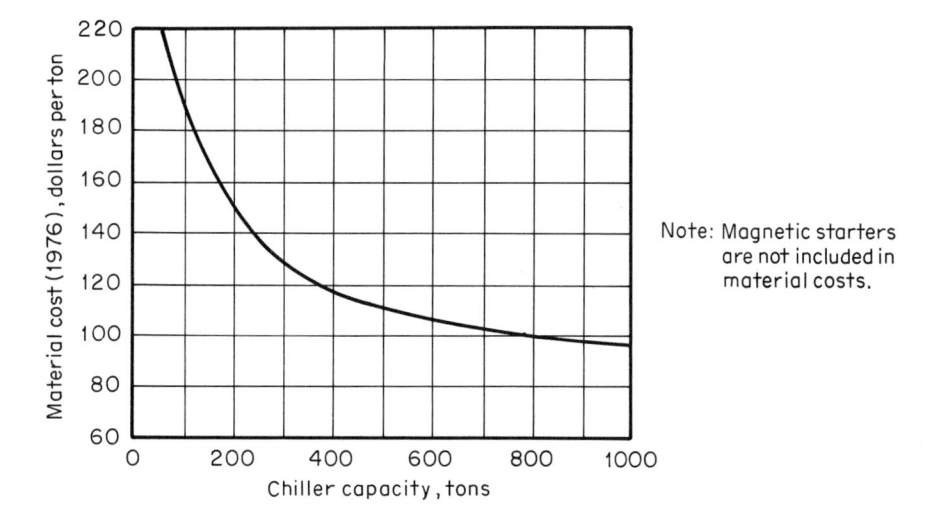

Note: Magnetic starters are not included in material costs.

2. Labor man-hours

Capacity, tons	Weight, lb	Man-hours
80	10,300	60
120	11,000	60
180	12,000	60
225	12,300	72
250	12,500	72
275	12,700	72
300	12,900	72
350	14,600	72
385	14,800	72

Capacity, tons	Weight, lb	Man-hours
435	17,400	96
490	19,600	96
555	20,000	96
620	21,300	96
675	21,600	96
700	22,300	96
765	22,800	96
900	36,000	144
1000	39,000	144

The boiling refrigerant vapor flows upward from the solution to the condenser, where it comes in contact with the coil surfaces. The coil is filled with condenser water, which is usually supplied by a cooling tower. The refrigerant vapor condenses and drops to the bottom of the condenser, from which it flows to the evaporator through a regulating orifice. This completes the operating cycle.

The efficiency of the cycle is improved by the use of a heat exchanger. Note that, for the given operating conditions, the concentrator has a temperature of 210°F, whereas the temperature of the absorber is about 105°F.

The heat exchanger is used to transfer heat from the hot solution leaving the concentrator to the lower-temperature solution going to the concentra-

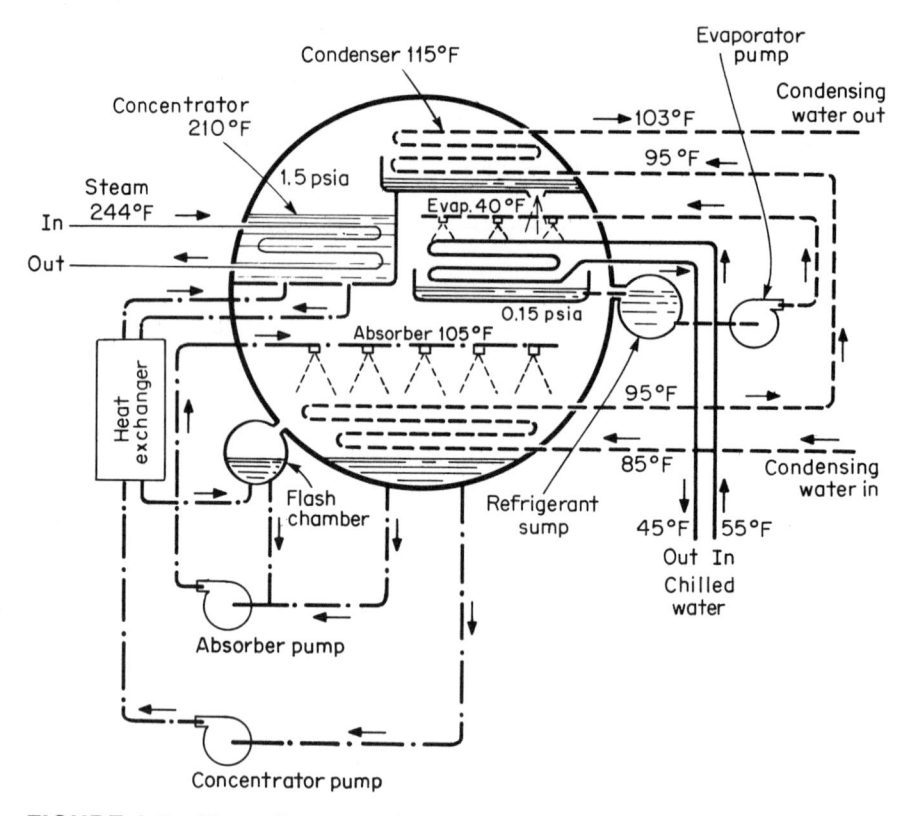

FIGURE 3-2 Flow diagram of one-shell absorption machine. (*Copyright © The Trane Company, 1965, The Trane Air Conditioning Manual. Used by permission.*)

tor. After passing through the heat exchanger, the concentrated solution enters the flash chamber. There, a small part of the water in the concentrated solution flashes, or evaporates, owing to the low pressure. This flashing cools the remaining solution. The flash vapors then move into the absorber, while the remaining solution flows to mix with solution being pumped to the absorber spray tree. The use of a heat exchanger results in lower steam consumption for the same amount of refrigerant evaporated from the concentrator; and less heat must be removed from the absorber by the cooling water.

Factory-packaged absorption chillers are available in capacities of 50 to 1500 tons. See Estimating Chart 3-3 for the estimating data for packaged absorption chillers.

CONDENSERS

The condensers in a refrigerating system remove, from the refrigerant vapor, the heat of compression and the heat absorbed by the refrigerant in the evaporator. The refrigerant is thereby converted back to the liquid phase at the condenser pressure and is available for reexpansion in the evaporator. The commonly used condensers may be classified on the basis of cooling medium as follows:

1. Water-cooled
2. Air-cooled
3. Air- and water-cooled (evaporative)

Many factors should be considered before a final decision is reached on the best method of condensing the refrigerant vapor. Some of these factors are indicated in Table 3-3.

Water-Cooled Condensers

The three common types of water-cooled condensers are:

1. The *double-pipe* or *double-tube* condenser, which consists of one or more assemblies of two tubes, one within the other. The refrigerant vapor is condensed in either the annular space or the inner tube.
2. The *shell-and-coil* condenser, in which the cooling water is circulated through one or more continuous or assembled coils contained within the shell.
3. The *shell-and-tube* condenser, which consists of large number of tubes installed inside a steel shell. The water flows inside the tubes, while the

ESTIMATING CHART 3-3 Packaged Absorption Water Chillers. (All Units Complete with Pumps, Purge System, and Control Panel, and Are Completely Piped.)

1. Material costs

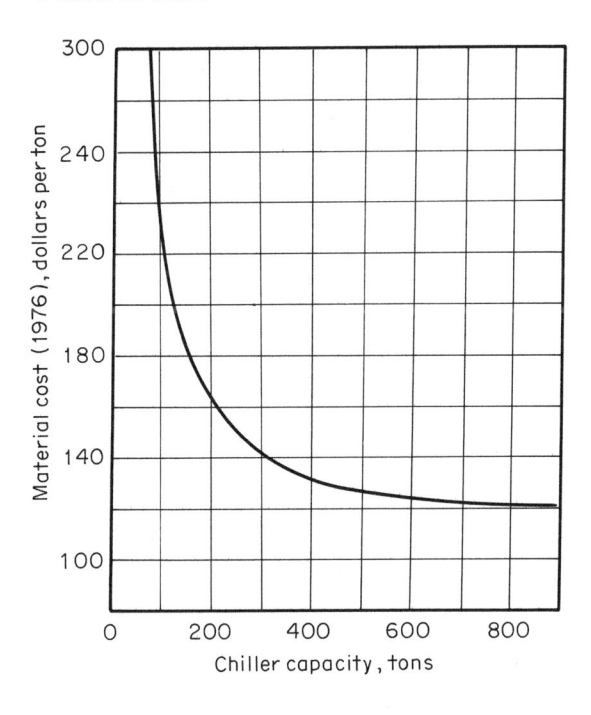

2. Labor man-hours

Capacity, tons	Weight, lb	Man-hours
100	8,600	64
115	9,400	64
130	10,100	64
150	11,400	64
180	12,200	64
200	13,400	80
230	14,300	80
260	16,200	80
315	18,100	104
385	20,600	104
465	23,000	104
520	25,600	144
590	29,000	144
650	31,600	144
725	33,000	144
850	41,200	160
935	44,800	160
1070	50,300	176

TABLE 3-3 Types of Condensers

Basis	Condenser Type		
	Water Cooled	Air Cooled	Evaporative Cooled
1. Space and location	Occupies the least amount of space except when a cooling tower is used. Located indoors, with cooling tower outdoors.	Usually located outdoors, sometimes indoors.	Requires less space than air-cooled condenser or water-cooled condenser with cooling tower. Located outdoors or indoors.
2. Unit size	Ranges from small to very large.	Up to 125 tons; larger capacities require multiple units.	Ranges from small to large.
3. Cooling medium and sources	Supplied with cooling water from a well or cooling tower or city water. Requires water treatment when used with cooling tower. Requires pumps and water piping large enough to handle total water flow.	Supplied with dry air by supply fans. Requires longer refrigerant lines.	Supplied with dry air and cooling water. Requires much less circulating water than water condenser with cooling tower. Uses a small water pump and water lines of smaller size and short runs. Requires water treatment.
4. Initial and life-cycle costs	High when cooling tower is included.	Lowest installation and maintenance cost. Higher power requirement per ton than evaporative condenser or water-cooled condenser.	Medium.

vapor flows outside the tubes. The vapor condenses on the outside surfaces of the tubes and drips to the bottom of the condenser, which may be used as a receiver for the storage of liquid refrigerant. Shell-and-tube condensers are used for practically all water-cooled refrigeration systems and are normally included in chiller packages. See Estimating Chart 3-4 for estimating data.

Air-Cooled Condensers

The air-cooled condenser is available with either propeller- or centrifugal-type fans for moving the air through the condenser coil. The propeller-fan type is generally used for outdoor applications, whereas the centrifugal-fan type, with ductwork to carry the air to and from the unit, is suitable for indoor applications. Either type can be built with a liquid subcooling circuit.

The main components of remote air-cooled condensers are a finned condensing coil, one or more fans and motors, and an enclosure. See Estimating Chart 3-5 for estimating data.

Air Condensing Units

The basic design considerations for these units are the same as for remote condensers, but the inclusion of the compressor makes it desirable in most cases to have the components enclosed in a cabinet. Thus, packaged units also contain controls, precharged line fittings or valves, and a liquid receiver (if separate from the coil). See Estimating Chart 3-6 for estimating data.

Evaporative Condensers

An evaporative condenser is a refrigeration-system component comprised of a coil in which refrigerant is condensed and a means for supplying a flow of air and water over its external surface. Heat is transferred from the condensing refrigerant inside the coil to the wet external surface and then into the moving airstream, principally by evaporation.

ESTIMATING CHART 3-4   Shell-and-Tube Water-cooled Condensers (ASME Constructed).

1. Material costs

2. Labor man-hours

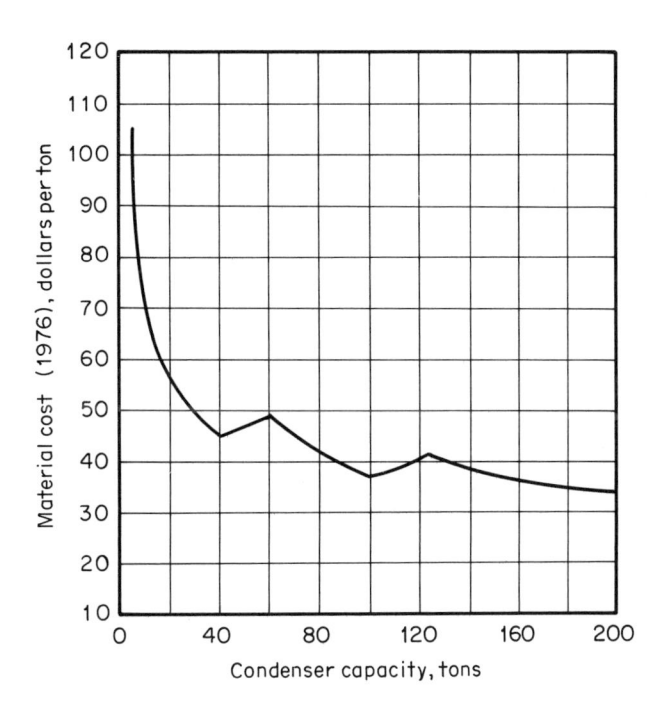

Capacity, tons	Length, ft	Weight, lb	Man-hours
5	3	85	8
10	4	145	8
15	4	160	8
20	6	260	12
25	6	290	12
30	6	360	12
40	6	395	12
50	6	500	18
60	6	590	18
75	6	670	24
100	8	775	24
125	8	1020	30
152	8	1200	30
175	8	1400	32
200	8	1600	32

Note: The curve fluctuates depending on a sharp increase in the weight of the condenser.

ESTIMATING CHART 3-5 Remote Air-cooled Condensers.

1. Material costs

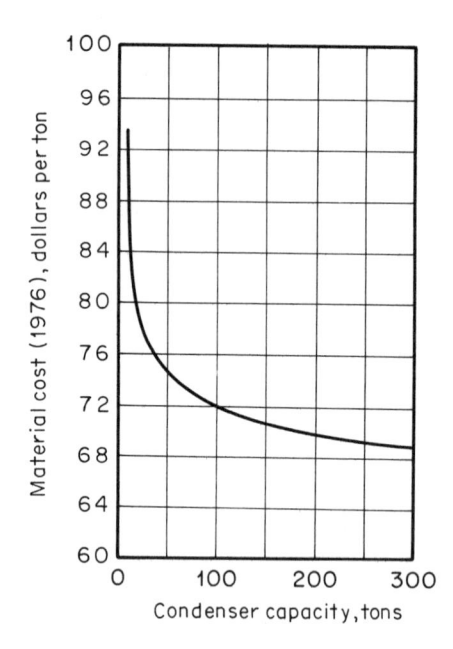

2. Labor man-hours

Capacity, tons	Weight, lb	Man-hours
10	525	8
15	690	8
25	1000	12
30	1300	12
45	1800	12
60	2100	12
75	2600	16
90	3300	16
120	4200	16
150	5100	24
180	6000	24
225	7200	24
300	9400	32

As shown in Fig. 3-3, refrigerant vapor from the compressor enters the top of the condensing coil and condenses to a liquid as it flows through the coil. Water is sprayed downward over the coil; the spray water falls into a water tank and is picked up by a pump and returned to the spray nozzles. A fan at the top of the unit draws air into the bottom of the casing and up through

ESTIMATING CHART 3-6 Air-cooled Condensing Units. (All Units Include Hermetic Compressors, Air-cooled Condensers, and Control Panels).

1. Material costs

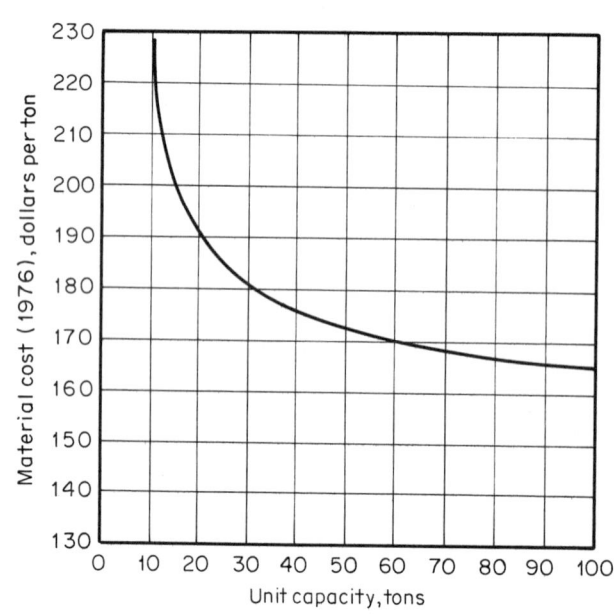

2. Labor man-hours

Capacity, tons	Weight, lb	Man-hours
10	1200	12
15	1675	12
20	1800	12
25	2200	12
30	2600	16
40	3400	16
50	4300	20
60	5100	20
75	5800	24
100	6500	24

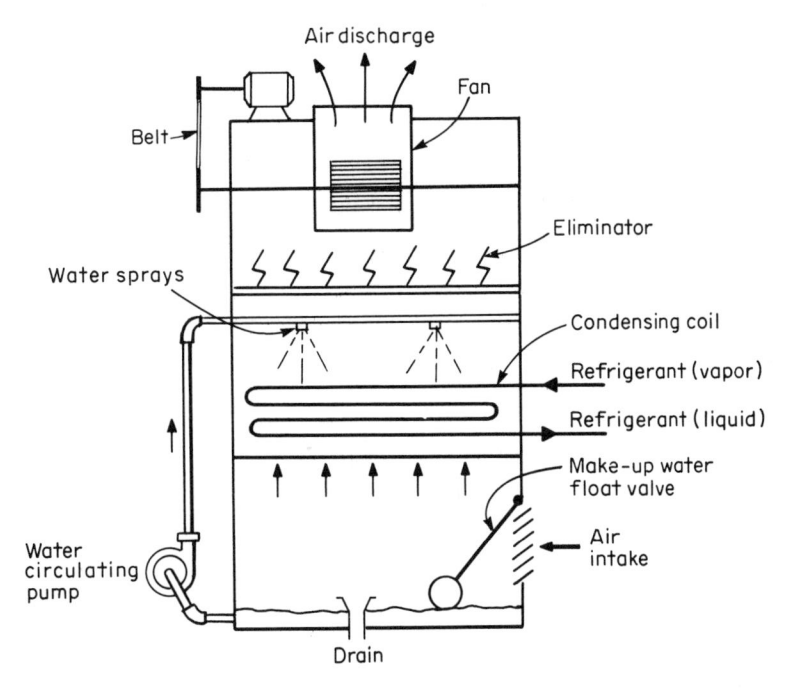

FIGURE 3-3 Evaporative condenser (draw-through type).

the spray-filled interior, discharging it out the top. Almost all evaporative condensers use a fan to either blow or draw air through the unit. See Estimating Chart 3-7 for estimating data.

COOLING TOWERS

A cooling tower is a steady-flow device that uses a combination of mass and energy transfer to cool water by exposing it to the atmosphere as an extended surface. The water surface is extended either by filling, which presents a film surface or creates drops due to splashing, or by spraying, which produces droplets. The airflow may be produced by mechanical means or by natural draft. Cooling towers are classified according to the method of moving air through the tower, as natural-draft, induced-draft, or forced-draft cooling towers. For estimating data, see Estimating Chart 3-8.

Natural Draft The natural-draft cooling tower is designed to cool water by means of air moving through the tower at the low velocities typical of open spaces during the summer. They are constructed of cypress or redwood and have numerous wooden decks of splash bars, installed at regular intervals from bottom to the top. Warm water from the condensers is flooded or sprayed over the distributing deck and flows by gravity to the water-collecting basin. This type of cooling tower is rarely used nowadays, unless low initial cost and minimum power requirements are primary considerations. The space requirements are much greater than for mechanical-draft cooling towers.

ESTIMATING CHART 3-7 Evaporative Condensers. (All Units Include Fan, Pump, and Motor.)

1. Material costs

2. Labor man-hours

Capacity, tons	Weight, lb	Man-hours
5	775	8
10	1500	12
15	1750	12
20	2150	12
25	2950	16
30	3400	16
40	4250	18
50	5400	20
60	6100	24
70	6950	24
75	7400	24
80	7900	28
90	8400	28
100	9500	32

Induced Draft An induced-draft cooling tower has a top-mounted fan that induces atmospheric air to flow up through the tower as warm water falls downward. An induced-draft tower may have only spray nozzles for water breakup, or it may be filled with slat-and-deck arrangements.

In a counterflow induced-draft tower, a top-mounted fan induces air to enter all four sides of the tower and flow vertically upward as the water cascades through the tower. The counterflow tower is particularly well

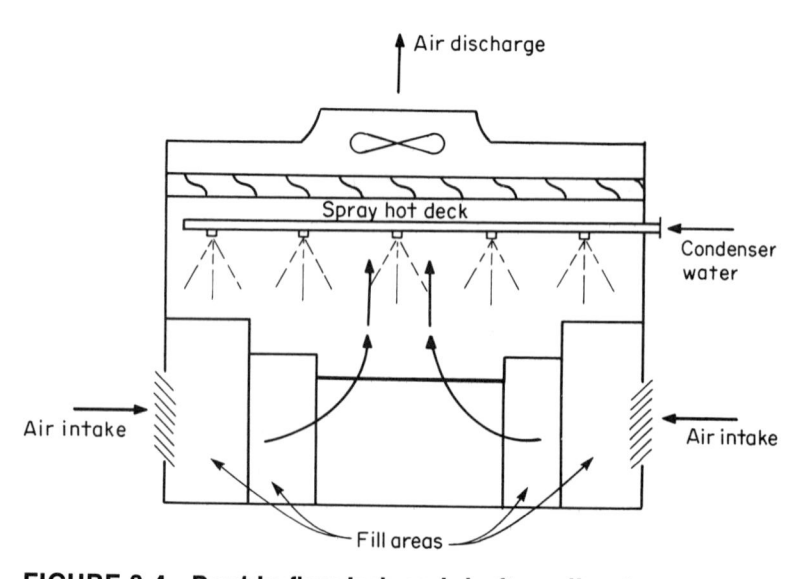

FIGURE 3-4 Double-flow induced draft cooling tower.

ESTIMATING CHART 3-8 Cooling Towers (Factory-Assembled, Galvanized-Steel Construction, Completely Fireproof).

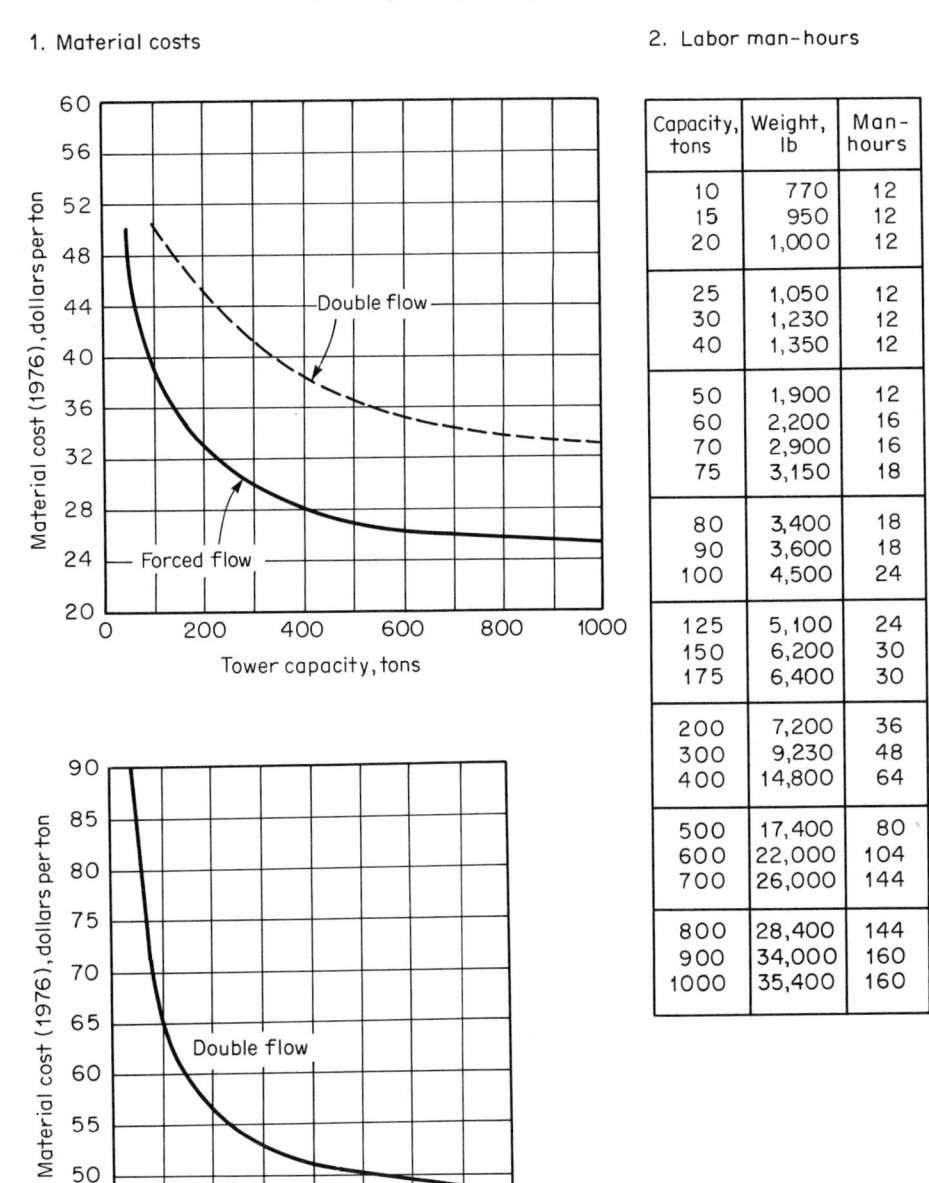

1. Material costs

2. Labor man-hours

Capacity, tons	Weight, lb	Man-hours
10	770	12
15	950	12
20	1,000	12
25	1,050	12
30	1,230	12
40	1,350	12
50	1,900	12
60	2,200	16
70	2,900	16
75	3,150	18
80	3,400	18
90	3,600	18
100	4,500	24
125	5,100	24
150	6,200	30
175	6,400	30
200	7,200	36
300	9,230	48
400	14,800	64
500	17,400	80
600	22,000	104
700	26,000	144
800	28,400	144
900	34,000	160
1000	35,400	160

adapted to a restricted space, as the discharge air is directed vertically upward, and the four sides require only minimum clearance for air intake. The primary breakup of water may be accomplished either by pressure spray or by gravity, from pressure-filled flumes.

The double-flow induced-draft tower (Fig. 3-4) has a top-mounted fan to induce air to flow across the fill material. The air is then turned vertically in the center of the tower. This type of induced-draft tower has two air intakes, on opposite sides of the tower.

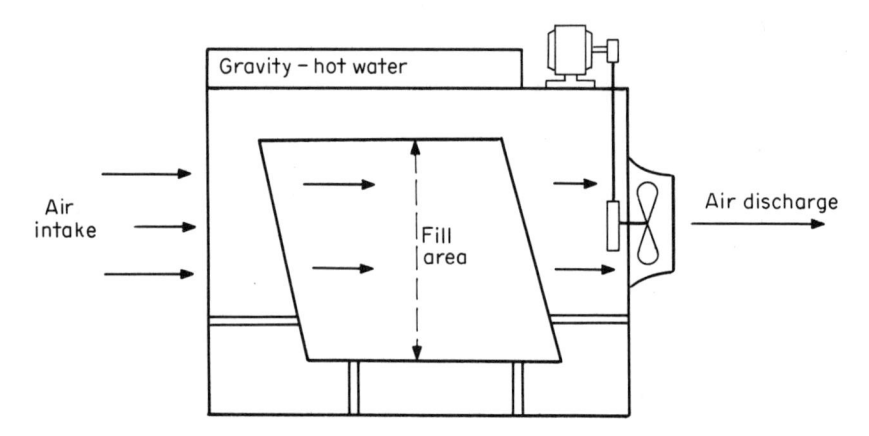

FIGURE 3-5 Cross-flow induced-draft cooling tower.

The crossflow induced-draft tower (Fig. 3-5) is a modified double-flow induced-draft tower. The fan in the crossflow tower draws air in through a single horizontal opening at one end and discharges the air at the opposite end.

Forced Draft

A fan is used to force air into the forced-draft cooling tower. In the usual installation, the fan shaft is in a horizontal plane; the air is forced horizontally through the fill, upward to the discharger, and out the top of the tower (see Fig. 3-6).

Factory-assembled mechanical-draft towers are constructed in single-fan modules having capacities up to 1000 tons. Greater capacities may be obtained in multiple-cell installations. The construction material varies from galvanized steel and plastic to preservative-treated wood. The fill materials are also quite varied; they include stainless steel, polyvinyl chloride, neoprene asbestos, and preservative-treated wood.

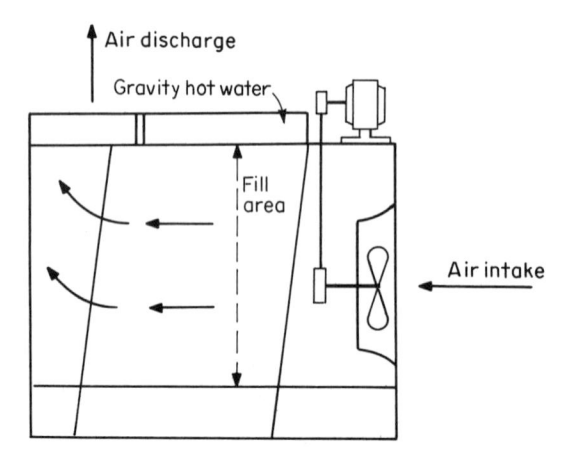

FIGURE 3-6 Forced-draft cooling tower.

4

HEAT- AND COOLING-GENERATION AUXILIARY EQUIPMENT

GENERAL

This chapter discusses heat- and cooling-generation auxiliary equipment:

- Water-treatment systems
- Pumps
- Hydronic specialties

Material cost charts and labor (man-hour) tables are given for each of these types of equipment. The material costs are presented for reference only and should not be considered reliable for estimating actual costs. Labor figures include the handling and erecting of the unit. For final hookup, see Chap. 8.

WATER-TREATMENT EQUIPMENT

Water treatment is an important aspect of the maintenance and operation of a recirculating water system. Generally, water treatment will effectively minimize corrosion and the formation of scale by removing dissolved oxygen, gases, and solids, and by maintaining the water in a slightly alkaline condition.

For *open recirculating cooling systems* (condenser water), the selection of the water treatment is affected to a considerable extent by the size of the system. The primary purposes of water treatment in an open recirculating system are the prevention of corrosion, the prevention of scale formation, and the elimination of organic growth. Thus, in small systems, treatment might consist of a controlled bleed (blowdown) to minimize scale formation and the maintenance of 200 to 500 ppm of sodium chromate for corrosion control, plus occasional application of sodium pentachlorophanate for the control of organic growth. In many cases, especially in larger systems, make-up water and operating conditions require a more complete water-treatment program.

For *closed recirculating systems* (hot water, chilled water, combined heating and cooling) with make-up water which is scale-forming at the operating temperatures, the use of softened make-up water is desirable. The system should be treated for corrosion control regardless of the quality of the water. Chromates are the most effective corrosion inhibitors for these systems. However, nitrite inhibitors may be used.

In hot-water heating systems, the concentration of chromate or nitrite inhibitor is normally higher than that carried in chilled-water systems. For example, chilled-water systems require a minimum of 200 ppm of sodium chromate or 500 ppm of sodium nitrite, whereas the usual hot-water heating systems (180 to 200°F) require a minimum of 2000 ppm of sodium chromate or 3000 ppm of sodium nitrite. A wide range of treatment procedures can be used for boiler water. In any particular case, the method selected must depend upon the composition of the make-up water, the operating pressure of the boiler, and the make-up rate.

A low-pressure steam heating boiler should have a very limited make-up requirement. Treatment with corrosion inhibitors such as a sodium-nitrite-borax combination or sodium chromate at higher concentration (on the order of 3000 ppm for sodium nitrite and 2000 ppm for sodium chromate) is generally sufficient. It is important to have a sufficiently high inhibitor concentration present during the summer lay up of such boilers, since the most serious damage can be caused during this time. In low-pressure systems in which steam is used for process, some make-up water is required, and the use of such inhibitors is ruled out. In these cases, it is necessary to remove dissolved oxygen from the feedwater through the use of a deaerator or feedwater heater. This is then followed by treatment to raise the pH value above 10.5, and treatment with sodium sulfite, to remove the last traces of dissolved oxygen.

The corrosion problem does not diminish as operating pressure and make-up rate go up, but the scale problem increases in importance. Scale is controlled by external softening and by internal treatment, usually with phosphates and organic dispersants, to precipitate residual traces of calcium.

The mechanical contractor is responsible for providing a complete chemical treatment system for the boilers, condenser-water system, chilled-water system, and hot-water heating system. These systems are complete in all respects and must include all piping, electrical work, controls, valves, hose bibbs, pumps, chemicals, test kits, etc. It is recommended that chemical-treatment-system manufacturers be contacted to determine the actual cost of a specific chemical treatment system. However, historical data indicate that for multi-million-dollar jobs, the cost of the water-treatment system might be approximately 0.2% of the total job cost. This percentage should only be used in schematic-type estimates. For bypass-type feeders of 2 to 10 gal capacity, 3 man-hours is a reasonable estimate for handling and erection.

PUMPS

In heating and cooling systems, water flow is normally produced by pumps. Centrifugal pumps are most commonly used in heating and cooling systems.

Centrifugal Pumps

The centrifugal pump is a machine that adds energy in the form of velocity to flowing water. All centrifugal pumps are built with the suction entrance directed toward the center of the pump. The suction entrance is sized to give the water a suitable velocity as it enters the impeller.

The water reaches the inlet edges of the impeller blades at a certain angle relative to the blades. The inlet edges of the blades are bent backward so that they face exactly in the direction of the water. This allows the blades to receive the water without a shock. During the flow along the impeller, from the impeller entrance to its exit, the water has a certain velocity relative to the impeller. The passage areas are designed so that this relative velocity is small, to minimize friction loss. At the impeller exit, an appreciable pressure has been produced as a result of the "centrifugal effect." Further, the water has a high absolute velocity, which is practically the velocity of the tips of the impeller blades.

A passage surrounding the impeller, commonly referred to as a *volute* or *diffuser*, receives high-velocity water from the impeller. It is shaped to reduce the velocity of the water gradually, with minimal friction loss. This efficient decrease in velocity results in an increase in pressure. Hence, the pressure at the diffuser exit is appreciably higher than the pressure at the impeller exit, because of the conversion of velocity into pressure.

In the heating, ventilating, and air-conditioning industry, the most commonly used type of centrifugal pump has a radial or *francis* type closed impeller and a volute casing, and is single-stage. The pumps are either close-coupled or flexible-coupled to the motor and generally are end-suction, horizontal or vertical split, or in-line mounted. Centrifugal-pump estimating data are given in Estimating Charts 4-1 through 4-5.

Pump formulas

$$\text{Pressure} = \frac{\text{head} \times \text{specific gravity}}{2.31} \quad (4\text{-}1)$$

where the pressure is in pounds per square inch, and the head is in feet.

$$\text{Brake horsepower} = \frac{\text{pump capacity} \times \text{head} \times \text{specific gravity}}{3960 \times \text{pump efficiency}} \quad (4\text{-}2)$$

where the pump capacity is in gallons per minute, and the head is in feet.

Water at 62°F weighs 8.33 lb/gal and is given the specific gravity 1.0.

ESTIMATING CHART 4-1 Vertical-Split, Close-Coupled In-line Centrifugal Pumps (Bronze-fitted or All-Iron Construction). (Prices Include Open Dripproof Motor (1750-pm), Standard Mechanical Seal, and Companion Flanges.)

1. Material costs

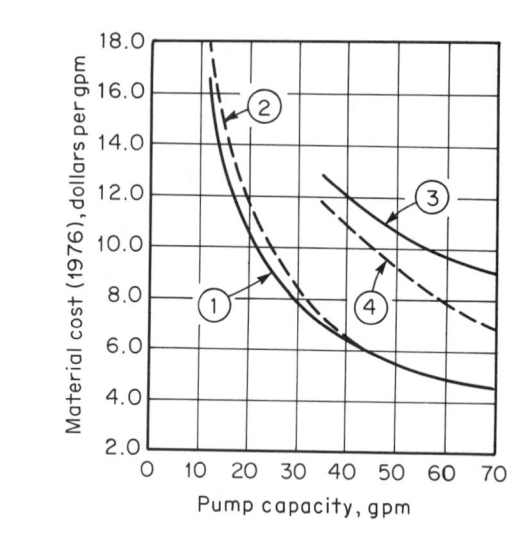

① Single phase – up to 30 ft head
② Three phase – 30 to 60 ft head
③ Single phase – up to 30 ft head
④ Three phase – 30 to 60 ft head

1 and 2 Motor size less than 1 hp
3 and 4 Motor size 1 hp and above

2. Labor man-hours

Capacity, gpm	Motor, hp	Pipe size, in.	Weight, lb	Man-hours
12	1/4	1	56	4
30	1/3	1-1/4	62	4
50	1/2	1-1/2	66	4
70	3/4	2	70	4
36	1	1-1/2	95	4.5
70	1-1/2	2	115	4.5

Figures 4-1 to 4-3 show typical hot-water heating, chilled-water, and condenser-water flow diagrams. These diagrams indicate the use of circulating pumps in the systems.

Pipe Sizes and Pump Capacities

Table 4-1 indicates the pipe sizes that should be used for particular pump capacities. These sizes are recommended as good practice based on fluid velocity in hydronic applications.

ESTIMATING CHART 4-2 Horizontal-Split, Close-coupled In-line Centrifugal Pumps (Bronze or All-Iron Construction). (Prices Include Open Drip-proof Motor, Companion Flanges, and Standard Mechanical Seal.)

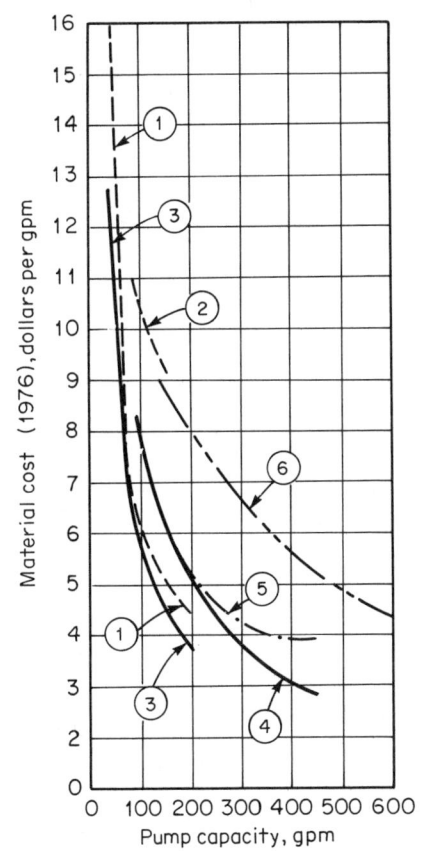

1. Material costs
2. Labor man-hours

Capacity, gpm	Motor hp	Pipe size, in	Weight, lb	Man-hours
40	3/4	1-1/2	155	6
75	1	2	195	6
160	1-1/2	3	195	6
80	2	2	230	6
200	2	4	225	6
100	3	2-1/2	265	9
90	5	2	215	6
150	5	3	275	9
140	7-1/2	2	270	9
350	7-1/2	4	410	12
240	10	3	355	12
450	10	4	365	12
140	15	2	455	12
300	15	4	455	12
190	20	2-1/2	505	12
400	20	4	485	12
250	25	3	565	16
450	25	4	585	16
400	30	3	700	16
600	40	3	735	16

Note:
Use the same material costs and labor man-hours for close-coupled end suction pumps with flanged suction and discharge.

① 1750 rpm, single phase, Up to 55 ft head
② 1750 rpm, single phase, 55 to 90 ft head
③ 1750 rpm, three phase, Up to 55 ft head
④ 1750 rpm, three phase, 55 to 90 ft head
⑤ 3500 rpm, three phase, 90 to 180 ft head
⑥ 3500 rpm, three phase, 180 to 380 ft head

Pump Capacities and Motor Horsepowers

Table 4-2 indicates the recommended motor horsepowers at various motor speeds for particular pump capacities and total heads. The following formula may be used to determine the motor size at a desired head and capacity:

$$\frac{\text{hp}_t}{\text{hp}_x} = \frac{(\text{gpm} \times H)_t}{(\text{gpm} \times H)_x} \tag{4-3}$$

where the subscript t refers to values obtained from the table; the subscript x refers to values for a specific pump; hp is the motor horsepower; and H is the total head in feet.

ESTIMATING CHART 4-3 Close-coupled, End-suction Centrifugal Pumps (Bronze-fitted Construction; Screw Suction and Discharge). (Prices Include Dripproof Motor and Standard Mechanical Seal.)

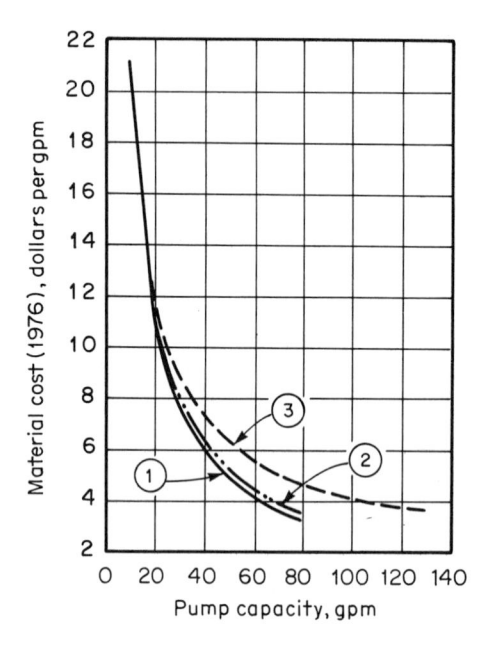

1. Material costs
2. Labor man-hours

Capacity, gpm	Motor hp	Suction size, in.	Weight, lb	Man-hours
10	1/4	1-1/4	48	4
20	1/3	1-1/4	45	4
30	1/3	1-1/4	53	4
10	1/2	1-1/4	51	4
40	1/2	1-1/2	58	4
60	1/2	2	59	4
20	3/4	1-1/4	50	4
80	3/4	2	60	4
30	1	1-1/4	56	4
40	1-1/2	1-1/2	63	4
100	2	2	77	4
130	2	2	77	4

① 1750 rpm, single phase, up to 30 ft head.
② 1750 rpm, three phase, up to 30 ft head.
③ 3500 rpm, single phase and three phase, from 30 to 110 ft head.

TABLE 4-1 Pipe Sizes and Pump Capacities

Pipe Size, in	Pump Capacity, gpm
1	Up to 15
1¼	5–20
1½	10–25
2	20–50
2½	30–75
3	55–135
4	115–275
5	200–500
6	330–800
8	700–1300
10	1250–1750
12	1600–2500

ESTIMATING CHART 4-4 Vertical-Split-Case, End Suction Base-mounted Centrifugal Pumps (1750 rpm; Bronze-fitted or All-iron Construction). (Pumps Include Flexible Coupling, Standard Mechanical Seal, and Fabricated Steel Base Plates; Prices Do Not Include Motors. For Motor Material Prices See Estimating Charts 4-6 and 4-7.)

1a. Material costs — UP to 55 ft Head – 1750 rpm

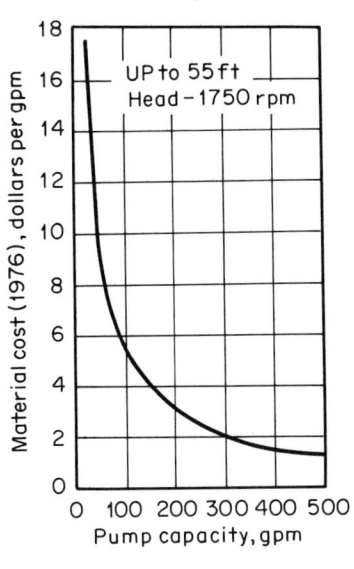

1b. Material costs — 55 to 110 ft Head – 1750 rpm

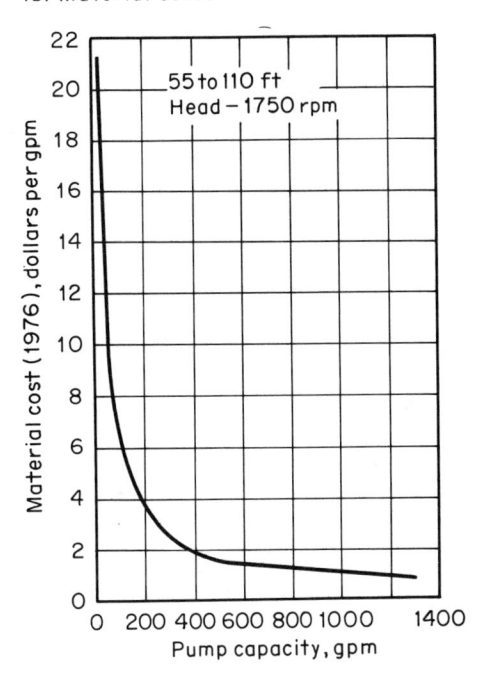

1c. Material costs — 110 to 160 ft Head – 1750 rpm

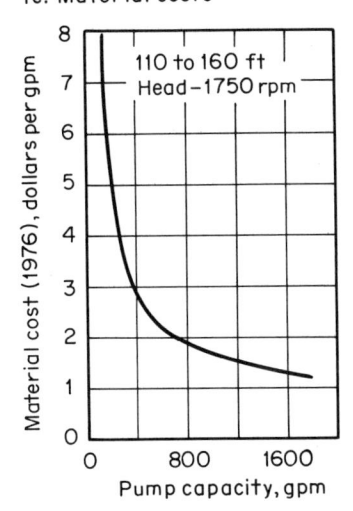

2. Labor man-hours
Refer to Estimating chart 4-5(2)

Notes applied to Estimating chart 4-4 (1c.)
For base-mounted centrifugal pump with enclosed double-suction impeller. Add to the material costs in this curve the following percentages.

Up to 500 gpm add 40%
500 to 800 gpm add 60%
800 to 1300 gpm add 50%
1300 to 1750 gpm add 65%

Example Determine the motor horsepower required for a pump delivering 500 gpm at a total head of 60 ft and a motor speed of 1750 rpm.

Solution From Table 4-2, at 500 gpm and a motor speed of 1750 rpm, the motor horsepower with a 90-ft head is 15 hp. Equation (4-3) gives

$$\frac{15}{\text{hp}_x} = \frac{500 \times 90}{500 \times 60}$$

Therefore, at a 60-ft head, the required motor horsepower is 10 hp.

ESTIMATING CHART 4-5 Vertical-Split-Case, End-Suction Base-mounted Centrifugal Pumps. (3500 rpm; Bronze-fitted or All-Iron Construction). (Pumps Include Flexible Coupling, Standard Mechanical Seal, and Fabricated Steel Base Plates; Prices Do Not Include Motors. For Motor Prices see Estimating Charts 4-6 and 4-7.)

1a. Material costs

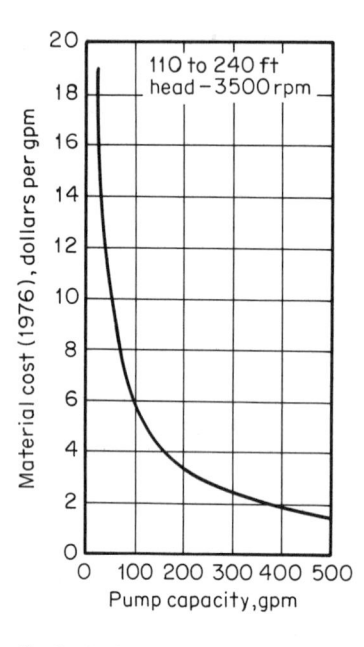

[Graph: 110 to 240 ft head – 3500 rpm; Material cost (1976), dollars per gpm vs Pump capacity, gpm (0 to 500)]

1b. Material costs

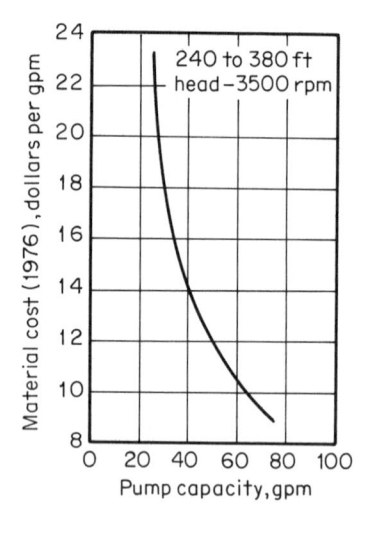

[Graph: 240 to 380 ft head – 3500 rpm; Material cost (1976), dollars per gpm vs Pump capacity, gpm (0 to 100)]

2. Labor man-hours
   For base-mounted centrifugal pumps with dripproof motors.

Suction size, in.	Capacity, gpm	Motor, hp	Weight, lb	Man-hours
1-1/2	25	3/4 2 5 10	132 176 195 317	6 6 6 12
2	50	1 3 7-1/2 15	160 206 215 343	6 6 6 12
2-1/2	75	1-1/2 5 10 20	172 260 327 468	6 9 12 12
3	135	2 7-1/2 10	186 387 566	6 12 16
4	275	3 10 15	216 411 667	6 12 16
5	500	5 15 20	290 507 731	9 12 16
6	800	20	571	16
8	1,300	25 40	773 1134	16 20
10	1,750	60	1428	24

For motor alignment: Up to 40 hp add 6 man-hours; 50 to 125 hp add 18 man-hours

Auxiliary Equipment | 71

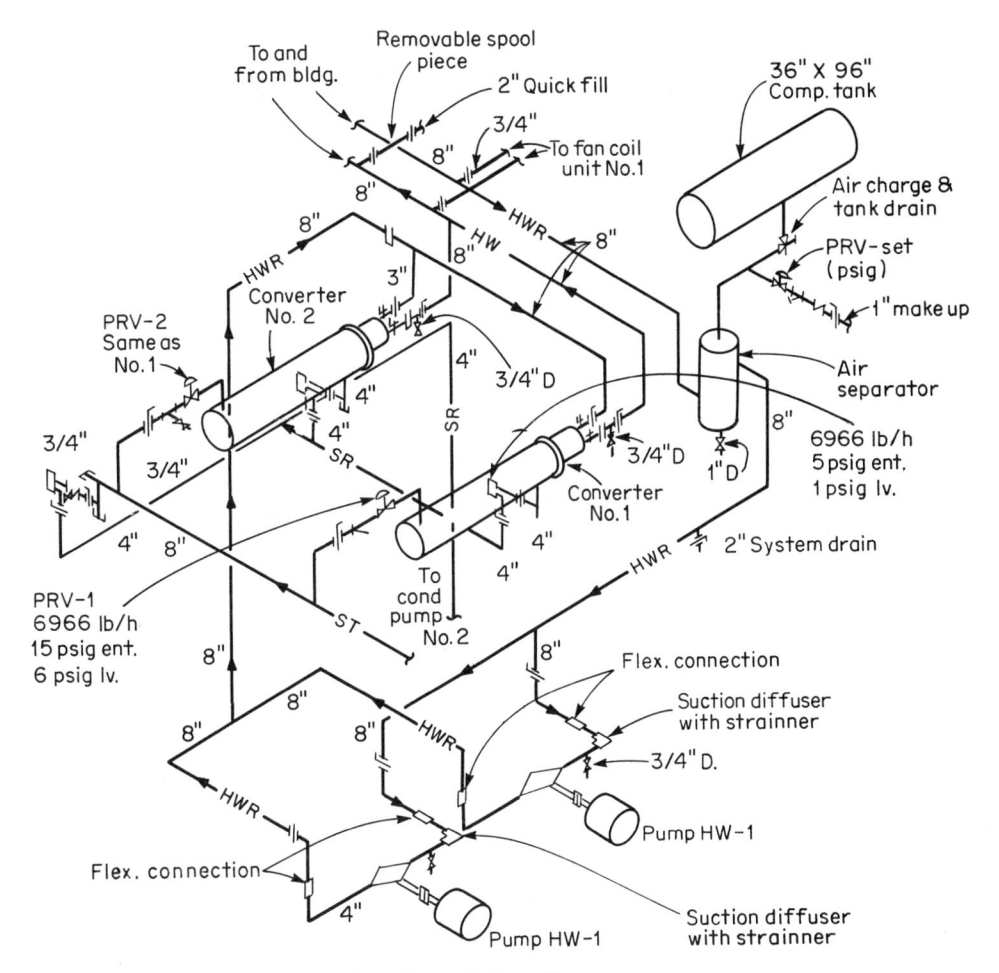

FIGURE 4-1 Hot-water heating piping diagram.

TABLE 4-2 Pump Capacities and Motor Horsepowers

Pump Capacity, gpm	Head, ft	Motor HP at 1150 rpm	Head, ft	Motor HP at 1750 rpm	Head, ft	Motor HP at 3500 rpm
Up to 15			Up to 24	¼–⅓		
20			25–100	⅓–2	100–450	2–10
25			30–110	½–3	110–380	3–15
50			35–160	¾–7½	160–380	7½–20
75			20–100	1–5	100–210	5–15
135			15–160	1–10	60–180	7½–15
275	20–30	2–3	20–160	2–15	120–220	15–30
500	10–50	3–7½	40–90	3–15		
800	20–40	5–10	40–160	10–30		
1250	20–60	10–20	55–160	25–60		
1750	30–75	20–40	80–180	60–100		
2500	25–75	20–50				

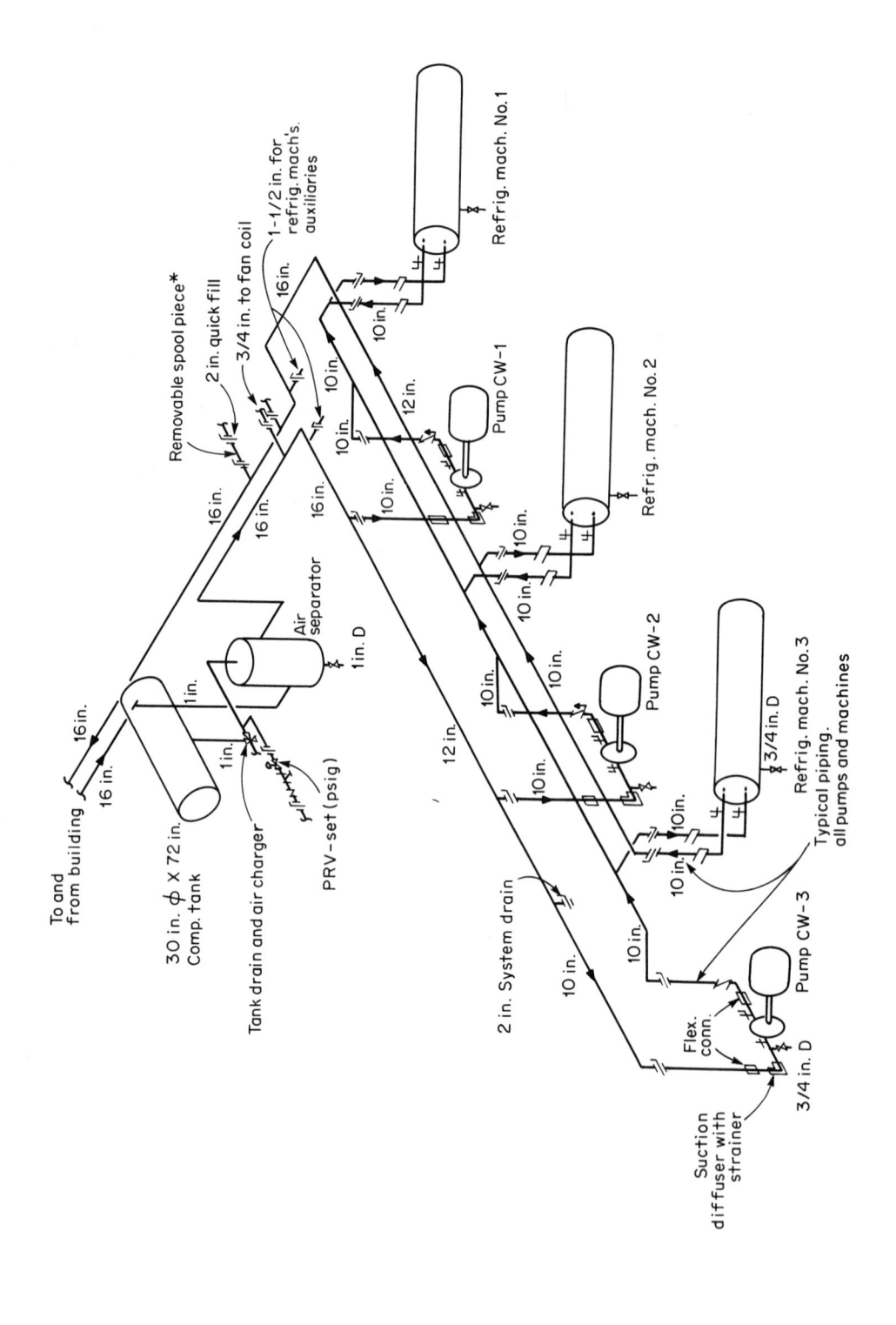

FIGURE 4-2 Chilled-water piping diagram. (*Spool is a piece of pipe flanged at both ends.)

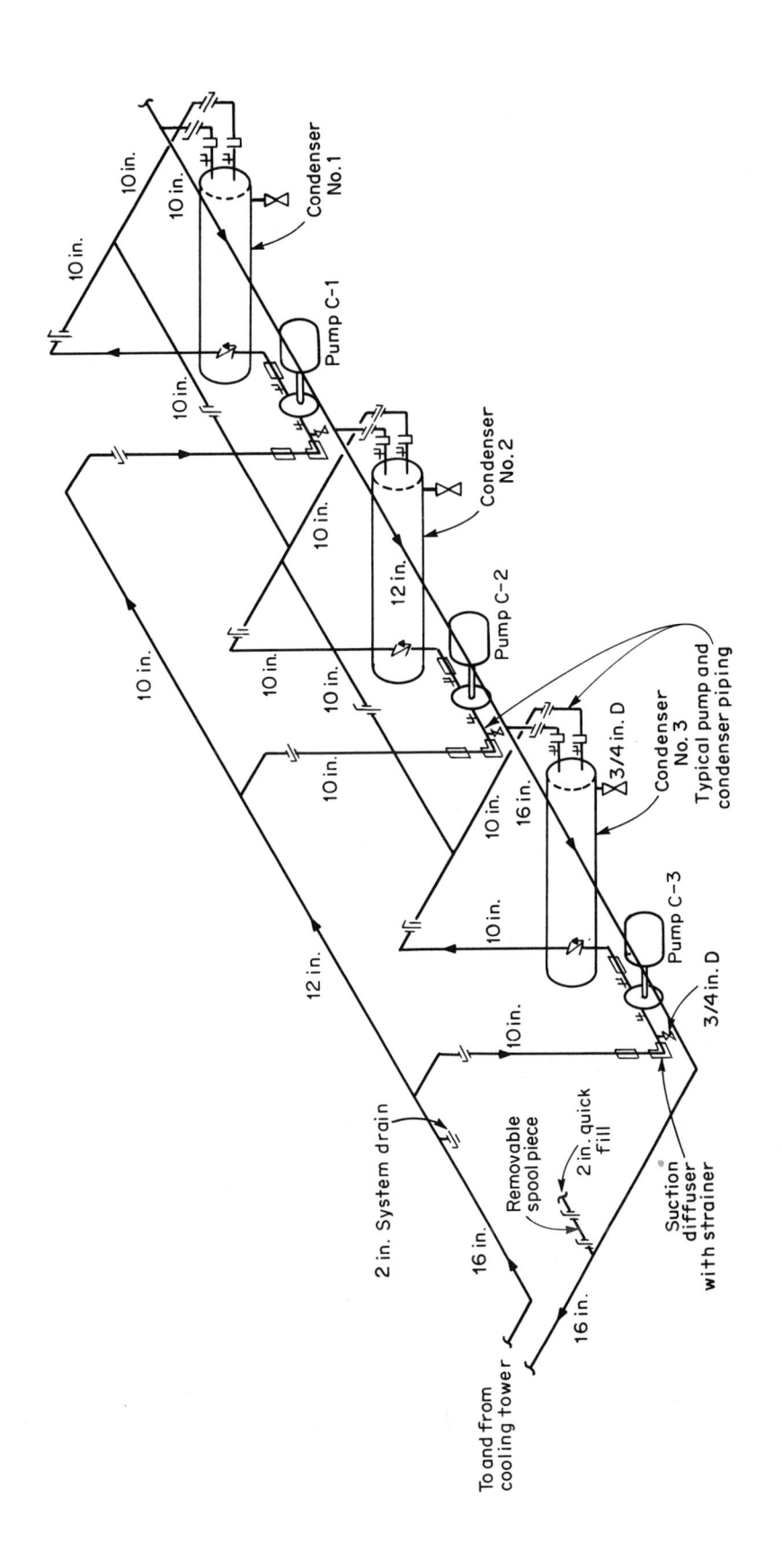

FIGURE 4-3 Condenser water piping diagram.

Standard Motor Horsepowers

The following are the standard motor horsepowers normally produced by electrical-equipment manufacturers:

Fractional hp	5	25	75
1	7½	30	100
1½	10	40	125
2	15	50	150
3	20	60	200

Material costs for dripproof and totally enclosed motors are given in Estimating Charts 4-6 and 4-7, respectively.

Material-Cost-Calculation Procedure

To see how to determine the material cost for a base-mounted centrifugal pump, consider a pump of capacity 500 gpm at a 120-ft head. The pump operates with a three-phase dripproof motor of 15 hp at 1750 rpm.

1. From the material-cost chart for 110 to 160 ft of head, find the material cost per gallon per minute at 500 gpm. It is $2.28/gpm [using Estimating Chart 4-4(1c)]. Therefore,

Material cost for pump = $2.28 × 500 = $1140.00

2. From the material-cost chart for three-phase dripproof motors, find the material cost per horsepower at 15 hp. It is $30/hp (see Estimating Chart 4-6c). Therefore,

Material cost for motor = $15 × 30 = $450.00

3. The total material cost (pump and motor) is

Total cost = $1140 + $450 = $1590.00

HYDRONIC SPECIALTIES

Expansion Tanks

In a closed recirculating system (hot-water heat, chilled water, combined heating and cooling), a given amount of air or gas space is required to accommodate water expansion and pressurization. Since water expands when heated and contracts when cooled in direct proportion to the temperature change, and since water is incompressible, a lack of expansion space means that any volume increase due to heating will cause an immediate and definite pressure increase. This is true even of chilled-water systems, which have limited expansion owing to their relatively narrow operating temperature range. An airtight expansion tank is the primary device used to

ESTIMATING CHART 4-6 Material Costs of Dripproof Motors for Base-mounted, Flexible-coupled Centrifugal Pumps. *(Standard Ball Bearing Motors.)*

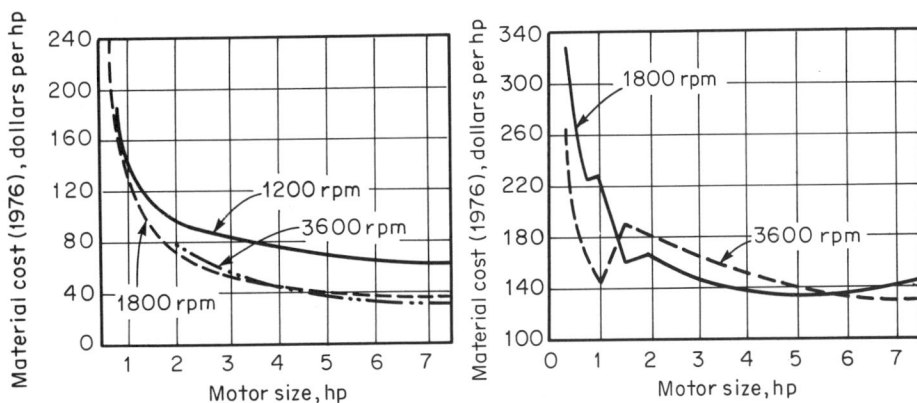

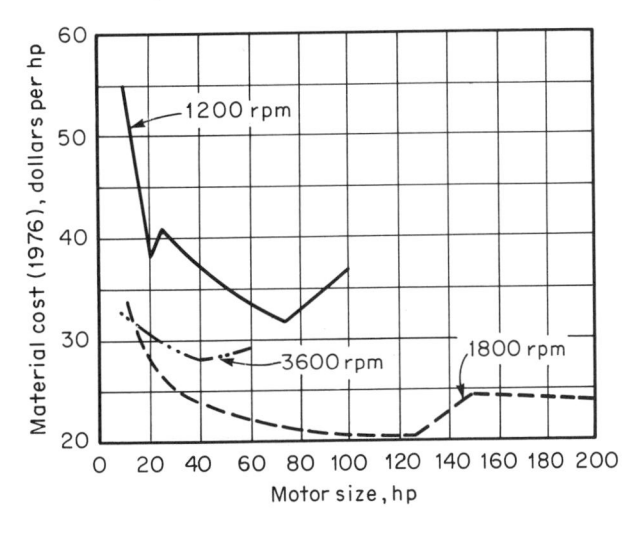

a. Dripproof motors – three phase, 60 Hz, 200, 208, or 230/460 V

b. Dripproof motors – single phase, 60 Hz, 115/230 V

c. Dripproof motor – three phase, 60 Hz, 200, 208, or 230/460 V

accommodate fluctuations in water volume within a closed system while maintaining a predetermined range of pressures, from the minimum cold-fill pressure up to (but not more than) the maximum working pressure of the system. The expansion tank acts as a spring on the system, keeping the pressure on it at all times.

The location of the pump relative to the expansion-tank connection determines whether the pump head is added to or subtracted from the system static pressure. This is because the expansion tank's point of connection in the system is the point of no pressure change for the system with regard to the pressures produced by pump.

ESTIMATING CHART 4-7 Material Costs of Totally Enclosed Motors for Base-mounted, Flexible-coupled Centrifugal Pumps. *(Standard Ball Bearing Motors.)*

a. Totally enclosed motors – three phase 60 Hz, 200, 208, or 230/460 V

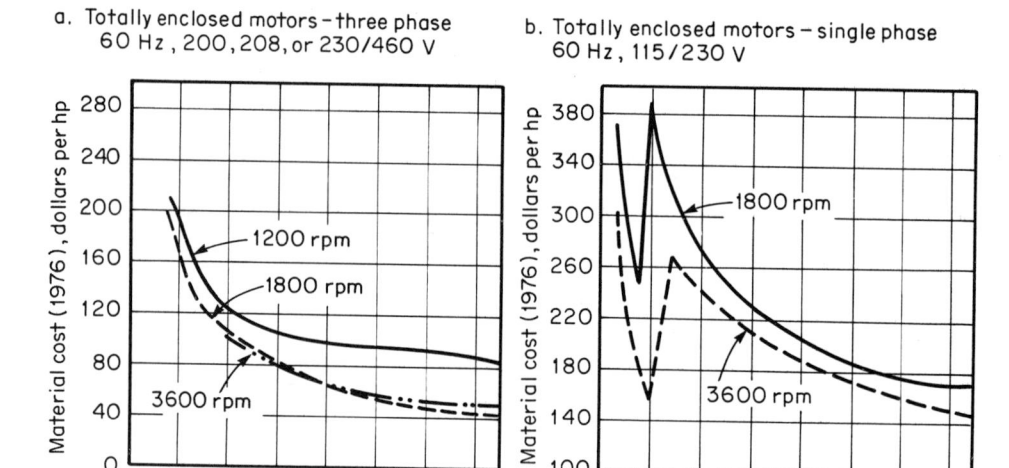

b. Totally enclosed motors – single phase 60 Hz, 115/230 V

c. Totally enclosed motors – three phase 60 Hz, 200, 208 or 230/460 V

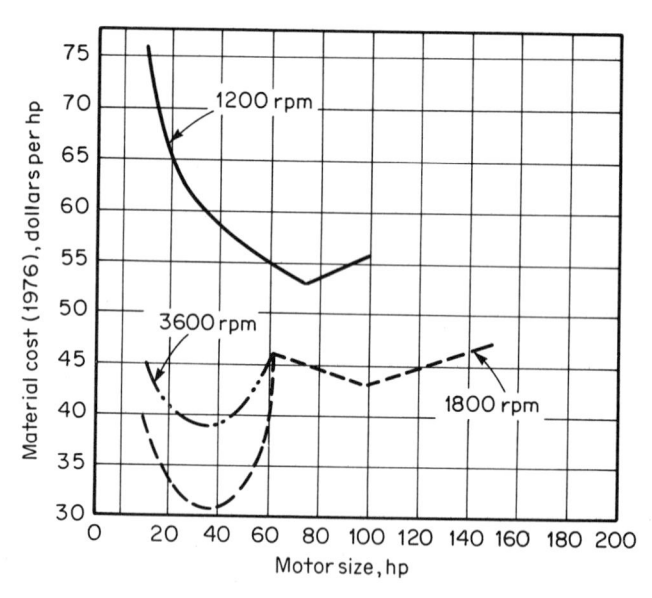

The point of connection of the expansion tank to a closed system is usually selected so that the pump discharges away from the expansion tank and the pump head adds to system pressures at all points.

Figure 4-4 shows one recommended pump location, relative to the expansion-tank connection. The tank is located close to the pump suction, the pump is discharging away from the boiler and the expansion tank, and the pump head appears as an increase in the pressure on the system.

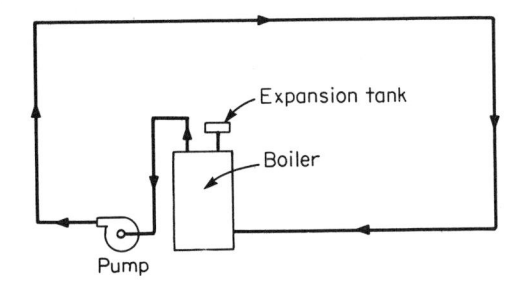

FIGURE 4-4 Recommended pump location for closed circulating water system.

Figures 4-1 and 4-2 show the locations of expansion tanks in hot-water heating and chilled-water systems. Estimating Chart 4-8 gives the estimating data for expansion tanks.

Air Separators

Air is roughly 80% nitrogen and 20% oxygen. Within a short time after the initial fill, the air in a properly operating closed system begins to lose oxygen through oxidation. Unless fresh water is added, the gas within the system is predominantly nitrogen. If water at its maximum air solubility level is heated and/or reduced in pressure, nitrogen will be released and the water will remain in a deaerated condition. The point of nitrogen separation should be the connecting point to the expansion tank, i.e., the point of no

ESTIMATING CHART 4-8 Expansion Tanks. (Tanks Include Airtrol Tank Fittings, Gauges, and Tapping.)

1. Material costs 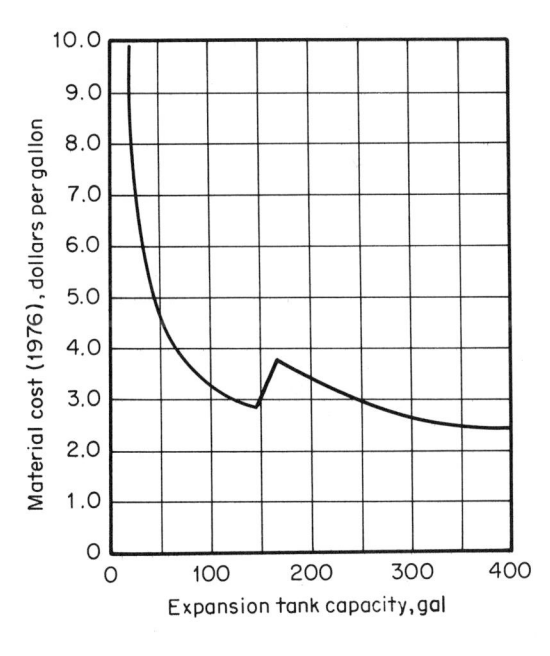
2. Labor man-hours

Capacity, gal	Dia. X length, in.	Weight, lb	Man-hours
15	13 X 35	69	3.0
24	13 X 51	90	3.0
30	13 X 62	104	3.0
40	16 X 53	122	3.0
60	16 X 77	160	3.0
80	20 X 68	203	4.5
100	20 X 82	234	4.5
120	24 X 72	288	4.5
144	24 X 83	319	6.0
163	30 X 60	375	6.0
202	30 X 72	485	6.0
238	36 X 84	495	6.0
270	30 X 96	680	6.0
306	30 X 108	760	9.0
337	36 X 84	795	9.0
388	36 X 96	805	9.0

ESTIMATING CHART 4-9 Tangential-flow Air Separators (Rol-Airtrol).

1. Material costs

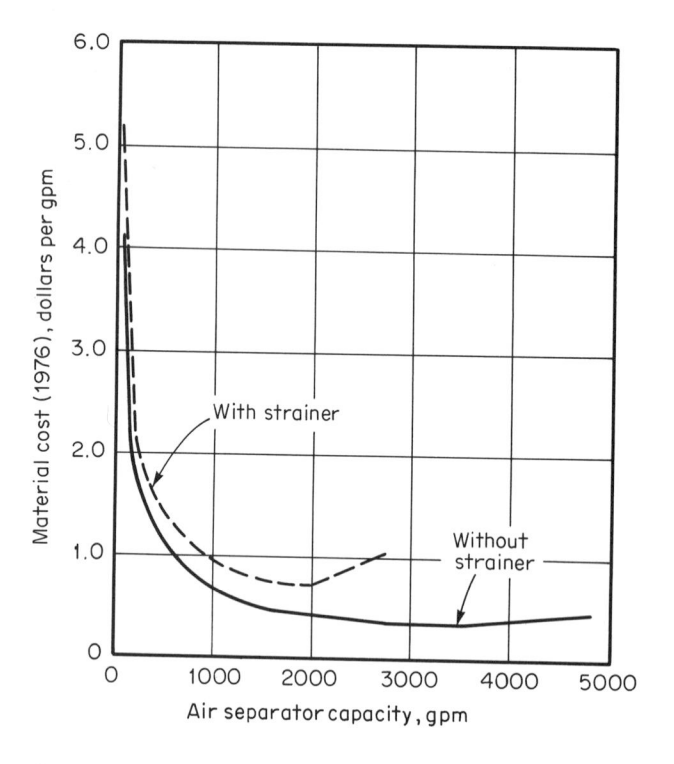

[Graph: Material cost (1976), dollars per gpm vs. Air separator capacity, gpm. Two curves shown: "With strainer" (dashed) and "Without strainer" (solid).]

2. Labor man-hours

a. With strainer

Capacity, gpm	Diameter X length, in.	Weight, lb	Man-hours
56	7 X 16	55	3
90	8 X 17	65	3
170	11 X 26	87	3
300	13 X 31	148	3
500	16 X 36	230	3
700	18 X 43	315	3
1,300	24 X 54	515	4.5
2,000	30 X 65	900	4.5
2,750	36 X 77	1,275	6

b. Without strainer

Capacity, gpm	Diameter X length, in.	Weight, lb	Man-hours
56	7 X 16	50	3
90	8 X 17	60	3
170	11 X 23	75	3
300	13 X 28	95	3
530	16 X 33	170	3
900	18 X 39	220	3
1,300	24 X 51	405	4.5
3,600	30 X 61	715	4.5
4,800	36 X 73	870	4.5

pressure change. An air separator should be installed at this point, so that air can be collected and raised into an expansion tank installed above (see Figs. 4-1 and 4-2).

The tangential-airflow-pattern air separator (Rol-Airtrol, with or without strainer) is commonly used in heating and cooling systems. Estimating Chart 4-9 indicates the estimating data for tangential-flow air separators.

5

HEAT-DISTRIBUTION EQUIPMENT

GENERAL   Heat should be introduced at, or directed toward, the local area of maximum heat loss, in order to offset or counteract this loss. In heating systems, heat-distribution units are used to supply heat to a space at the areas of greatest heat loss within that space; such areas include windows, cold walls, exposed floor slabs, and doorways. The heating medium may be steam, hot water, gas, oil, or electricity.

Many types of heat-distribution units are used in heating systems. Usually, these units are of the following types:

1. *Natural-convection units*, including cast-iron radiators, convectors, baseboard, and finned-tube radiation.

2. *Forced-convection units*, including unit heaters, duct reheat coils, and central heating-ventilating units (see Chap. 7) as well as certain types of fan-coil units and unit ventilators, normally used where the service is heating (see Chap. 6).

3. *Radiation units*, including infrared heaters and panel systems. These units transfer some heat by convection.

Heat-distribution units may be used in combination within a system to produce optimum results and to take full advantage of the characteristics of each type. However, units with different performance characteristics should not be installed in the same zone, e.g., cast-iron radiators should not be mixed with fin-tube baseboards.

To estimate material costs and labor man-hours, use the cost charts and man-hour tables that follow. The man-hour figures include handling and erecting the unit (see Chap. 8 for final hookup). Material costs are based on 1976 prices and are given for reference only.

NATURAL-CONVECTION UNITS

Radiators, convectors, baseboard, and finned-tube units are the types of natural-convection heat-distributing units commonly used in steam or low-temperature water heating systems and some medium-temperature water systems. These types of units emit heat by a combination of radiation to the space and convection to the air within the space. In general, natural-convection units are installed in the areas of greatest heat loss, i.e., under windows, along cold walls, and at doorways and stairways.

The output ratings of these units are expressed in Btu's per hour, MBH (1 MBH = 1000 Btu/h), or in square foot of EDR. (An EDR, or equivalent direct radiation, is the amount of heat emitted from one square foot of heating surface. It is equal to 240 Btu/h for steam and 150 Btu/h for hot water.)

Radiators (Small-Tube Type)

Radiators are generally constructed of sectional cast-iron radiation of the column, wall-tube, large-tube, or small-tube type. Small-tube-type radiators, with a length of only 1¾ in per section, are the only type now being manufactured, because these units occupy less space than the older column and large-tube units. Estimating Chart 5-1 gives the estimating data for small-tube radiators.

Convectors

Factory-assembled convectors, consisting of a ferrous or nonferrous heating element and an enclosure or cabinet, are widely used. They are made in a variety of sizes and types, including freestanding (Fig. 5-1), wall-hung, and recessed. Normally, the enclosures have outlet grilles and inlet openings, or vice versa. Convectors are usually operated with gravity-circulated

ESTIMATING CHART 5-1 Small-Tube Radiators.

1. Material costs for free-standing radiators
2. Labor man-hours

Number of tubes	Rating based on steam Sq.ft. EDR	Rating based on steam Btu	Dimensions, (L X W X H), in.	Man-hours
4	1.6	384	7 X 4.5 X 19	2
	1.8	432	7 X 4.5 X 22	2
	2.0	480	7 X 4.5 X 25	2
6	2.3	552	10.5 X 8 X 19	2
	3.0	720	10.5 X 8 X 25	2
	3.7	888	10.5 X 8 X 32	2

Note: Add 10% for recessed-cabinet-type radiators

Heat-Distribution Equipment | 81

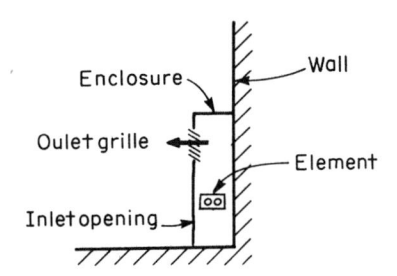

FIGURE 5-1 Free-standing convector

air. The air enters the enclosure below the heating element, is heated in passing up through element, and leaves the enclosure through the outlet grille located above the heating element. The heating medium may be steam, hot water, or electricity.

Estimating Charts 5-2 and 5-3 give the estimating data for cabinet convectors. The following example describes the use of Estimating Chart 5-2.

Example Find the material cost for a convector 36 in long, 24 in high, and 6 in wide.

Solution Use the 6-in-width enclosure-material chart. Find the material costs per foot of enclosure length for an element and an enclosure 24 in high at the 36-in enclosure length. The costs are $9.43 for the element, and $15.37 for the enclosure. Therefore,

Convector cost = 3 ft × $(9.43 + 15.37) = $74.40.

Baseboard Units

Baseboard units are designed for installation along the bottoms of walls. They may be made of cast iron with the front face directly exposed to the room, or with a finned-tube element in a sheet-metal enclosure. They operate with gravity-circulated room air.

Cast-Iron Radiant Baseboard (Steam or Hot Water)

These units are provided with air openings at the top and bottom to permit the circulation of room air over the extended heating surfaces on the backs of the units (see Fig. 5-2). They are made of cast-iron sections in 12-, 18-, and 24-in lengths. For lengths greater than 2 ft, sections are joined together with push nipples and tie bolts. Lengths up to and including 6 ft, in 6-in increments, are factory assembled. Longer lengths are shipped in two or more subassemblies for easy assembling on the job. These units are available in 7- and 10-in heights.

An estimated material cost, as of 1976, is $2.76/ft^2 of EDR, or $7/ft for 7-in-high panel, and $10/ft for 10-in-high panel. The labor required for handling and erection is 0.4 man-hours/ft.

ESTIMATING CHART 5-2 Free-standing Cabinet-type Convectors (Steam or Hot Water). Elements Are Copper Tubes and Nonferrous Fins (18-gauge Steel Enclosure for Tops and Fronts; 20-gauge for Backs and Ends, including Outlet Grilles).

1. Material costs

    a. 4 in. width enclosures

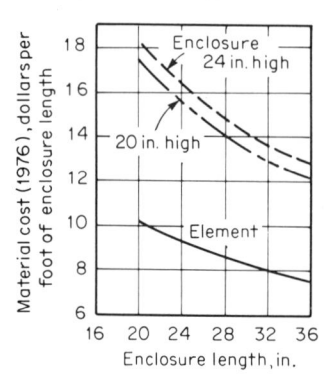

    b. 6 in. width enclosures

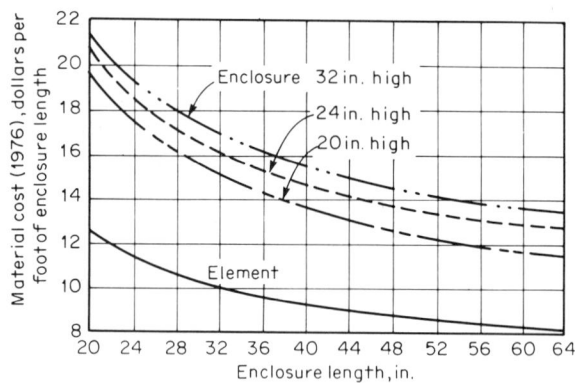

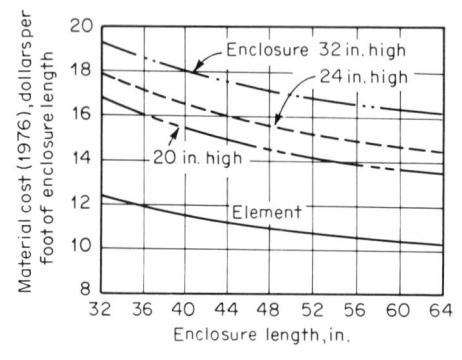

    c. 8 in. width enclosures

    d. 10 in. width enclosures

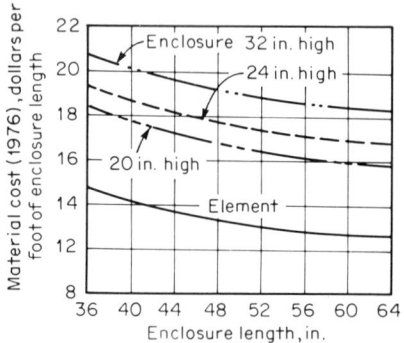

2. Labor man-hours

Sq. ft. EDR (steam)	Dimensions (L X W X H), in.	Man-hours
12.0	24 X 4 X 20	3.0
17.0	28 X 4 X 24	3.0
20.0	32 X 6 X 20	3.5
33.5	36 X 6 X 24	3.5
58.5	48 X 8 X 24	4.0
73.5	56 X 8 X 32	4.0
92.0	56 X 10 X 24	4.5
101.5	64 X 10 X 32	4.5

ESTIMATING CHART 5-3 Electric Cabinet Convectors (Single Phase; 208, 240, or 277 volts; with All Basic Controls).

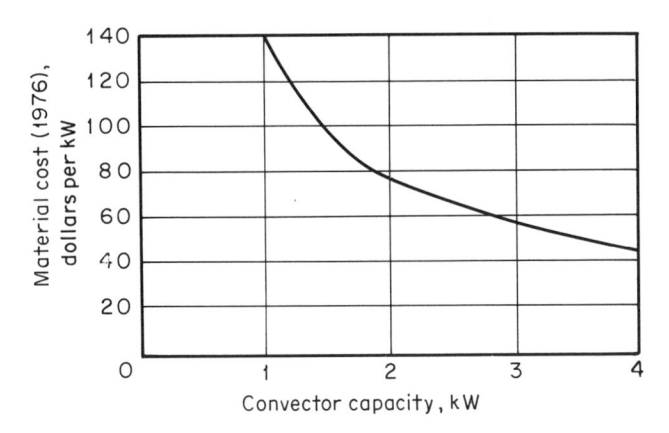

1. Material costs

[Graph: Material cost (1976), dollars per kW vs Convector capacity, kW]

2. Labor man hours: Use the same labor units for steam or hot water convectors.

Finned-Tube Baseboard

These units have a nonferrous finned-tube heating element that is concealed by a long, low, sheet-metal enclosure or cover (up to 10 in high). The heating medium may be steam, hot water, or electricity.

Steam or hot-water finned-tube baseboard. Factory-assembled units include copper tubing, aluminum fin elements, and 8-in-high steel enclosures. See Estimating Chart 5-4 for estimating data.

Electric baseboard heaters. Single-phase 120/208- and 240/277-V units include heater, enclosure, and basic control. Each linear foot of element has a capacity of 250 W. For labor, estimate 0.30 man-hours/ft. For material costs, see Estimating Chart 5-5.

Finned-Tube Units

These units are fabricated from metal tubing with metal fins bonded to the tube, and they are generally installed in an enclosure or cover. They operate with gravity-circulated room air.

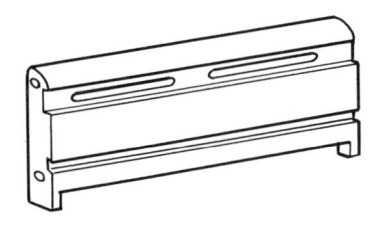

FIGURE 5-2 Cast-iron radiant baseboard unit.

ESTIMATING CHART 5-4 Finned-Tube Baseboard (Steam or Hot Water).

1. Material costs

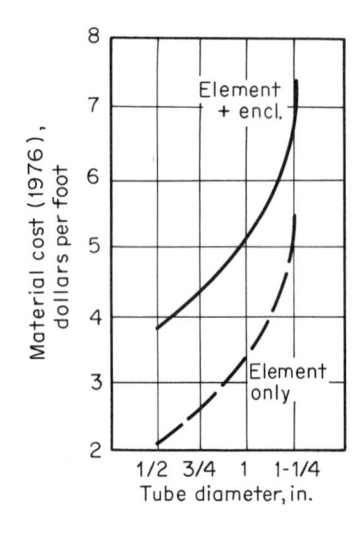

2. Labor man-hours

Tube diameter, in.	Man-hours per foot
1/2	0.35
3/4	0.35
1	0.40
1-1/4	0.40

Finned-tube elements are available in several tube sizes (either steel or copper), with fins of various sizes, spacings, and materials (steel or aluminum). They can be installed in one, two, or three tiers, as required. The elements can be used with either steam or hot water.

Commercial units are provided with enclosures in a variety of architectural designs. Utility units are provided with an expanded-metal cover for minimal protection.

For estimating data, see Estimating Chart 5-6.

FORCED-CONVECTION UNITS

Forced-convection heat-distribution units are usually used where heating-capacity requirements and the volume of heated space are too large to be handled by natural-convection units. Unit heaters and duct reheat coils are discussed in this section.

ESTIMATING CHART 5-5 Electric Baseboard Heaters.

1. Material costs

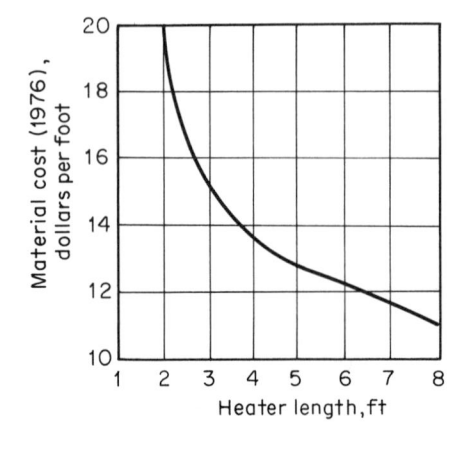

ESTIMATING CHART 5-6 Finned-Tube Units.

1. Material costs

   a. Elements 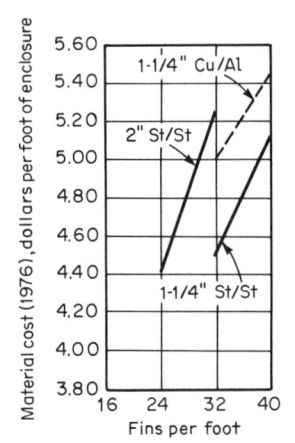
   b. Enclosures (slope type) without damper or back panel 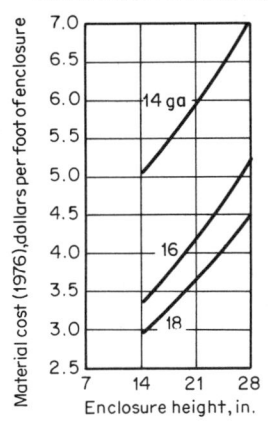
   c. Enclosures (round front type) without damper or back panel.

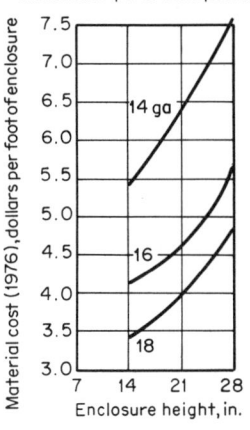

   St/St – steel pipe and steel fins
   Cu/Al – copper tube and aluminum fins

   d. Dampers and back panels

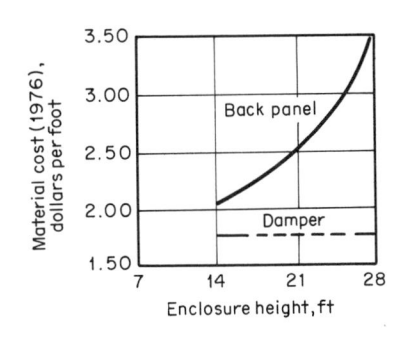

   Use for commercial-type enclosures:

   14" high enclosure for 1 tier
   21" high enclosure for 2 tiers
   28" high enclosure for 3 tiers

   For multi-tiers the unit material cost for elements – number of tiers X price per foot of one tier element

   e. Enclosures (flat-top type)

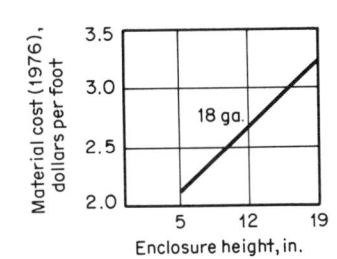

   Use for utility type enclosures:

   5" high enclosure for 1 tier
   12" high enclosure for 2 tiers
   19" high enclosure for 3 tiers

2. Labor man-hours

No. of tiers	Enclosure type	Man-hours per foot
1	Commercial	0.60
2		0.80
3		1.00
1	Utility	0.40
2		0.60
3		0.80
1	Bare	0.30
2		0.45
3		0.60

TABLE 5-1 Types of Unit Heaters

By Heating Medium	By Type of Fan	By Arrangement of Elements
Steam	Propeller	Drawthrough
Hot water	Horizontal blow	Blowthrough
Gas-fired	Downblow	
Oil-fired	Centrifugal	
Electricity	Cabinet (floor mounted)	
	Cabinet (suspended)	

Unit Heaters

The principal function of a unit heater is to heat a space. The major components of a unit heater are a fan and motor, a heating element, and an enclosure provided with an air inlet and a directional outlet. The mounting height and heat coverage can vary widely, depending on the type of directional outlet, existing obstructions and drafts, and the exposure areas. These factors require special attention during unit selection. Unit heaters may be classified in a number of different ways, as indicated in Table 5-1.

ESTIMATING CHART 5-7 Steam or Hot-Water Unit Heaters (propeller-fan type).

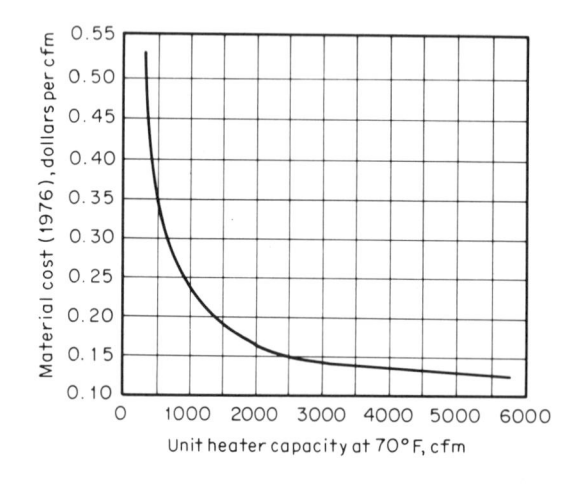

1. Material costs
2. Labor man-hours

cfm at 70°F	Capacity, MBH Steam	Capacity, MBH Water	Weight, lb*	Man-hours
280	17.4	3.5	38	4
545	38.7	19.1	42	4
815	60.5	34.6	57	4
1100	68.2	40.4	61	4
1215	87.6	55.7	73	4
1535	96.0	62.5	84	4
1760	125.7	84.9	102	4
2380	172.0	115.3	145	4
2800	185.0	125.0	147	4
3300	229.7	161.7	190	5
4100	256.3	181.1	195	5
4480	324.0	234.3	245	6
5660	355.5	264.9	250	6

*Weights for horizontal units.

Heat-Distribution Equipment | 87

Unit heaters are normally used for heating commercial and industrial structures, as well as corridors, lobbies, vestibules, and similar auxiliary spaces in all types of buildings.

Propeller-Fan Unit Heaters

Steam or hot-water standard horizontal or vertical types. All units include a nonferrous heating coil, fan, motor, casing, and louver diffuser. See Estimating Chart 5-7 for estimating data.

Gas unit heaters (horizontal). All units include a heat exchanger, burner, fan, motor, casing, louver diffuser, and control. See Estimating Chart 5-8 for estimating data.

Electric unit heaters. All units include an electric heater, fan, motor, casing, louver diffuser, and controls. For labor, use the same man-hour figures as for steam or hot-water propeller unit heaters. For material costs, see Estimating Chart 5-9.

Cabinet Unit Heaters (Centrifugal Fans)

See the material on fan-coil units in Chap. 6.

ESTIMATING CHART 5-8 Gas Unit Heaters (propeller-fan type).

1. Material costs
2. Labor man-hours

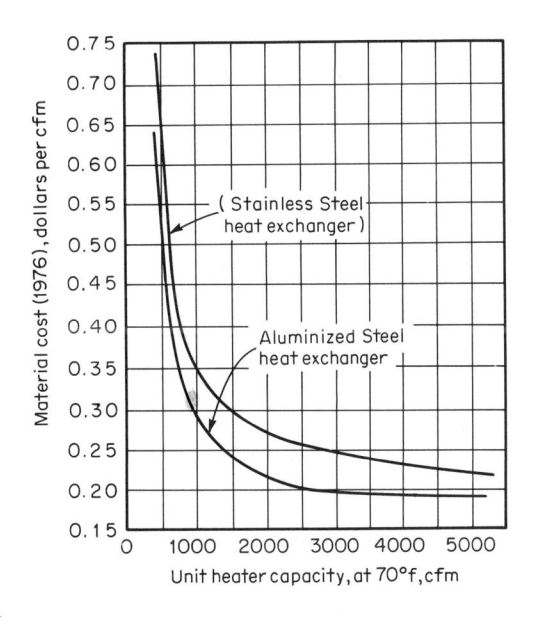

cfm at 70°F	Capacity, MBH Input	Capacity, MBH Output	Weight, lb *	Man-hours
400	30	24	100	4
650	50	40	100	4
1000	75	60	137	4
1300	100	80	165	5
1650	125	100	200	5
1950	150	120	240	5
2300	175	140	265	6
2600	200	160	285	6
2950	225	180	313	6
3300	250	200	340	6
3900	300	240	440	8
5200	400	320	530	8

Note: For oil-fired deduct 20%.

*Weights for steel heat exchangers.

ESTIMATING CHART 5-9 Electric Unit Heaters (propeller-fan type).

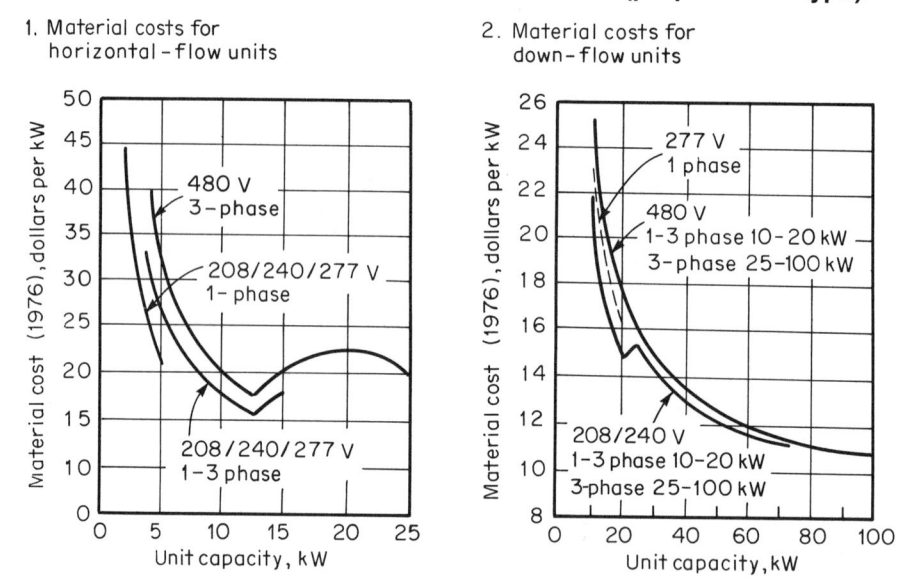

1. Material costs for horizontal-flow units
2. Material costs for down-flow units

Duct Reheat Coils

Duct reheat coils are used to reheat preconditioned air in a cooling system as a form of zone control for areas of unequal loading. A heating coil is normally inserted in the duct system, downstream of the cooling coil. The medium for heating may be hot water, steam, or electricity.

Steam or hot-water reheat coils. The standard coil is ⅝-in OD copper tubing, with 11 aluminum fins per inch. One or two rows may be used for a hot-water coil, but only one row for a steam coil. Prices are quoted for 12-in-fin-width coils at various finned lengths. See Estimating Chart 5-10 for estimating data.

Electric duct reheat coils. These units include an electric coil for either single-phase or three-phase operation, frame, overheat protection, and built-in relay. See Estimating Chart 5-11 for estimating data.

RADIANT HEATING

Panel heating and infrared heating are the most commonly used radiant heating systems. The panel heating system supplies heat through a relatively large, low-temperature source, whereas the infrared system has a relatively small, high-temperature source.

Heat-Distribution Equipment | 89

ESTIMATING CHART 5-10 Steam or Hot-Water Duct Reheat Coils.

1. Material costs

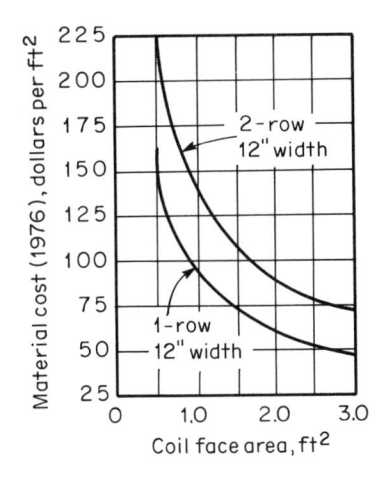

2. Labor man-hours

Length X width, in.	Face area, ft²	Weight, lb 1 row	Weight, lb 2 row	Man-hours
6 X 12	0.50	4.5	6.7	3
9 X 12	0.75	6.0	9.2	3
12 X 12	1.00	7.6	11.9	3
15 X 12	1.25	8.8	14.0	3
18 X 12	1.50	10.2	16.4	3
21 X 12	1.75	11.5	18.6	4
24 X 12	2.00	12.7	20.8	4
27 X 12	2.75	14.0	23.0	4
30 X 12	2.50	15.2	25.1	4
33 X 12	2.75	16.5	27.3	4
36 X 12	3.00	17.7	29.5	4

ESTIMATING CHART 5-11 Electric Duct Reheat Coils.

1. Material costs

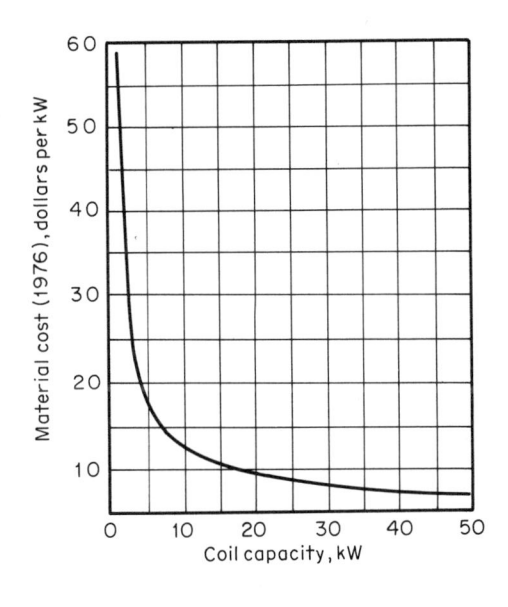

2. Labor man-hours

Width X height, in.	cfm	kW	Man-hours
6 X 8	150	1	3.0
8 X 8	200	2	3.0
10 X 8	250	3	3.0
12 X 8	300	4	3.0
12 X 12	400	5	3.0
14 X 12	500	7.5	3.0
18 X 12	600	10	3.0
20 X 12	750	15	3.0
24 X 12	800	20	4.0
28 X 12	1000	25	4.0
36 X 12	1200	30	4.0
28 X 18	1400	35	4.0
24 X 24	1600	40	4.5
36 X 18	1800	45	4.5
28 X 24	1900	50	4.5

Panel Heating In this system, a heated panel transfers heat to a room by radiation to objects and other surfaces in the room, and by convection to the room air. In panel heating, the heat-emitting surfaces may be ceilings, floors, and/or walls. The heating medium may be low-temperature hot-water heating, warm air, or electricity. The various forms of panel heating are:

1. Metal ceiling panels (prefabricated)
2. Piping embedded in ceilings, walls, or floors
3. Air-heated floors
4. Electric heating cables or prefabricated electric heating panels

In the United States, metal ceiling panels are most commonly used in panel heating and cooling systems.

Prefabricated Metal Ceiling Panels (Heating and Cooling Panels)

These panels are normally used in a system that provides heating and cooling. Forced ventilation air is required in such a system, especially in the summer. When metal panels are used for heating only, ventilation may or may not be required, depending upon local codes. The types of metal ceiling panels normally used in suspended perforated acoustical ceilings are:

1. Lightweight aluminum panels, usually 12 × 24 in, that are attached in the field to ½-in galvanized steel pipe coils.
2. Copper coils bonded to aluminum modular panels; the module may be as large as 36 × 60 in.

Metal-ceiling-panel systems are classified as combination air-water systems, i.e., two-pipe, three-pipe, or four-pipe systems, where the heating medium circulated through the coils is hot water, chilled water, or a combination.

The modular panels are normally furnished and installed by the general contractor. However, the mechanical contractor is responsible for the installation of piping loops and the final hookup. If the ceiling panels attached to galvanized steel coils are being used, the mechanical contractor is normally responsible for the installation of the coils and piping loops and the hookup (see Chap. 8).

Piping or Electric Cables Embedded Beneath Surfaces (Snow Melting)

Snow may be melted from driveways, walkways, and airport runways by circulating a nonfreezing liquid or solution through pipes embedded beneath the surface of the area to be protected. The heating medium normally used in such a snow-melting system is glycol solution, steam, or hot water. Electric heating cables are also used for this purpose.

Where steam is the basic heat source, a heat exchanger is usually employed to provide the heating water, to prevent condensate freezing, and to prevent the excessive thermal stresses that would be present if steam were used directly. Expansion tanks must be used in the various liquid-filled heating systems, to accommodate liquid expansion.

Liquid snow-melting systems include components such as heat exchangers, pumps, piping, coils, and controls. To estimate the cost of a system, the estimating procedure given in Chap. 1 should be followed. For large areas, the total cost may run from $4.50 to $5.00 per square foot of surface area for a complete snow-melting system. These costs are given for reference only, and should not be used for final estimates.

Infrared Heaters[1]

Infrared heaters are used to provide localized supplementary outdoor heat during cold weather and for snow melting. They may be located at department-store windows, under building entrance canopies, at bank tellers' drive-up windows, and at loading docks, grandstands, and hangars. Infrared heaters may be electric, gas-fired, or oil-fired. They consist of an infrared source or generator operating in a temperature range of 500 to 5000°F. Reflectors are usually used to control the distribution of radiation within specific patterns.

Electric infrared heaters. Electric infrared heaters are available with either metal-sheath, quartz-tube, or quartz-lamp heating elements located at the focus of two optically correct reflectors. The unit casing is of 18-gauge steel construction. See Estimating Chart 5-12 for estimating data.

[1]The *ASHRAE Handbook—1975 Equipment Volume* has been used as a reference, with the permission of ASHRAE.

ESTIMATING CHART 5-12   Electric Infrared Heaters.

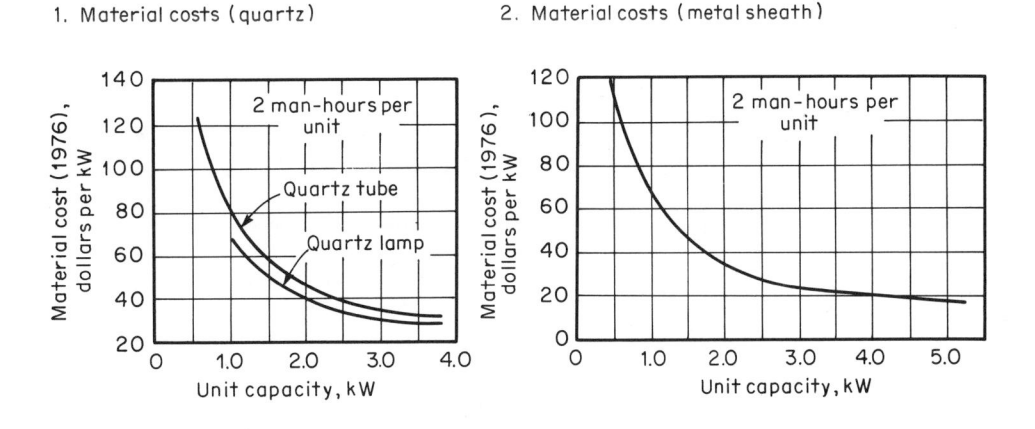

1. Material costs (quartz)
2. Material costs (metal sheath)

ESTIMATING CHART 5-13 Indirect Gas- or Oil-Fired Infrared Heaters.

1. Material costs

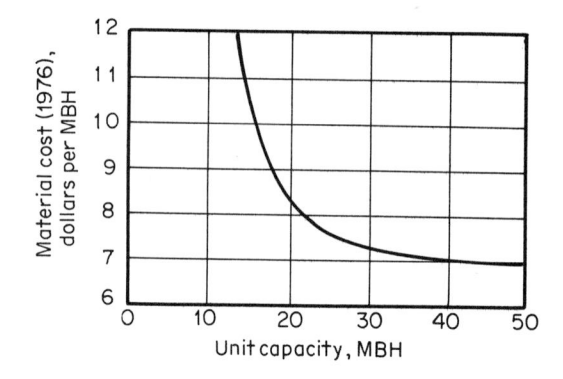

[Graph: Material cost (1976), dollars per MBH vs. Unit capacity, MBH. Curve descends from about 12 at 15 MBH to about 7 at 50 MBH.]

2. Labor man-hours

Capacity, MBH	Man-hours
10–60	3
60–120	4

Gas infrared heaters. These heaters are available in four basic types:

Indirect infrared heaters are internally fired, and the radiating surface is located between the hot gases and the load. Combustion takes place within the radiating elements, which may be metallic or nonmetallic tubes or panels. The heaters operate with surface temperatures up to 1200°F. These units generally require fume exhaust units (educators).

Direct-fired refractory infrared heaters are constructed of an open refractory surface; the infrared energy is produced by hot gases or flame striking the surface. Temperatures range from 1650 to 2800°F.

Porous refractory infrared heaters are constructed of a porous (or drilled) ceramic or metallic screen. The refractory material is enclosed, except for one major surface facing the load. A gas-air mixture enters the enclosure and flows through the refractory material to the exposed surface, and combustion occurs on this surface. Temperatures range from 1650 to 1800°F.

Catalytic-oxidation infrared heaters are somewhat similar to the porous refractory type, with the exception that the refractory material is glass wool and the radiating surface is a catalyst that causes oxidation to proceed without visible flames.

Oil infrared heaters. These units are similar to the gas-fired indirect units. Oil-fired units are vented.

Estimating Chart 5-13 gives the estimating data for indirect gas- or oil-fired infrared heaters. These units include a metallic-tube refractory, burner, and educator.

6

COOLING-DISTRIBUTION EQUIPMENT

GENERAL The function of cooling-distribution equipment is to introduce air into a conditioned space to obtain the desired indoor comfort conditions, i.e., temperature, humidity, noise, and cleanliness control.

The cooling-distribution equipment discussed in this chapter is:

1. *Forced-convection terminal units*, including unit ventilators, fan-coil units, and induction units.

2. *Air-diffusing units*, including supply-air outlets, return- and exhaust-air inlets, and terminal control units or boxes for all high-pressure air systems.

To estimate material and labor costs, use the cost charts and man-hour tables that follow. The man-hour figures include handling and erecting the unit (see Chaps. 8 and 9 for final hookup). Material costs are based on 1976 prices and are given for reference only.

FORCED-CONVECTION TERMINAL UNITS

Unit ventilators, fan-coil units, and induction units are the types of forced-convection cooling-distribution units commonly used in cooling systems. Some of these units can also be used for heating and ventilating.

Fan-Coil Units and Unit Ventilators

The fan-coil unit is a room terminal which is used in all-water systems and sometimes in air-water systems. Fan-coil units may be used with or without outside air where the service is heating only. Units that use 100% circulated air must rely upon infiltration as a source of ventilation air. Cooling and humidification are provided by circulating chilled water through a finned coil in the unit. Heating is provided by supplying hot water through the same coil or a separate coil, using a two-, three-, or four-pipe hydronic system. Electric heating or a separate steam coil may also be used.

The basic components of a fan-coil unit are an enclosure, finned-tube coil, fan-and-motor section, filter, condensate pan, manual fan-speed control, and automatic control of water flow, fan speed, or both (see Fig. 6-1). In addition, the unit may contain an auxiliary heating coil, usually of the electric, steam, or hot-water type. The fan-coil enclosure is equipped with a return-air inlet and discharge-air outlet. Units having provisions for introducing ventilation air are normally equipped with a damper capable of introducing up to 25% outside air through a wall opening.

Types of Fan-Coil Units

Room fan-coil units are available in vertical wall-mounted models or horizontal ceiling-mounted models, in nominal sizes of 200, 300, 400, 600, 800, and 1200 cfm.

Vertical Wall-Mounted Models

Vertical wall-mounted models are generally installed along an outside wall. Three types are available.

Type 1: models having no provision for introducing ventilation air. These models can be applied to recirculate 100% of the air within the space, with no provision within the space for positive ventilation. Type 1 units, commonly referred

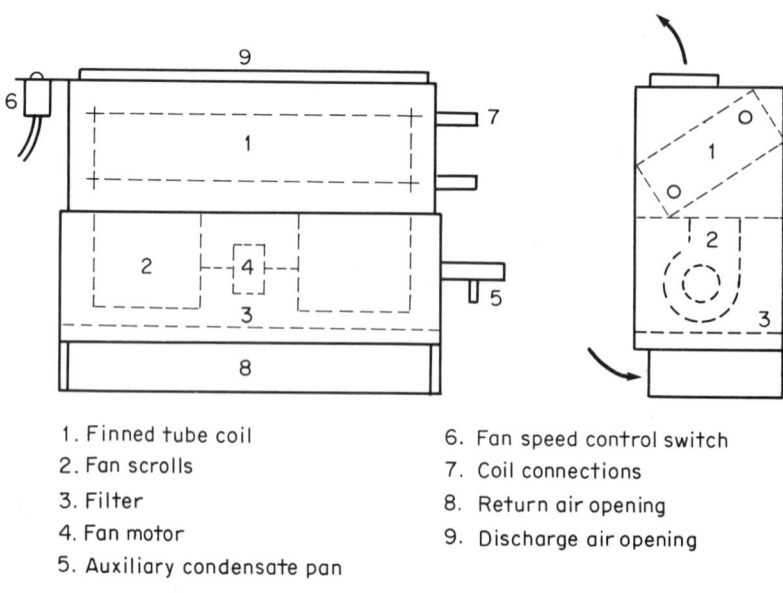

1. Finned tube coil
2. Fan scrolls
3. Filter
4. Fan motor
5. Auxiliary condensate pan
6. Fan speed control switch
7. Coil connections
8. Return air opening
9. Discharge air opening

FIGURE 6-1 Typical fan-coil air conditioner. (*Reprinted by permission from ASHRAE—1973 Systems Handbook.*)

ESTIMATING CHART 6-1 Fan-Coil Units. (Units include cabinet, fan, motor, coil(s), filter, grille, and drain pan.)

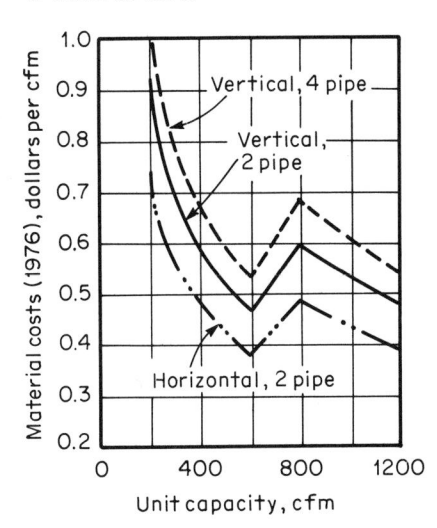

1. Material costs

2. Labor man-hours

Capacity, cfm	Dimensions (L X W X H), in.	Man-hours Vertical	Man-hours Horizontal
200	34 X 9 X 25	5	7
300	42 X 9 X 25	5	7
400	50 X 9 X 25	5	7
600	62 X 9 X 25	5	7
800	60 X 12 X 28	6	8
1000	72 X 12 X 28	6	8
1200	84 X 12 X 28	6	8

to as *fan-coil units* or *cabinet unit heaters*, are particularly adapted to space-heating applications. Type 1 units can also be applied in spaces where a separate duct system is used to supply ventilation air. In this case, the ventilation air is provided from a central heating-ventilating unit. For estimating data, see Estimating Chart 6-1.

Type 2: models having provisions for introducing up to 25% outdoor air. These models have a damper capable of introducing up to 25% outdoor air through a wall louver. (A *louver* is an air device consisting of sloping vanes which permit air to pass through but prevent the transfer of water.) Type 2 units can be used in spaces requiring heating, ventilating, and cooling. For estimating data, see Estimating Chart 6-1.

Type 3: models having provisions for introducing up to 100% outdoor air (as unit ventilators). These models have a damper to permit the introduction of up to 100% outdoor air during summer and winter months, to provide a complete cycle of heating, ventilating, ventilation cooling, or mechanical cooling, as required. Type 3 units are commonly referred to as *unit ventilators* or *heavy-duty fan-coil units*. They are available in sizes ranging from 200 to 1500 cfm, with automatic dampers and through-the-wall outdoor air intakes for up to 100% outside ventilation air. Sometimes these units can be constructed with a direct-expansion cooling coil; the condensing unit can be furnished as an integral part of the unit-ventilator assembly (see Estimating Chart 6-2), or it can be remotely located. The heating section may be a gas-fired heating furnace.

ESTIMATING CHART 6-2 Vertical-type Self-Contained Unit Ventilators (Heating and Cooling, Cooling Only, or Heating Only.)

1a. Material costs for unit with water-cooled condenser with self-contained thermostat (hot water heating)

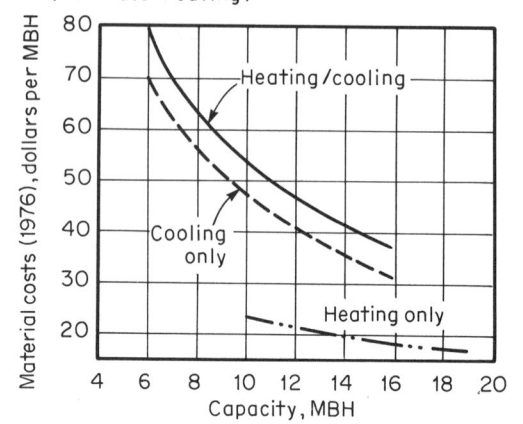

2a. Labor man-hours

cfm		MBH		Man-hours
Cooling	Heating	Cooling	Heating	
160	137	6.0	10.0	5
255	210	9.4	13.7	5
355	295	12.7	15.8	5
430	380	16.0	19.0	5

Dimensions: 43 in. X 9 in. X 26 in.

Horizontal Ceiling-Mounted Models

Horizontal ceiling-mounted models are similar to the three types of vertical models in construction and operation. With horizontal units, ventilating air is generally ducted to the unit plenum or a corridor supply. Horizontal units are frequently used where heat is not needed along outside walls, as in warm climates. These units are also desirable because they do not occupy any floor space. See Estimating Chart 6-1 for estimating data.

Induction Units

Induction units are normally used to distribute air for heating and cooling. They are normally installed at a perimeter wall under a window, but units designed for overhead installation are also available. During the heating season, the floor-mounted induction unit can function as a convector during off hours, with hot water to the coil and without primary air supply.

Induction units are similar to fan-coil units, but with air circulation provided by central high-pressure primary air handling a part of the load, instead of by a fan in each cabinet. The high-pressure air is supplied to the unit plenum. In an air-water induction unit (see Fig. 6-2), this primary air flows through the induction nozzles and induces secondary air from the room to and over the secondary water coil. The secondary air is either heated or cooled at the coil, depending on the room requirements and the season. The primary air and secondary air are mixed and discharged to the room. A drain pan is provided to collect the condensed moisture resulting from unusual latent loads of short duration.

A lint screen is normally provided at the inlet to the secondary coil. In addition, the unit is equipped with a balancing damper to allow adjustment

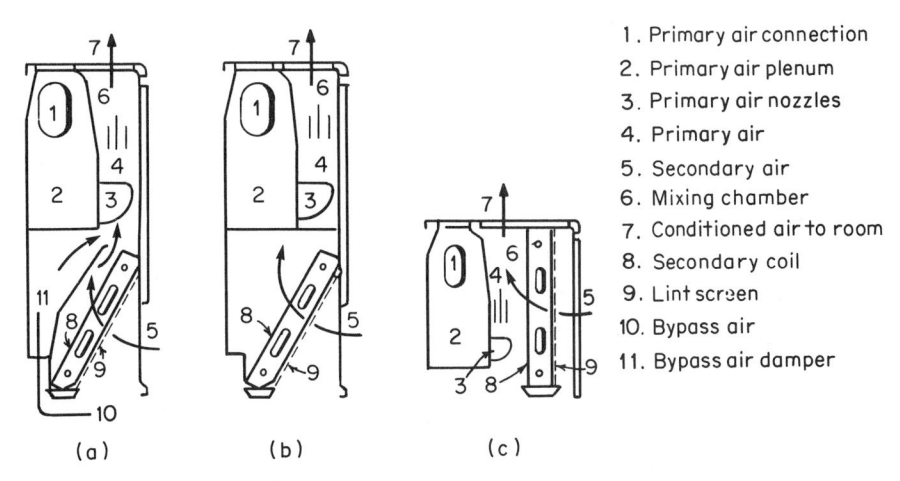

FIGURE 6-2 Typical air-water induction unit. (*a*) **Air bypass damper unit;** (*b*) **water control unit;** (*c*) **low-height water control unit.** (*Reprinted by permission from* ASHRAE—1973 Systems Handbook.)

1. Primary air connection
2. Primary air plenum
3. Primary air nozzles
4. Primary air
5. Secondary air
6. Mixing chamber
7. Conditioned air to room
8. Secondary coil
9. Lint screen
10. Bypass air
11. Bypass air damper

of primary air quantity. Induction units are installed in custom enclosures designed to blend with the architectural treatment, or in standard cabinets provided by the unit manufacturer.

For air-water induction-unit estimating data, see Estimating Chart 6-3.

Induction units can be used in all-air systems as reheat-type units or boxes (see Fig. 6-3), where individual reheat is accomplished by heating the secondary air. They may also be used as low-temperature reheat-type units, where the reheat coil is located in the primary airstream to provide for a very low supply-air temperature at high load conditions, and a higher supply-air temperature for lower load conditions. See Estimating Chart 6-4 for estimating data.

Induction units can also be used in air-water high-pressure dual-duct systems as dual-duct induction units or boxes, where the mixing boxes are designed to accommodate the full heating load and part of the cooling load.

ESTIMATING CHART 6-3 Air-Water Floor-mounted Induction Units. (Units Include Cabinet, Secondary Coil, and Inlet Screen.)

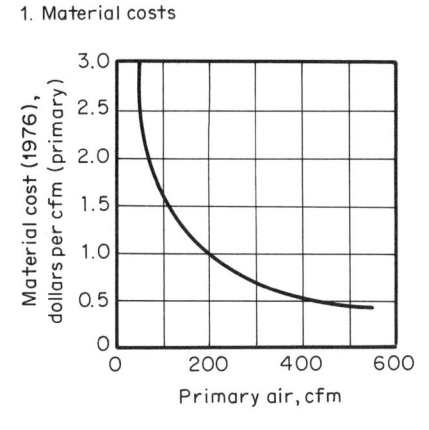

1. Material costs
2. Labor man-hours

Primary air, cfm	Secondary air, cfm	Dimensions (L×W×H), in.	Man-hours
50-250	90-510	36 × 9 × 22	3.5
200-450	360-860	48 × 9 × 22	4.0
300-500	515-985	60 × 9 × 22	4.0

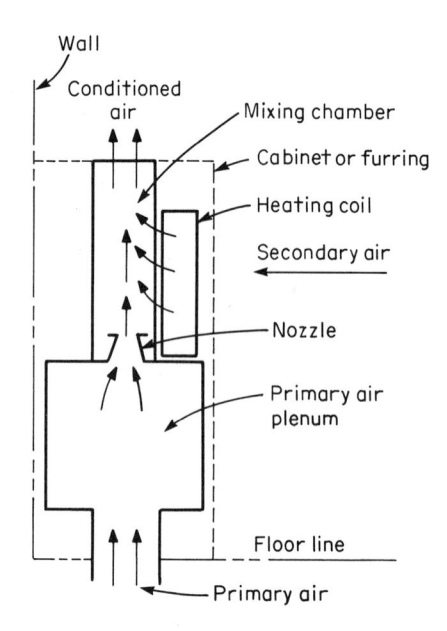

FIGURE 6-3 All-air induction unit. (*Reprinted by permission from ASHRAE—1973 Systems Handbook.*)

The mixing boxes contain induction nozzles and a secondary water coil which receives cooled secondary water to extend the unit capacity during the summer. During the off season, the secondary water system may be shut down, and the system may then be operated as a dual-duct system. During the winter, hot water may be provided to the secondary coil to permit gravity heating during off hours; i.e., the unit functions as a convector.

TERMINAL CONTROL BOXES FOR ALL-AIR HIGH-PRESSURE SYSTEMS

Terminal control boxes are used in all-air high-pressure systems to provide all or some of the following: pressure reduction, volume control, temperature modulation, air mixing, and sound attenuation. These boxes may be classified as on pages 99 to 101 (see Estimating Chart 6-5 for estimating data).

ESTIMATING CHART 6-4 All-Air Ceiling-mounted Induction Boxes with Hot-Water Reheat Coils.

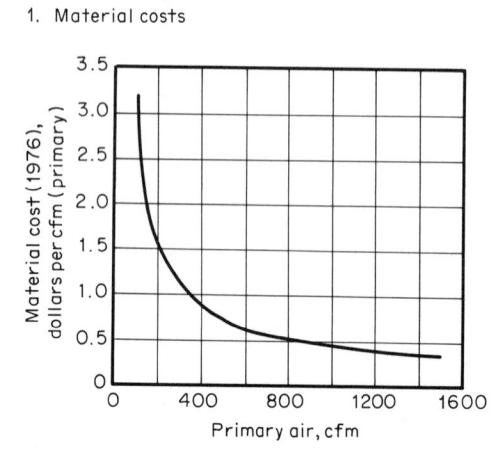

1. Material costs

2. Labor man-hours

Primary air, cfm	Dimensions (L × W × H), in.	Inlet diam., in	Man-hours
100 – 300	50 × 18 × 8.5	5	2.5
300 – 500	50 × 22 × 8.5	6	2.5
500 – 750	56 × 26 × 10.5	8	3.0
750 – 1500	59 × 38 × 12.0	10	3.0

Pressure-reducing Air Valves

Pressure-reducing air valves consist of series of operated vane sections mounted within a rigid casing, and jacketed to reduce air leakage between valve and duct as much as possible. They are generally equal in size to the low-pressure branch rectangular duct connected to the valve discharge. This arrangement provides minimum pressure drop with the valve fully opened.

ESTIMATING CHART 6-5 Terminal Control Boxes.

1. Material costs

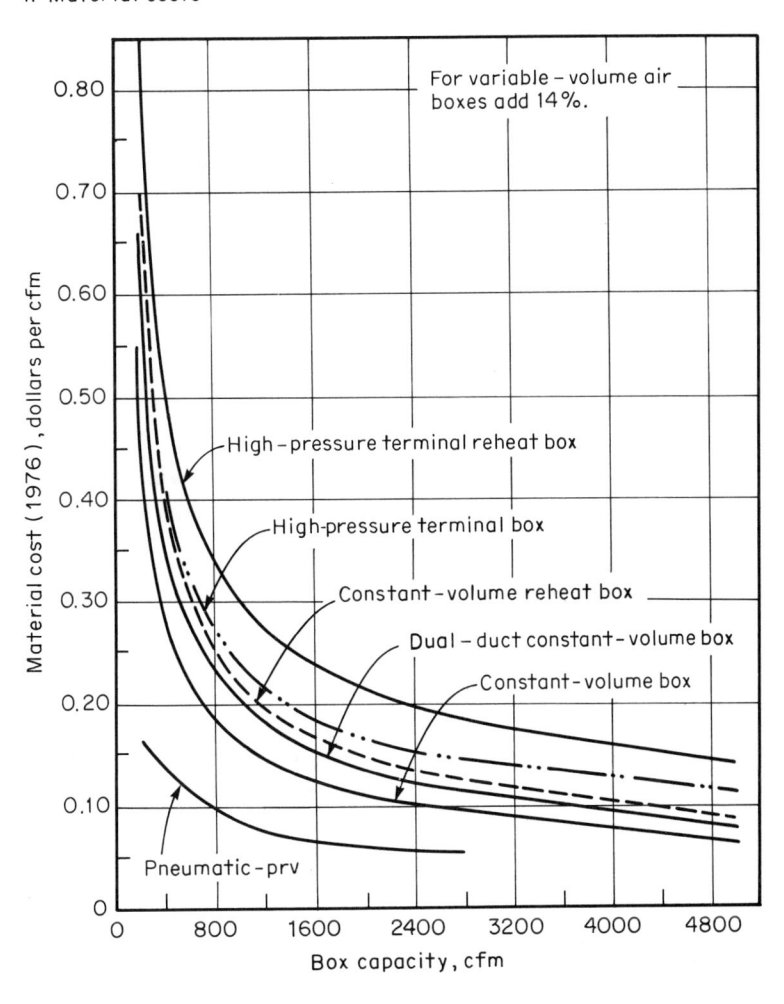

[Chart: Material cost (1976), dollars per cfm vs. Box capacity, cfm. Curves shown for: High-pressure terminal reheat box, High-pressure terminal box, Constant-volume reheat box, Dual-duct constant-volume box, Constant-volume box, Pneumatic-prv. Note on chart: "For variable-volume air boxes add 14%."]

2. Labor man-hours

Capacity, cfm	Man-hours
200 – 1000	2.50
1200 – 2200	3.00
2400 – 3600	3.50
3800 – 5000	4.00

Note: Terminal control boxes are normally provided by the sheet-metal contractor.

The valve inlet side is normally connected to a high-pressure trunk duct. Pressure reduction is produced by the partial closing of the valve and results in high pressure drop through the valve. This action generates noise, which must be attenuated in the low-pressure discharge duct. Volume control is obtained by adjustment of the valve, either manually or by a pneumatic or electric control motor, actuated by a pressure regulator or thermostat. (A pneumatic device is one that is operated by a supply of compressed air.)

High-Pressure Single-Duct Boxes

The high-pressure single-duct box normally consists of a pressure-velocity-reducing air valve, or a damper and a sound-attenuation chamber lined with thermal- and sound-insulating material, equipped with baffles for sound reduction. With boxes of this type, lining is not required in the low-pressure discharge duct. Volume control is obtained by adjustment of the box valve or the box damper, either manually or by a pneumatic or electric control motor, actuated by a thermostat or volume regulator. Air may be discharged through a single rectangular opening suitable for low-pressure branch-duct connection or through a supply outlet connected directly to the unit. Several outlets may be connected to the box by flexible ducts (see Fig. 6-4).

Sometimes a hot-water, steam, or electric reheat coil may be equipped with these boxes to accomplish individual reheat. Boxes so arranged are commonly referred to as *high-pressure single-duct reheat boxes*. A high-pressure single-duct box may be equipped with a means of flow control to overcall a thermostat to reset a pressure-velocity air valve or a dampering device to the desired change in airflow; a box with this arrangement is known as a *variable-air-volume box*.

Dual-Duct Mixing Boxes

The dual-duct mixing box is constructed in the same way as the high-pressure single-duct box, but it has two pressure-velocity-reducing air valves. The box is supplied with warm and cold air from warm and cold air decks. In response to a room thermostat, the valves can be actuated to new positions to regulate the flow of warm and cold air in accordance with room requirements.

When individual modulating damper operators are used to regulate the flows of warm and cold air, it is recommended that the size of the damper opening be varied to effect both pressure reduction and volume control with the same damper. When a single modulating damper operator is used to regulate the flows of both warm and cold air, a pressure-reducing damper or constant-volume controller is needed in the box to reduce pressure and limit airflow. When a large-capacity dual-duct box is used, the low-pressure discharge duct may require acoustic lining.

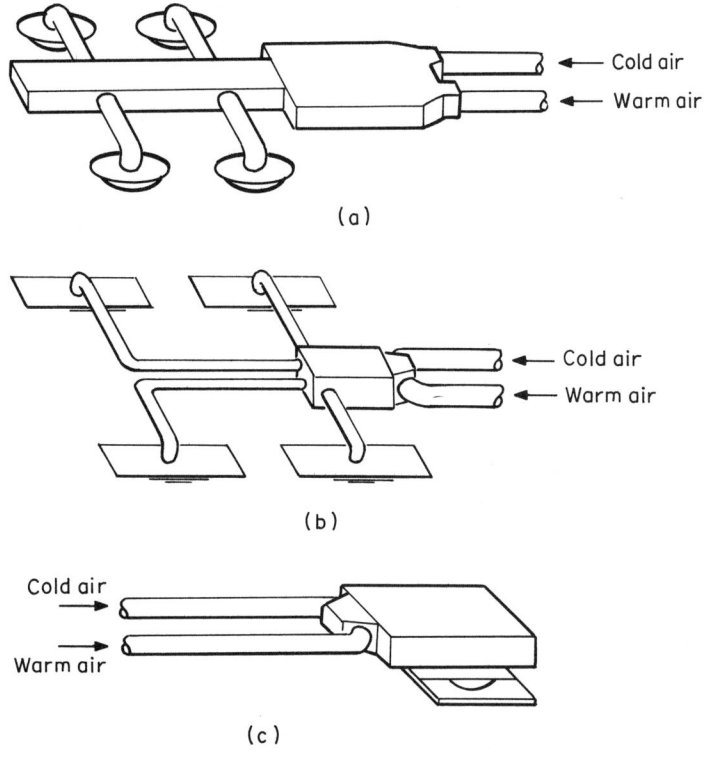

FIGURE 6-4 Dual-Duct Mixing Box. *(a)* High-pressure type (discharges air to a low-velocity duct which supplies a series of round diffusers by flexible ducts); *(b)* mechanical constant-volume type (supplies a series of square diffusers by flexible ducts connected to the box); (c) mechanical constant-volume type (supplies a diffuser directly connected to the box).

Constant-Volume Boxes

The function of a constant-volume box is to maintain a constant volume flow. Constant-volume boxes normally include one of the following volume-control devices:

1. A mechanical volume-control device actuated by static pressure in the primary air-duct system without an outside source of power.

2. A pneumatic volume regulator and selector devices actuated by outside pneumatic pressure and an internal-calibration pressure drop.

Boxes of type 1 may be equipped with a means for automatically resetting the mechanical constant-volume regulator to a different control point within the range limit of the control device. Boxes with this arrangement are known as *variable constant-volume boxes*, *single duct* or *dual duct*, and may be used with a reheat coil.

SUPPLY-AIR OUTLETS

The function of supply-air outlets is to correctly distribute supply air throughout a given space, to maintain proper comfort conditions. A specific type and shape of supply-air outlet is usually selected to satisfy the architectural and engineering requirements of a specific job. The types of supply-air outlets commonly used in air-distribution systems are:

1. Ceiling diffusers (circular, half-round, square, and rectangular)
2. Slot linear diffusers (wall to wall)
3. Perforated (square and rectangular)
4. Supply registers and grilles (side, wall, and floor)
5. Air-outlet light fixtures with troffer air boots (single and double)

Supply-air outlets generally incorporate all or some of the following basic functions:

1. Breaking the flow of supply air into several layers.
2. Reducing the air velocity and the temperature differential between the room air and supply air. This can be achieved by inducing room air into the primary airstream within the air outlet. Outlets with high induction characteristics are usually constructed so that the moving airstream creates a low-pressure area immediately beyond the rim of the air outlet, and deflects the airstream back to the ceiling, causing thorough mixing of room and supply air.
3. Eliminating the prime causes of drafts. This can be achieved by the slow, decreasing air motion set up within the entire radius of diffusion, which also equalizes the temperature and humidity.

Supply-air-outlet performance is usually influenced by these factors: mounting height, hot or cold temperature differential, ceiling construction, exposed ductwork, rate of air change, and air approach to the air outlet as determined by the proper use of accessories.

A diffuser consists of concentric elements, usually cones or pyramids, arranged so as to deflect the air through a large angle and create local high-ratio mixing. Diffusers come in round, square, rectangular, and strip-linear shapes. They can be selected to produce one-way, two-way, three-way, four-way, or circular distribution patterns. Adjustable-pattern diffusers are also available for use in both heating and cooling. In general, diffusers give more nearly draftless distribution than side-wall grilles.

Supply grilles usually have independently movable vertical and horizontal bars, the latter for upward deflection. A supply register is a grille with a volume-regulating device attached.

Grilles and diffusers are equipped with accessories permitting volume control and, in the case of grilles, additional directional control. Diffusers often have equalizing grids at the junctions of their necks with the supply duct, to ensure uniform distribution of air across the diffuser (see Fig. 6-5).

Diffusers are selected for noise level and ceiling height, which affects the jet velocity, pattern of distribution, and throw (radius of diffusion). Some manufacturers list throws at more than one terminal velocity. For example, the *maximum* radius of diffusion is assured if an average velocity between 25 and 35 ft/min is maintained and the maximum throw does not exceed 1½

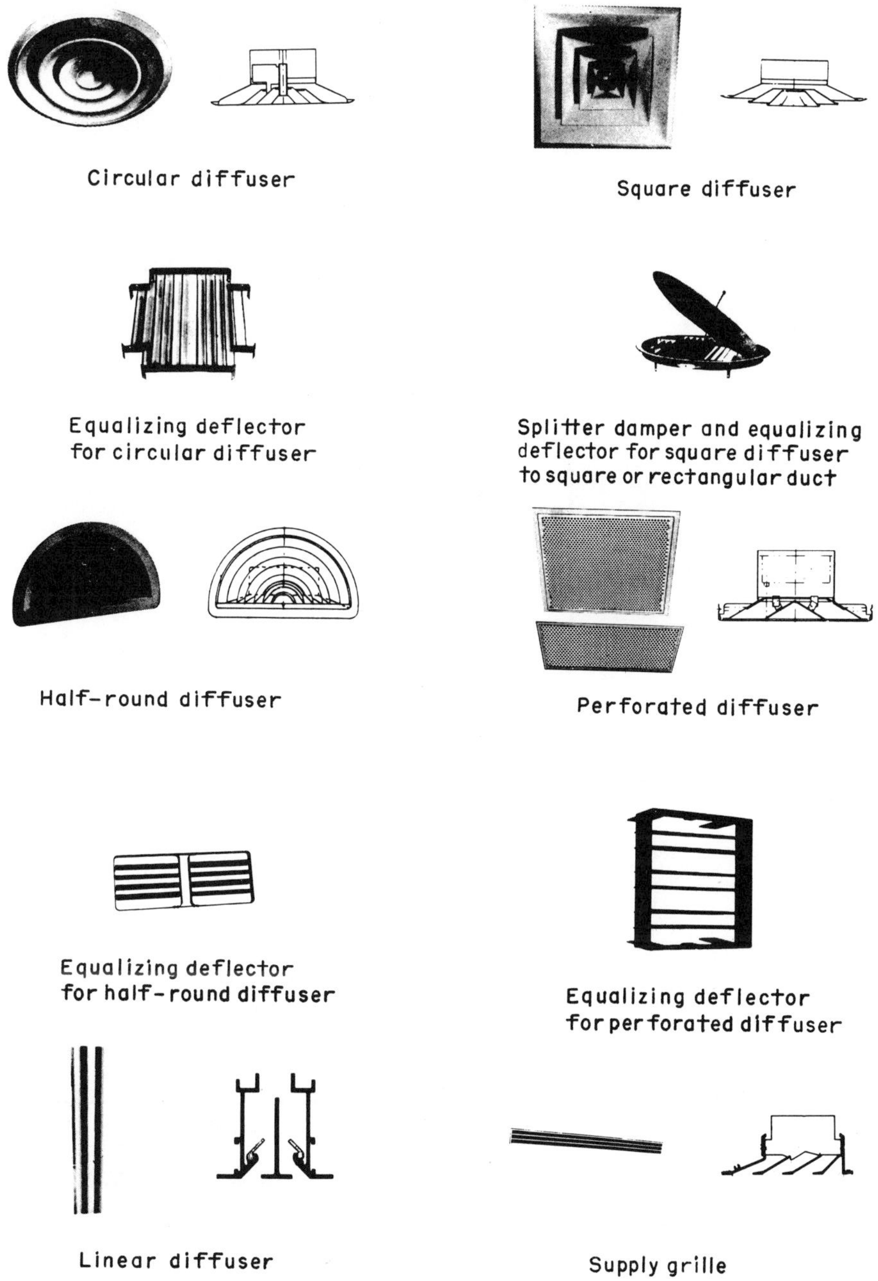

FIGURE 6-5 Diffusers and accessories. *(Reproduced by permission of Anemostat Products Division, Dynamics Corporation of America.)*

times the mounting height; the minimum radius of diffusion is assured if an average velocity between 25 and 50 ft/min is maintained and the stated minimum distance between the diffuser and a room obstacle is observed. Supply grilles are selected for throw, distribution pattern, drop (as affected by velocity), and velocity (as affecting noise control).

ESTIMATING CHART 6-6 Supply-Air Outlets (Diffusers).

A. Round and square diffusers

1. Material costs

2. Labor man-hours

Up to 12" neck size	1/2 man-hour
14" to 20" neck size	3/4 man-hour
24" to 32" neck size	1 man-hour

Note: Diffusers are normally provided by sheet metal contractor.

B. Linear diffusers

Material costs per foot		Man-hours per foot
1 Slot	$10	1/4
2 Slot	$12	1/4
3 Slot	$15	1/3
4 Slot	$20	0.4

C. Troffer air boot

Material cost each		Labor man-hours each
2' X 2' (Single inlet)	$18	3/4
2' X 4' (Single inlet)	$22	1
2' X 2' (Double inlet)	$20	3/4
2' X 4' (Double inlet)	$24	1

Refer to Estimating Chart 6-6 for estimating data for supply-air outlets (diffusers).

RETURN AND EXHAUST INLETS

Return-air inlets may be connected to a duct, or they may be simple vents which transfer air from one area to another. Exhaust-air inlets remove air directly from a building and, therefore, are always connected to a duct. Return and exhaust inlets usually have fixed bars or louver grilles at angles to break the line of sight into the duct. Where continuity of the surface effect is considered important, stamped lattice-pattern grilles are used. Registers are often used where local control is desired, but they should not be used for the basic balancing of high-pressure-drop systems because of unacceptable noise levels.

Return and exhaust grilles and registers are selected on the basis of velocity, which depends on location and mounting height. Large-capacity inlets

ESTIMATING CHART 6-7 Registers and Grilles (Supply and Return).

1. Material costs

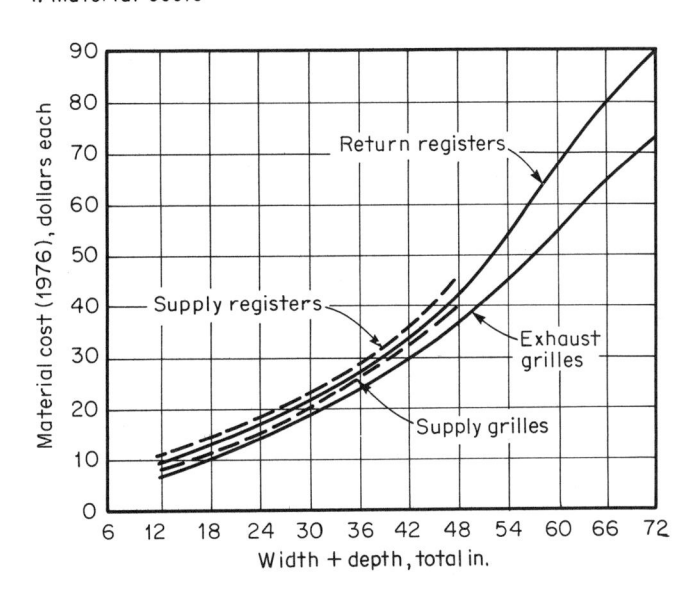

2. Labor man-hours

Width + depth, total in.	Man-hours (each)
Up to 30	1/2
32 to 48	3/4
54 to 72	1

NOTE: Registers and grilles are normally provided by sheet metal contractor.

should not be located in corners or other restricted spaces, as excessive air motion may result. Exhaust-air inlets are, in most cases, located in ceilings or high on side walls. Return-air inlets for cooling systems can be located in ceilings or high on side walls, provided they do not short-circuit the primary airstream to any appreciable degree. Return-air inlets for warm-air heating systems must be located near the floor and distributed so as to compel the air to traverse the entire room.

Refer to Estimating Chart 6-7 for estimating data for registers and grilles (supply and return).

7

AIR-HANDLING EQUIPMENT

GENERAL Almost every building requires some kind of air-handling equipment for human comfort. A well-designed and effective heating, ventilating, and air-conditioning system normally requires the use of properly selected air-handling equipment that can be easily installed and maintained. This chapter discusses the basic types of air-handling equipment used for heating, ventilating, and air conditioning. This equipment consists of

1. Air-handling units, including air-conditioning units and heating-ventilating units

2. Heat-recovery systems

3. Fans

4. Air-treatment equipment, including humidifiers and sound traps

5. Dampers

To estimate material and labor costs, use the material cost charts and man-hour tables that follow. The man-hour figures include handling and erecting the unit (see Chaps. 8 and 9 for final hookup). The material costs are based on 1976 prices and are given for reference only.

CENTRAL AIR-HANDLING UNITS

Central air-handling units were originally used with forced warm-air heating and ventilating systems. It was found that cooling and dehumidification equipment, when added to these units, produced satisfactory air conditioning where heat gains were relatively uniform throughout the conditioned space. These units can also be used where heating and cooling loads vary within the space served. In such cases, the space must be divided into sections or zones, and the central system supplemented by additional equipment and controls.

A central air-handling unit normally has centrally located equipment, with the air distributed through ducts. It may be installed in a basement or service area of a commercial building, and in the truss space or on the roof of a factory. It can be located adjacent to a central cooling and heating plant or at a considerable distance from it, and use circulated chilled water, hot water, or steam.

Central air-handling units can be used in the following applications:

1. Spaces with uniform loads, such as auditoriums, theaters, the public areas of most buildings, and the interior areas of commercial buildings (also exterior areas, with supplementary perimeter units, if required)

2. Areas with short occupancy, such as cafeterias, supermarkets, and exhibition areas

3. Spaces requiring precision controls, such as particular rooms (within a larger building) with stringent requirements for cleanliness and temperature control

4. Multiple systems for large areas, such as hangars, factories, and some large stores, where the units are often located in the truss space, against the outside wall, or on the roof

5. Primary source of conditioned air for other systems, such as with induction units, fan-coil units, and panel systems

The basic types of central air-handling units (heating, ventilating, and air-conditioning units and heating-ventilating units) will now be discussed.

HEATING, VENTILATING, AND AIR-CONDITIONING UNITS

Heating-ventilating-air-conditioning units are available in blowthrough and drawthrough configurations, for either indoor or outdoor installation. The basic unit normally consists of the following (see Fig. 7-1):

Mixing plenum or box. This is particularly useful when a mixture of recirculated (return) air and substantial quantities of outdoor air are needed to meet system requirements. The air mixing box is an effective aid in overcoming air stratification. Normally, outside air and recirculated air are drawn through the mixing-box dampers. Recent studies indicate that parallel-blade dampers can be used more effectively than opposed-blade dampers for mixing boxes. If parallel blades are used, each damper should be mounted so that its partially opened blade directs its airstreams toward the other damper, to assure maximum mixing and proportioning of the air before it enters the main airstream.

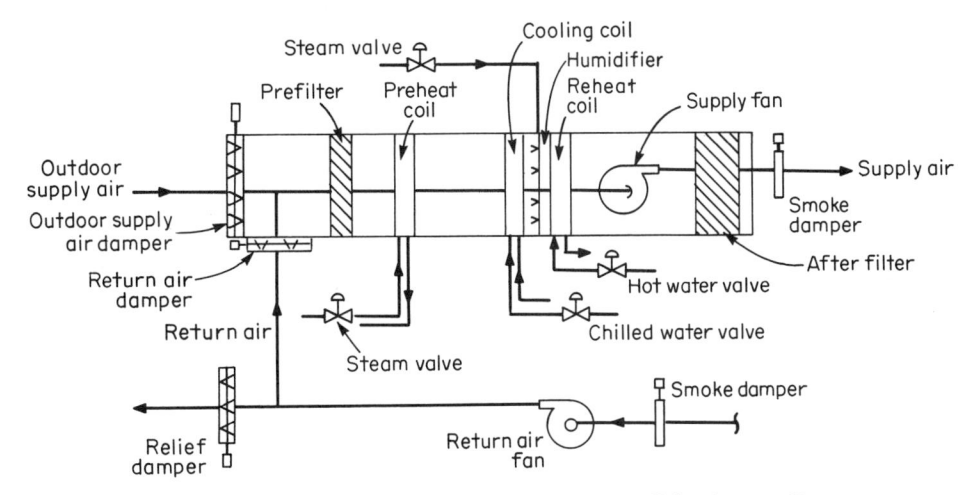

FIGURE 7-1 Central heating–ventilating–air-conditioning unit.

Filter section. Filters may be of the automatic or manual-roll type, with disposable or cleanable media, of the throwaway or cleanable flat panels, of the disposable bag type, electrostatic, or a combination of these types. Disposable-type filters are made of fibrous media (dry or viscous impingement), whereas the cleanable-type filters are made of metal screens. Normally, the filter section is located directly upstream of the mixing box and is referred to as *prefilter*. Some applications, such as hospitals, require additional high-efficiency filter sections, located upstream of the coil sections, to assure a high degree of air cleanliness; such filters are referred to as *after* or *final filters*. The overall performance of the unit depends heavily upon the filter. It must be properly installed (no bad connections) and regularly maintained.

Preheat-coil section. A preheat coil is normally used in cold climates to heat the air mixture by not more than 35°F; it should not operate in outdoor temperatures above 35°F, so as not to overheat the mixture. The preheat coil should have a wide fin spacing and be accessible for easy cleaning. The heating medium is usually either steam or electricity.

Humidifier section. A humidifier is often required to obtain a desired relative humidity. Moisture can be added to the air by a grid-type steam humidifier or, in some cases, by a pan-type humidifier with a heating coil or mechanical atomizer. The location of the humidifier is important in preventing stratification of moist air in the system. (Effective locations for various types of humidifiers are given later in this chapter.)

Cooling and dehumidifying section. Cooling coils may be chilled-water coils or direct-expansion (D-X) coils. Coils are manufactured with various fin spacings, face areas, lengths, and numbers of rows of tubes in the face and in the direction of airflow. In all finned coils, some air passes through without being treated by contact with the fins or tubes and becomes part of the room load. In order to reduce this bypass, a deeper coil with recirculating sprays or an air washer might be used.

Cooling coils normally remove the sensible and latent heat loads, and a second mixing action occurs in air passing through them. The cooling medium may be either chilled water or refrigerant.

Reheat-coil section. Reheat coils are normally used to reheat the cooled air below room temperature, to prevent occupied interior areas from being overcooled when their heat gains are relatively constant. Reheat coils are also used to provide supplemental heat in winter for warm-up, ventilation, and humidity control. The heating medium may be hot water, steam, gas, oil, or electricity.

Bypass section. A bypass damper may be used when the internal loads are reduced. If it is included in the system, the room thermostat opens the bypass damper and permits return air at design conditions to enter the bypass section. At the same time, the face damper on the cooling section closes and reduces the flow of cooled air. The temperature of the mixture is thus raised to compensate for the reduction of the sensible heat in the room, and overcooling is prevented. Bypass dampers may be used when the outside-air percentage is small. They cannot be used when ventilation requirements are very high; in this case, a reheat coil and a humidifier must be installed to control the relative humidity in the room.

Fan section. Supply fans are either centrifugal fans or axial-flow fans. Centrifugal fans may be specified as:

- Single or double width
- Single or double inlet
- Forward- or backward-inclined
- Flat-plate or airfoil

These fans are commonly used in packaged air-handling units.

Axial fans are either vane, axial, or in-line centrifugal. They are normally used for high-pressure, dual-duct, or built-up air-handling units.

Supply fans are normally selected according to the amount of airflow, the outlet velocity, and the static pressure. The supply-fan design static pressure may vary by up to 12 in of water. HVAC units may be designed for:

1. Low pressure, with velocities below 2,000 ft/min and duct static pressures of 2 in of water or less

2. Medium pressure, with velocities greater than 2,000 ft/min or duct static pressures up to 6 in of water

3. High pressure, with velocities greater than 2,000 ft/min or duct static pressures above 6 in of water and up to 10 in of water

HVAC units may be factory-assembled units (up to 50,000 cfm) or built-up units (with larger capacities or for special installation requirements). These units may be classified as single-duct, multizone, or dual-duct units.

For heating and cooling unit estimating data, see Estimating Chart 7-1.

Unitary Air-Conditioning Units

HVAC units are also available as unitary or self-contained units. They are factory-assembled in a single enclosure, including a refrigeration compressor, direct-expansion coil, filters, fan, heating coil, and air- or water-cooled condenser. These units are arranged to supply outside air for ventilation and are frequently designed and styled for installation within the conditioned space.

Unitary-type HVAC units can be grouped in the following classifications:

Indoor Self-Contained Air-Conditioning Units

These units are installed directly in the conditioned space and use either a discharge plenum or ductwork for air distribution, as required by the application. Normally, they are floor-mounted units. Refer to Estimating Chart 7-2 for estimating data.

Rooftop Units These units are self-contained in weatherproof housings, and are installed directly above the conditioned space. Accessory equipment, such as water, steam, gas, oil, or electric heating coils and steam or electric humidifiers, is available, so that rooftop units may be used for year-round air conditioning. Rooftop units have become increasingly popular, especially for low-budget jobs. Refer to Estimating Chart 7-3 for estimating data.

Window Units These units are similar to the self-contained unit ventilator discussed in Chap. 6. They are available in capacities from ½ to 2½ tons and are used to cool small spaces. Window units are designed for minimal installation cost. Such units can be moved from one location to another. Refer to Estimating Chart 7-4 for estimating data.

ESTIMATING CHART 7-1 Packaged Heating and Cooling Units. (Prices do not include motors. For motor prices see Estimating Charts 4-6 and 4-7).

1a. Material costs for units with chilled-water coils and flat filters.

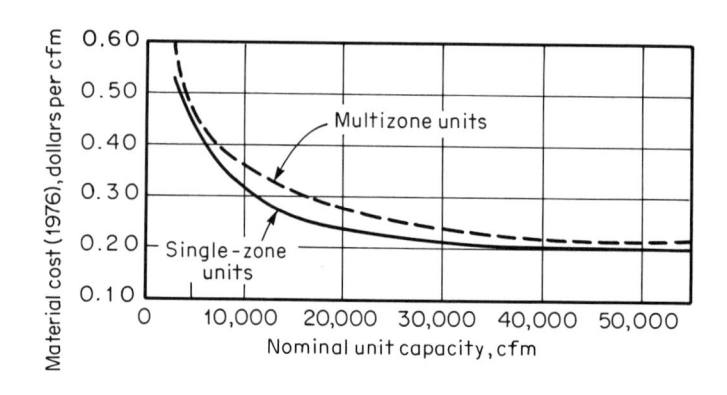

1b. Material costs for units with chilled-water coils and roll filters.

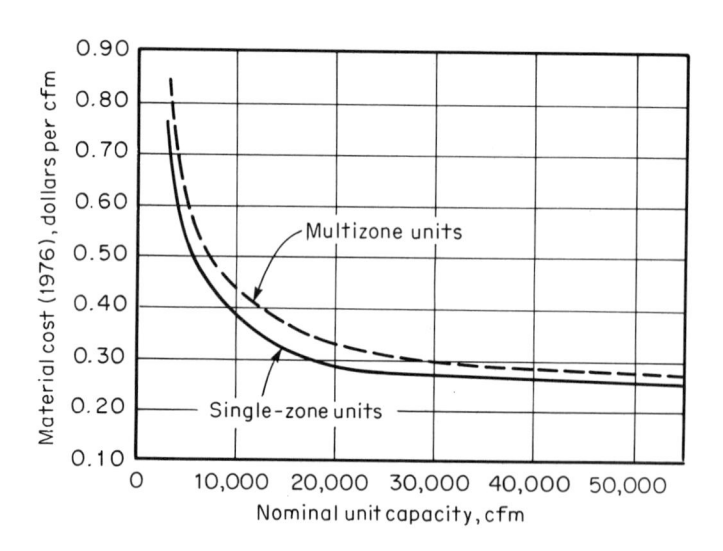

Heat Pumps

A heat pump usually provides both cooling and heating from within a single unit. It can be used to cool air in summer in the conventional way, and to supply winter heat by removing heat from a low-temperature source and pumping it to a higher-temperature space. Heat pumps may be classified as follows:

Air-to-Air Heat Pump. In the air-to-air heat pump, air is used as a source of heat during the heating cycle; the air takes its heat from the condenser. During the cooling

ESTIMATING CHART 7-1 Packaged Heating and Cooling Units. (Prices do not include motors. For motor prices see Estimating Charts 4-6 and 4-7). *(Continued)*

1c. Material costs for units with chilled-water coils and automatic fitters.

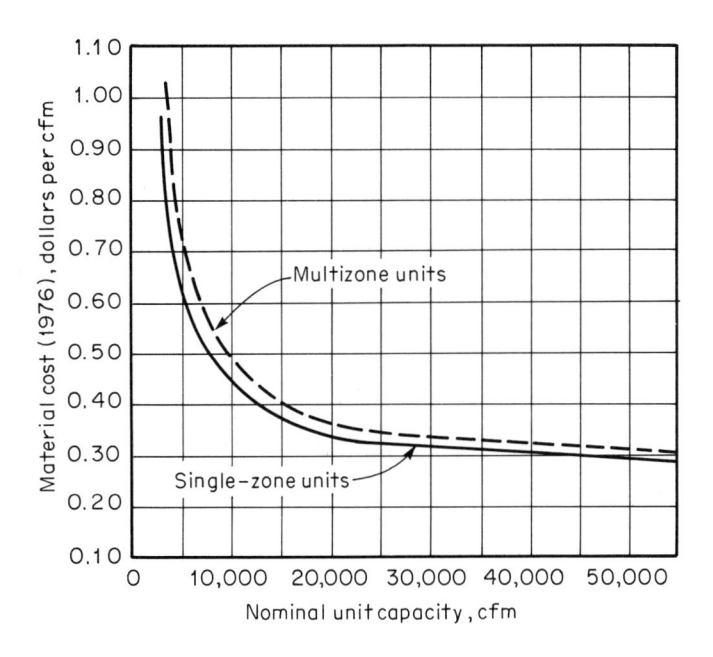

[Graph: Material cost (1976), dollars per cfm vs. Nominal unit capacity, cfm (0 to 50,000). Curves shown for Multizone units (dashed) and Single-zone units (solid).]

2. Labor man-hours for all heating and cooling units.

Nominal capacity, cfm	Man-hours Floor-mounted	Man-hours Ceiling-mounted
2,750	16	24
3,750	16	24
6,600	24	32
8,700	24	32
12,500	32	48
16,000	40	56
19,000	48	64

Nominal capacity, cfm	Man-hours Floor-mounted	Man-hours Ceiling-mounted
23,500	56	72
27,000	64	80
35,000	72	88
47,000	80	96
55,000	80	96

cycle, air is used to cool the space; the air rejects its heat to the outside. The path of the refrigerant is reversed by means of eight two-way valves, as shown in Fig. 7-2.

Water-to-Air Heat Pump. In the water-to-air heat pump, air is used as a source of heat, and water is employed to transfer the heat from the condenser and chiller. The refrigerant circuit is fixed; refrigerant flows from the compressor to the

ESTIMATING CHART 7-2 Vertical-type Indoor Self-contained Air-conditioning Units. (Units include all components, hot-water coil, and basic controls.)

1. Material costs

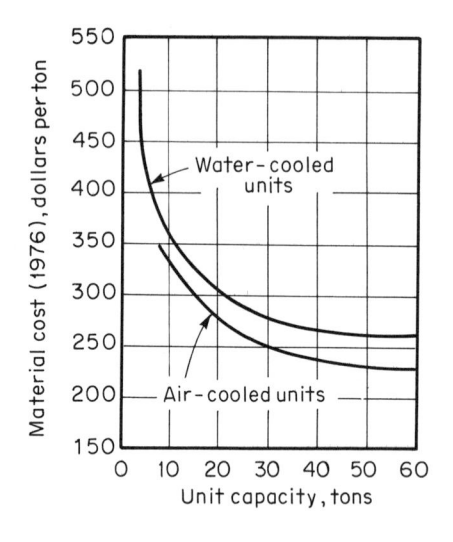

2. Labor man-hours

Nominal capacity		Weight, lb*	Man-hours
tons	cfm		
3	1 200	510	12
5	2 000	685	12
7.5	3 000	800	12
10	4 000	1010	16
15	6 000	1590	16
20	8 000	2130	16
25	10,000	2560	20
30	12,000	2950	20
40	16,000	3750	20
50	20,000	4900	24
60	24,000	5640	24

*Weights for water-cooled units.

ESTIMATING CHART 7-3 Packaged Rooftop Air-conditioning Units (air-cooled). (Units include all components and basic controls.)

1. Material costs

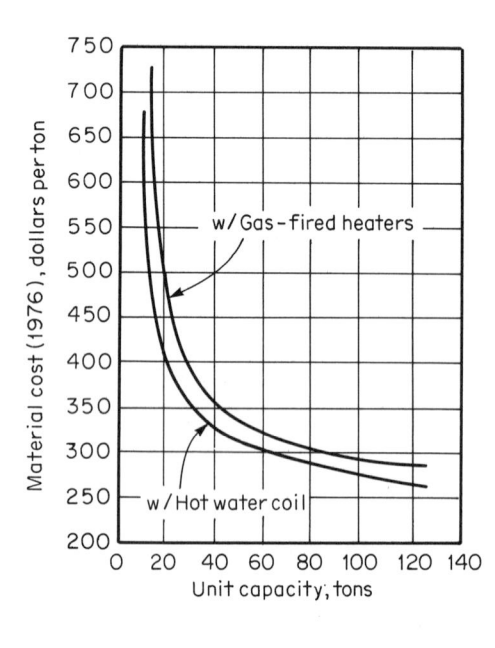

2. Labor man-hours

Nominal capacity		Weight, lb	Man-hours
tons	cfm		
10	4,000	3,900	24
15	6,000	4,200	24
20	8,000	5,800	24
25	10,000	6,400	32
30	12,000	6,800	32
35	14,000	7,100	32
40	16,000	9,100	40
50	20,000	10,000	40
60	24,000	11,500	40
75	30,000	14,000	48
90	36,000	17,000	56
100	40,000	20,000	64
125	50,000	24,000	72

Air-Handling Equipment | 115

ESTIMATING CHART 7-4 Air-conditioning Window Units (Slide-out Chassis).

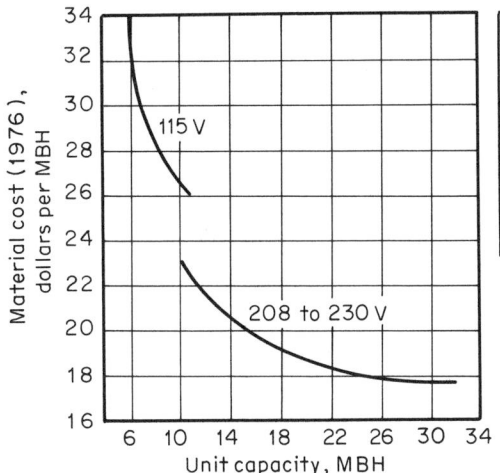

1. Material costs

2. Labor man-hours

MBH (cooling)	Dimensions, (H X W X D), in.	Man-hours
6.0 to 8.0	14 X 25 X 17	2.0
9.0 to 14.6	16 X 26 X 21	2.0
18.6 to 21.0	18 X 27 X 29	2.5
25.0 to 32.0	20 X 28 X 35	3.0

condenser, through the expansion valve and chiller, and back to the compressor, as shown in Fig. 7-3.

Water-to-air heat pumps may be connected with a two-pipe closed-loop system. Cooling is accomplished by each unit in the conventional manner, by supplying cool air to the individual module and rejecting the heat thus removed to the two-pipe system. All the heat gathered by the two-pipe system is rejected to an evaporative cooler, usually mounted on the roof. If some modules call for heat, the individual units switch into the heating cycle (by means of reversing valve), and derive their heat from the two-pipe water loop; i.e., they obtain heat from the condenser water of other units. When all units are in the heating cycle, a boiler must be added to provide 100% heating capability. The water loop is usually maintained between 70 and 90°F and therefore needs no piping insulation.

Refer to Estimating Chart 7-5 for heat-pump estimating data.

Split-System Air-Conditioning Units

In this arrangement, the unit is split into an indoor blower-coil section and outdoor condensing units (see Fig. 7-4). For condensing-unit material costs and labor man-hours, see Chap. 3. For blower-coil units, see Estimating Chart 7-6.

HEATING-VENTILATING UNITS

Heating-ventilating (HV) units are similar to central HVAC units, except for the lack of a cooling section. The heating medium may be steam, hot

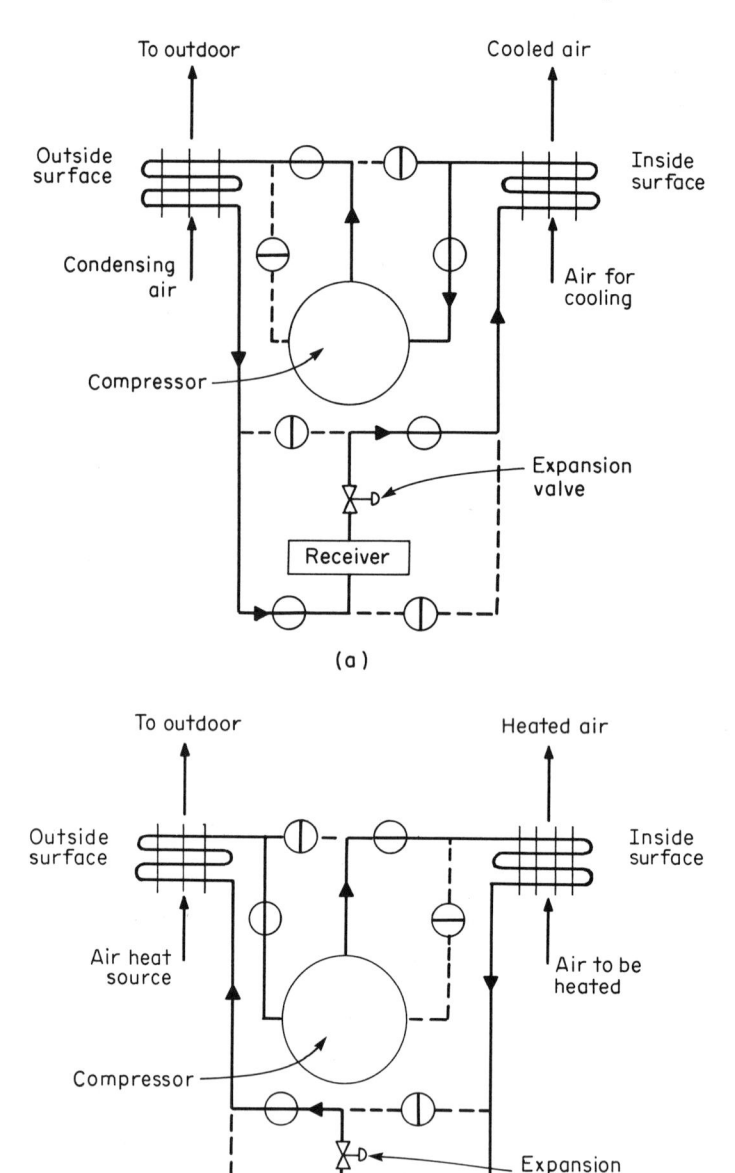

FIGURE 7-2 Air-to-air heat pump. (*a*) Cooling cycle; (*b*) heating cycle.

water, gas, oil, or electricity. They are used to supply heat and ventilation air to a space as required. These units are referred to as *make-up-air* units when used in spaces requiring 100% outside air. Refer to Estimating Chart 7-7 for estimating data.

Sometimes these units are referred to as *warm-air furnaces* when they are equipped with fuel-oil (or gas) burners and heat exchangers (indirect-fired units), or with electric heating elements. They are available in horizontal, upflow (high-boy), and downflow (counterflow) configurations. They may also be equipped with cooling equipment to provide year-round air conditioning. Heating-ventilating units may be used in residential, commercial, and industrial applications. Refer to Estimating Chart 7-8 for furnace estimating data.

HEAT-RECOVERY SYSTEMS

When exhaust-air quantities are large, a reheat-recovery coil can be located downstream of the exhaust fan, in addition to the preheat coil located at the HV unit. In this arrangement, the exhaust air transfers heat to a circulated fluid (normally glycol solution); then the hot fluid circulates in the tubes of the preheat coil. An outside airstream passes through the fins and tubes of this coil and can be heated to limited termperatures. This cycle may be referred to as a heat-recovery cycle.

Instead of the coil arrangement, heat recovery may be provided by an air-to-air heat-recovery wheel. A heat-recovery wheel recovers 60 to 90% of the total energy from the exhaust airstream before it is vented to the atmosphere, and transfers this energy to the incoming outside air. These heat-recovery systems can be used in heating and cooling systems.

In central cooling plants, the chiller can be equipped with an additional condenser (a *heat-recovery* condenser) which extracts heat from the chilled liquid and rejects some of that heat, plus the energy of compression, to a warm-water circuit for reheat or heating. Such a chiller is normally referred to as a *chiller with double-bundle condensers*.

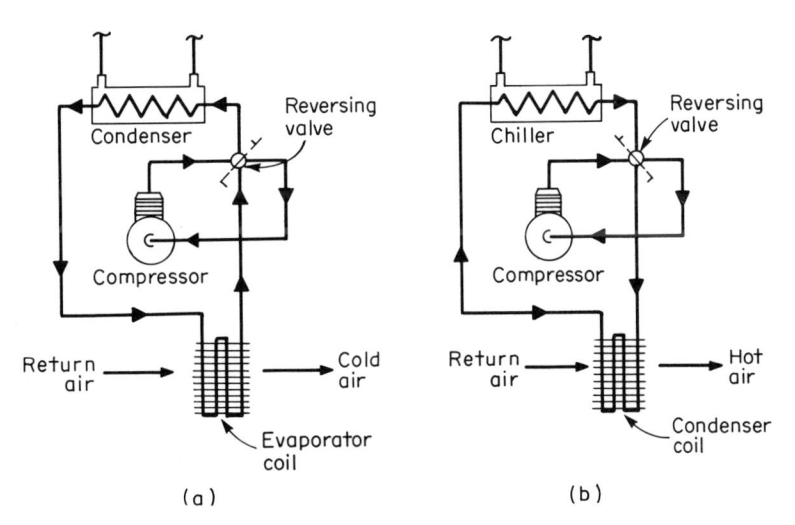

FIGURE 7-3 Water-to-air heat pump. (a) Cooling cycle; (b) heating cycle.

ESTIMATING CHART 7-5 Heat Pump Units. (Units include all components and basic controls.)

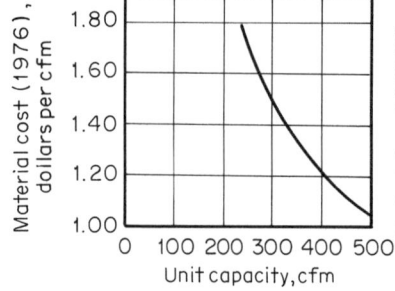

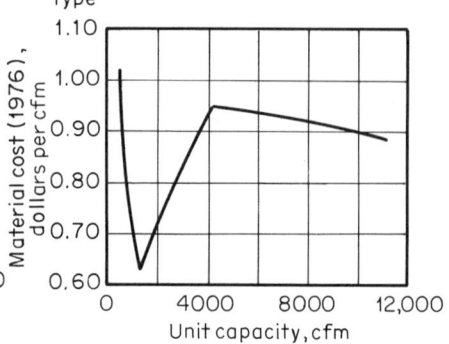

1. Material costs

   a. Cabinet floor-mounted water-to-air type

   b. Ceiling-mounted water-to-air type

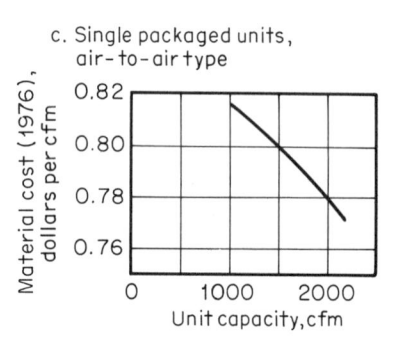

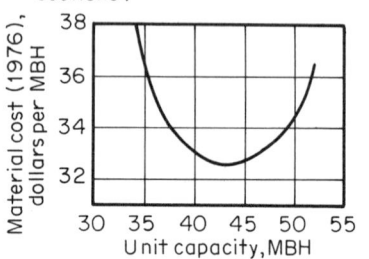

   c. Single packaged units, air-to-air type

   d. Air-cooled split-system units (prices include indoor and outdoor sections)

2. Labor man-hours

cfm	MBH (cooling)	Man-hours
250	7.5	6
350	12	6
500	14	6
650	19	6
900	20	6
1300	33	8

cfm	MBH (cooling)	Man-hours
1,750	44	12
2,200	55	12
4,200	124	16
6,800	184	20
11,200	304	24

For split system add 6 man-hours for air-cooled condensers.

Heat-recovery systems offer low heating costs and reduce the capacities required for heat-generation equipment. However, these systems require carefully designed control systems if they are to make the most use of the free heat recovered and maintain the proper temperature and humidity in all parts of the building. Estimating Chart 7-9 indicates the material costs

Air-Handling Equipment | 119

and labor man-hours for *rotary air-to-air heat-recovery wheels (heat exchangers)*. Such units include a galvanized steel casing, wheel, and variable-speed motor.

FANS

A fan is a machine that creates a pressure difference and causes an airflow through the rotation of an impeller. The impeller imparts both static and kinetic energy to the air; their proportions depend on the fan type.

Fans are used in HVAC systems either to supply air to a space or to exhaust air from a space. They are classified generally as centrifugal fans or axial-flow fans, according to the direction of airflow through the impeller.

Definitions

The following definitions apply to all fans.

1. *Rating.* A statement of performance for one operating condition. Includes fan size, speed, capacity, pressure, and horsepower.

2. *Capacity.* Measured in cubic feet per minute (cfm) at fan capacity. Directly related to the speed and the total pressure against which the fan operates. Resistance to airflow in a duct increases the pressure and reduces the capacity. Normally, the outlet volume is substantially equal to the inlet volume.

3. *Total pressure.* The rise in pressure from inlet to outlet; the sum of the static pressure and the velocity pressure.

4. *Velocity pressure.* Corresponds to the average velocity of the air at the outlet.

5. *Static pressure.* The total pressure less the velocity pressure.

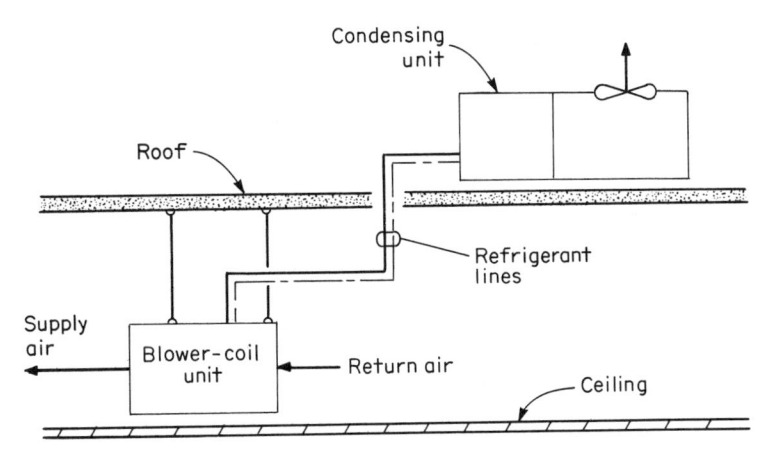

FIGURE 7-4 Split-system air-conditioning units.

ESTIMATING CHART 7-6 Split-system Air-conditioning Units (Blower-Coil Units). (Prices do not include motors. For motor prices see Estimating Charts 4-6 and 4-7).

1a. Material costs for heating and cooling units with D-X coils and flat filters.

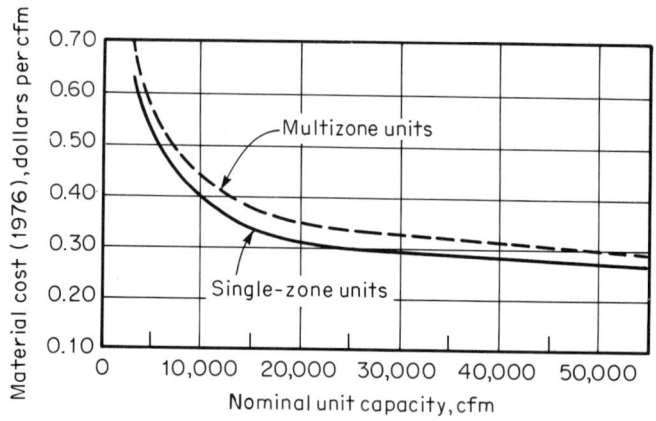

1b. Material costs for packaged heating and cooling units with D-X coils and roll filters.

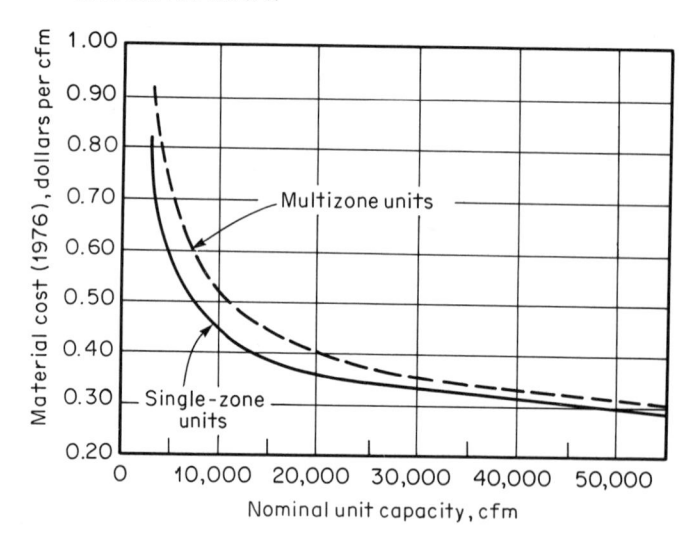

6. *Power output.* Based on the volume and total pressure and expressed in horsepower, as

$$hp = 0.0001575 \times p \times Q \tag{7-1}$$

where Q = air volume, cfm
p = total pressure, in of water

7. *Power input.* The measured horsepower delivered to the fan shaft.

ESTIMATING CHART 7-6 Split-system Air-conditioning Units (Blower-Coil Units). (Prices do not include motors. For motor prices see Estimating Charts 4-6 and 4-7). *(Continued)*

1c. Material costs for packaged heating and cooling units with D-X coils and automatic filters.

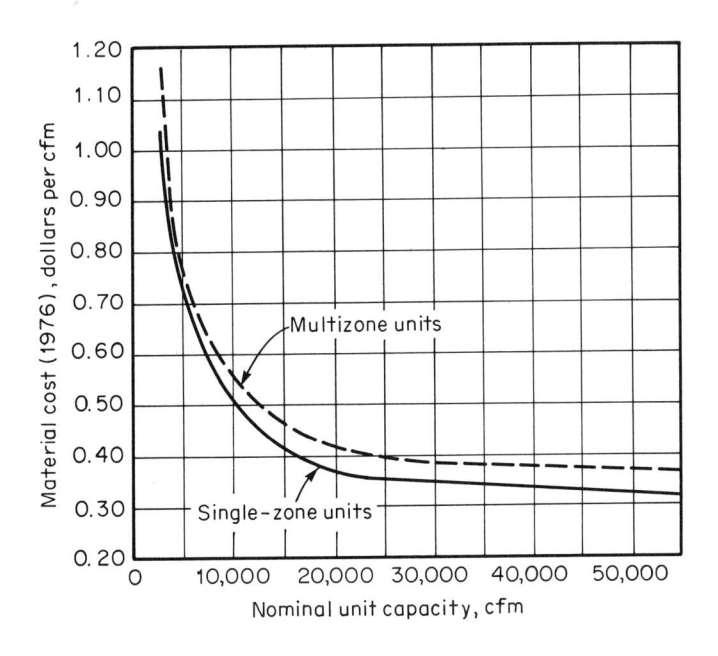

[Graph: Material cost (1976), dollars per cfm vs. Nominal unit capacity, cfm. Shows curves for Multizone units and Single-zone units.]

2. Labor man-hours for blower coil units, see man-hours table (7-1-2) for units with chilled water coils

8. *Mechanical efficiency.* The ratio of power output to power input.

9. *Static efficiency.* The mechanical efficiency multiplied by the ratio of static pressure to total pressure.

Centrifugal Fans

Centrifugal fans may be classified according to the impeller blade type, as follows:

1. *Radial blade.* This blade is straight and radial to the axis (see Fig. 7-5a). It is efficient for handling abrasive dust and exhausting fumes containing dirt, grease, or acids. The blade efficiency is 60 to 75%, but slightly lower for protective-coated blades.

2. *Forward-curved blade.* The outer edge of this blade curves forward in the direction of fan-wheel rotation (see Fig. 7-5b). It is good for high-pressure applications, and is often used where space is limited. It moves large quantities of air at low wheel speeds; the curved blade increases the

ESTIMATING CHART 7-7 Packaged Heating-Ventilating Units. (Prices do not include motors. For motor prices see Estimating Charts 4-6 and 4-7).

1. Material costs

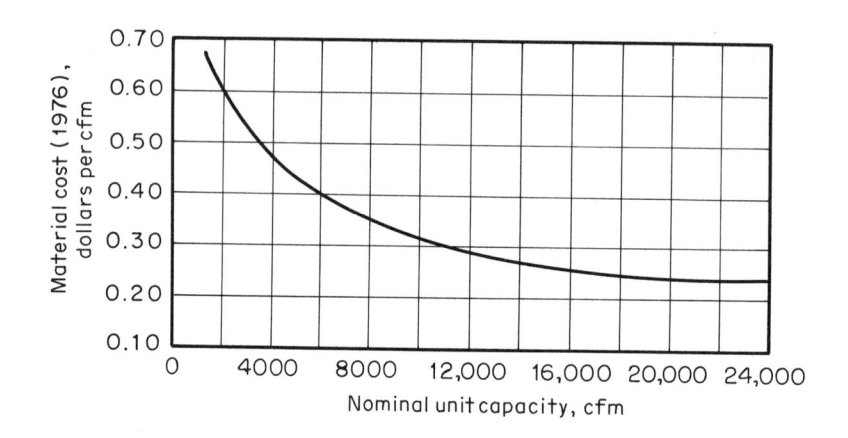

2. Labor man-hours

Nominal capacity, cfm	Man-hours Floor mounted	Man-hours Ceiling mounted
1,200	12	20
1,700	12	20
2,500	12	20
3,500	16	24
4,500	16	24
5,500	16	24
7,000	24	36
8,500	24	36
10,000	32	48
15,000	40	56
20,000	48	64
25,000	56	72

air velocity. The housing diffusion space converts velocity to static pressure. It is the least efficient blade.

3. *Backward-inclined blade.* The outer tip inclines to the rear, away from the direction of fan-wheel rotation (see Fig. 7-5c). Its efficiency is somewhat higher than that of the forward-curved blade.

4. *Airfoil blade.* This blade has the highest efficiency of all centrifugal-fan designs. It is often used for high-pressure applications. This type of blade

moves large quantities of air at high speed with low noise level (see Fig. 7-5d). Refer to Estimating Chart 7-10 for estimating data for centrifugal fans with airfoil blades.

ESTIMATING CHART 7-8 Packaged Furnaces.

1. Material costs

a. Gas-fired furnaces (heating only)

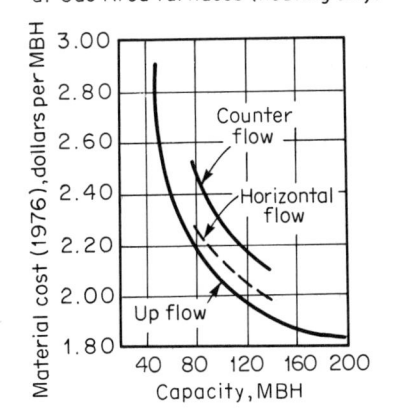

b. Gas-fired furnaces (heating and cooling)

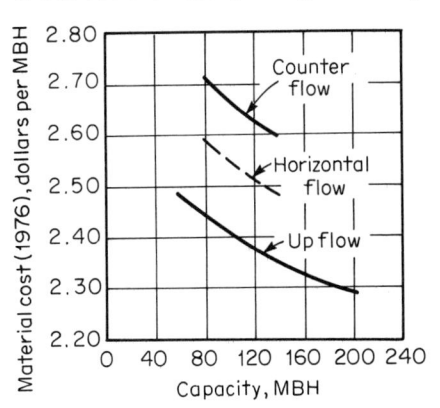

c. Electric furnaces (heating only)

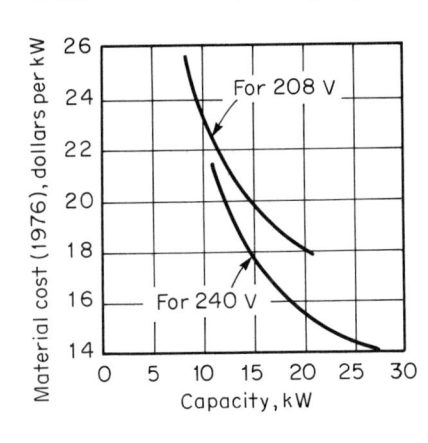

d. Oil-fired furnaces (heating only)

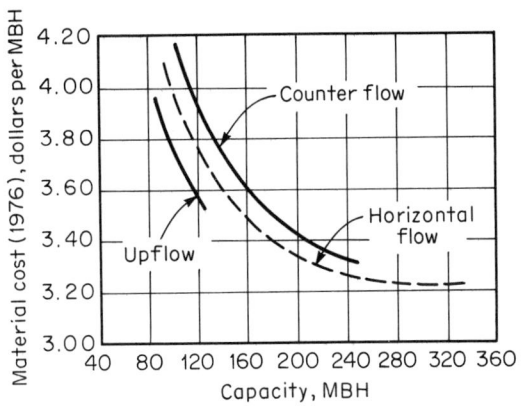

2. Labor man-hours

MBH	Man-hours
50–80	6.0
85–120	6.0
125–180	8.0

MBH	Man-hours
185–205	8.0
250–335	10.0

ESTIMATING CHART 7-9 Rotary Air-to-Air Heat-Recovery Wheels (heat exchangers).

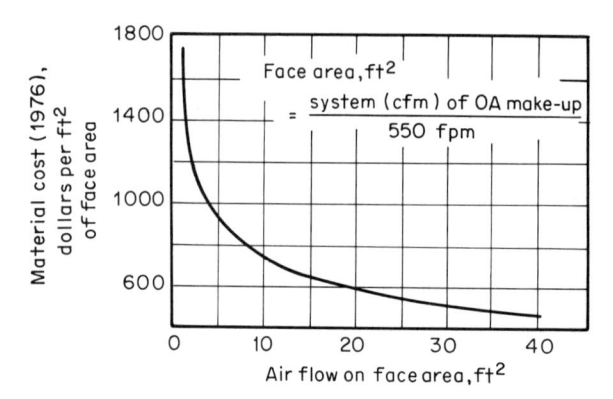

1. Material costs

Face area, ft² = system (cfm) of OA make-up / 550 fpm

2. Labor man-hours

Face area, ft²	Wheel dia., ft	Weight, lb	Man-hours
1.84	2-1/2	240	6
3.30	3	440	8
8.13	5	875	12
13.15	6	1750	18
21.10	8	2200	24
31.00	10	2600	26
46.05	12	3450	32

Centrifugal fans may also be either single-width–single-inlet or double-width–double-inlet types. The centrifugal fan normally consists of a steel housing, impeller wheel, impeller blades, shaft, bearings, and motor. The fan may have a direct drive or a belt drive.

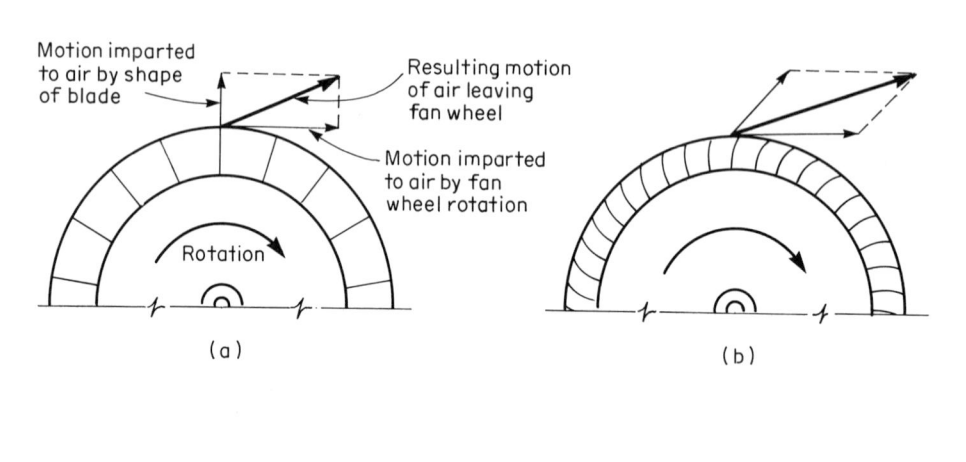

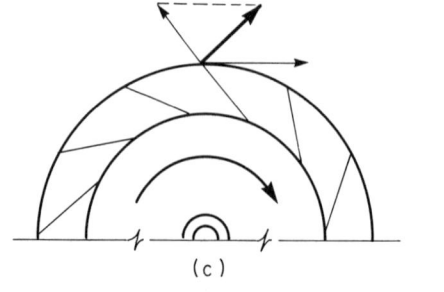

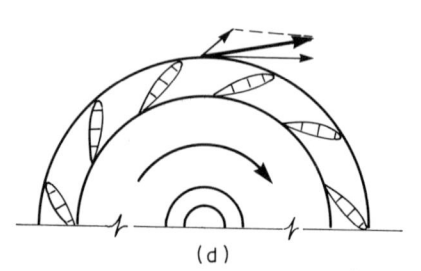

FIGURE 7-5 Types of centrifugal fan blades. (*a*) Radial; (*b*) forward-curved; (*c*) backward-inclined; (*d*) airfoil (forward-inclined).

Air-Handling Equipment | 125

ESTIMATING CHART 7-10 Centrifugal Fans with Airfoil Blades. (Prices do not include motors. For motor prices see Estimating Charts 4-6 and 4-7.)

1. Material costs

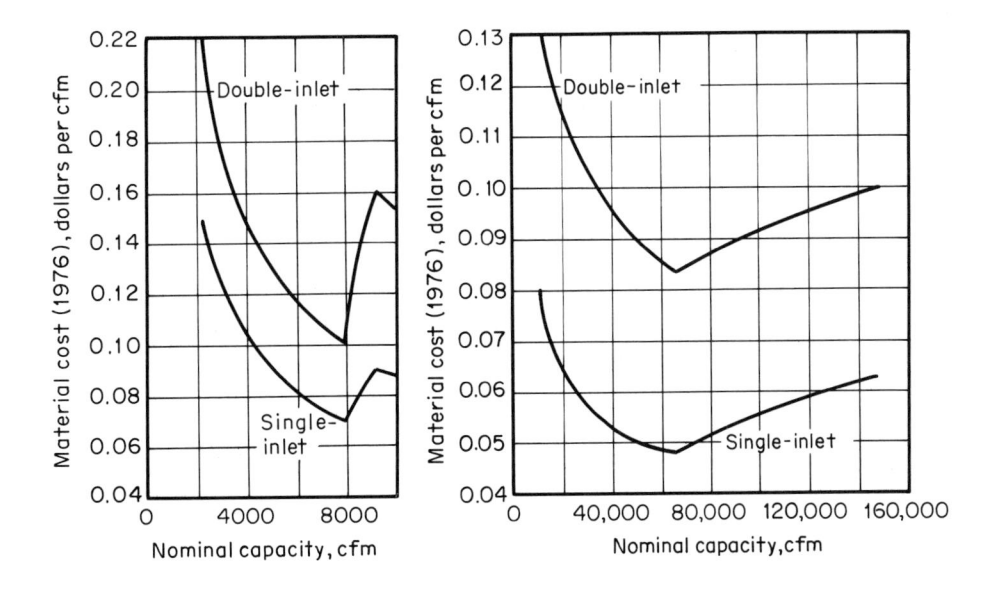

2. Labor man-hours

Nominal capacity, cfm	Wheel diameter, in.	Man-hours Single inlet	Man-hours Double inlet
2,000	12-1/4	4	6
3,000	15	5	6
5,000	18-1/4	6	8
6,500	20	6	8
8,000	22-1/4	8	12
9,000	24-1/2	8	12
12,000	27	8	12
15,000	30	10	16
18,000	33	10	16
25,000	36-1/2	12	16

Nominal capacity, cfm	Wheel diameter, in.	Man-hours Single inlet	Man-hours Double inlet
30,000	40-1/4	16	24
37,000	44-1/2	16	24
45,000	49	20	32
55,000	54-1/4	24	32
67,000	60	32	40
81,000	66	40	48
99,000	73	48	56
122,000	80-3/4	60	72
148,000	89	72	96

The rotation of the impeller wheel may be either clockwise or counterclockwise. The discharge of the fan is determined by the direction of the line of air discharge and its relation to the fan shaft. There are many configurations for the discharge of the fan, including

- Top or bottom horizontal
- Upblast or downblast
- Top or bottom angular

Axial Fans Axial fans may be classified according to the impeller type, as follows:

1. *Propeller fan.* Mainly for moving air from one non-air-conditioned space to another, this fan is normally used without ductwork. Two types, roof and wall-mounted, are available (see Fig. 7-6). Small and high-speed units are usually direct drive; large and low-speed units are belt driven, for variable-speed capability and vibration absorption. They have one or more blades or disk wheels mounted on the motor shaft, usually within a panel mounting ring or plate. The mechanical efficiency increases with speed, and maximum efficiency is reached when the unit is wide open. Wide blades run slowly but quietly; narrow blades are more efficient but noisier. Automatic or motorized louvers are most frequently used on wall-mounted types. Single-blade, automatic backdraft eliminators are used with smaller roof fans. Refer to Estimating Chart 7-11 for estimating data.

2. *Tube-axial fan.* A heavy-duty propeller fan, normally used without inlet or outlet vanes (see Fig. 7-7a). The air is discharged in helical motion. These fans are constructed with up to eight wheel blades, which are wider than those of propeller fans. They produce static pressures up to 3 in with mechanical efficiencies of 50 to 80 %.

3. *Vane-axial fan.* This fan normally has a surrounding cylinder, similar to those of tube-axial fans but with air guide vanes to straighten the helical air-discharge pattern (see Fig. 7-7b). The vanes are located either before or after the fan wheel; they convert spin velocity pressure into useful static pressure. Some units produce static pressures up to 15 in with mechanical efficiencies up to 90% when equipped with airfoil blades and vanes. Refer to Estimating Chart 7-12 for estimating data.

Tube-axial and vane-axial fans cost less than centrifugal fans, but have high noise level.

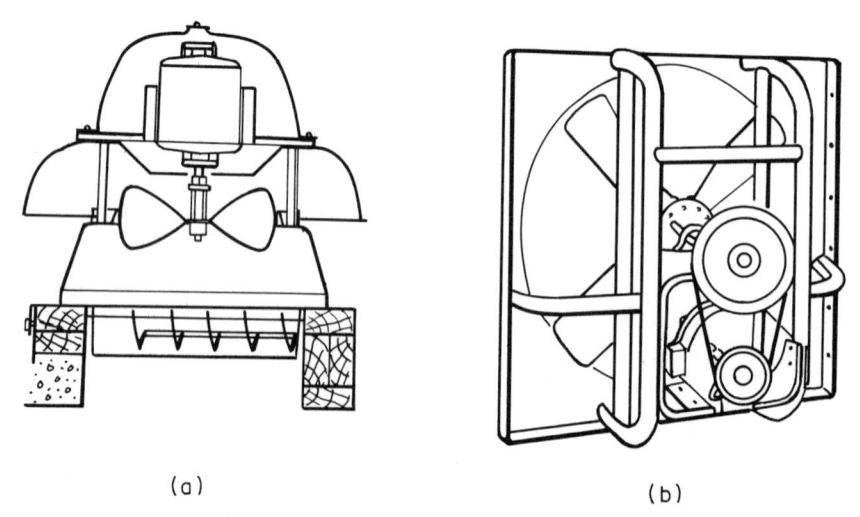

FIGURE 7-6 Propeller fans. (*a*) Roof type; (*b*) V-belt driven.

ESTIMATING CHART 7-11 Propeller Fans. (Prices do not include motors. For motor prices see Estimating Charts 4-6 and 4-7.)

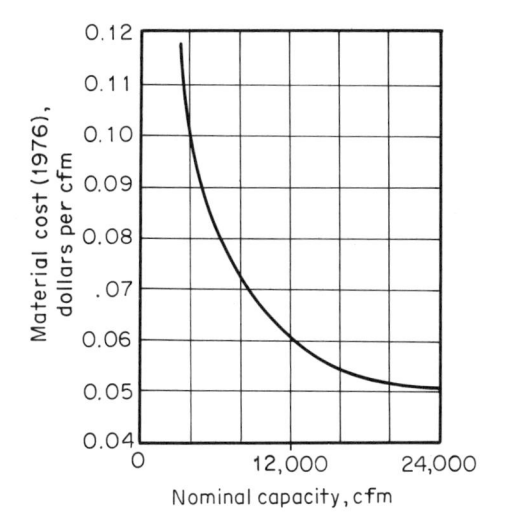

1. Material costs

2. Labor man-hours

Nominal capacity, cfm	Wheel diameter, in	Man-hours
3,000	16	4
5,000	20	4
7,500	24	6
12,000	30	6
20,000	36	8

4. *In-line centrifugal fan.* This fan is recommended for all air conditioning, including conventional low-velocity, high-pressure, or double-duct systems. Its straight-through airflow eliminates scrolls, elbows, plenum boxes, and offset connections. The fan and motor are usually combined in a single drive package. Conversion vanes eliminate air turbulence and give a smooth discharge. The fan may be mounted in the ceiling, floor, or wall, with the motor mounted independently. Sometimes this fan can also be mounted on the roof, as a power roof ventilator. Refer to Estimating Chart 7-13 for estimating data.

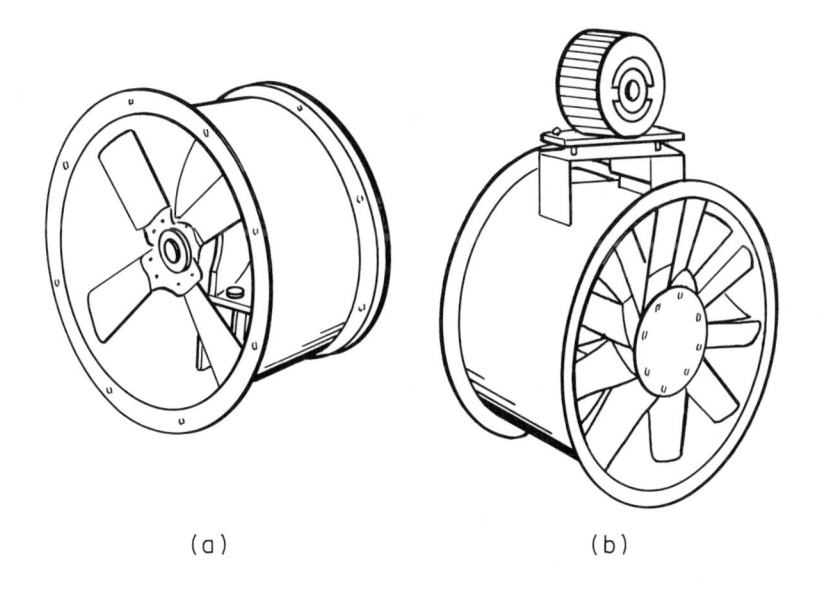

FIGURE 7-7 Axial Fans. (a) Tube-axial (direct driven); (b) Vane-axial.

ESTIMATING CHART 7-12 Vane-axial Fans. (Prices do not include motors. For motor prices see Estimating Charts 4-6 and 4-7.)

1. Material costs

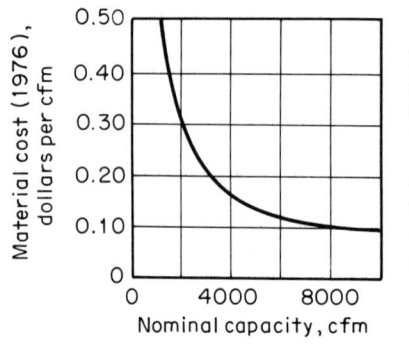

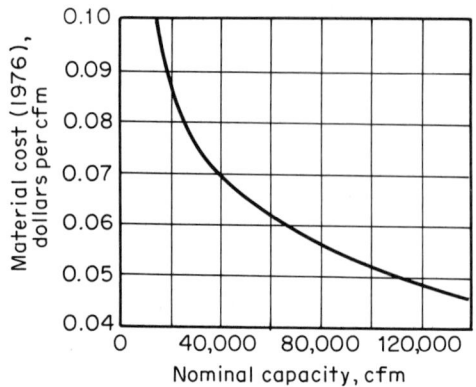

2. Labor man-hours

Nominal capacity, cfm	Wheel diam., in.	Man-hours Floor mounted	Man-hours Ceiling mounted
1,200	16	4	6
1,500–2,750	19	4	6
5,000–6,000	21	6	8
8,000–10,000	27	6	8
10,800–13,000	30	8	12
13,500–20,000	36	8	12
23,000–27,000	44	12	16
30,000–33,000	48	12	16
34,000–36,000	54	16	24
37,000–45,000	60	24	32
53,000–60,000	72	32	40
75,000–100,000	78	40	48
125,000	84	48	56

AIR-TREATMENT EQUIPMENT

Humidifiers Humidifiers may be used to obtain a desired indoor relative humidity for occupant comfort or process requirements. They are commonly used in central air-handling systems.

Heated-water-pan type. Units of this type may be heated by an electric, steam, or hot-water coil. They may be attached directly to the underside of a duct or installed remote from the duct; in the latter case, the unit is generally provided with a fan.

Steam-grid type. Direct-steam units are available in a wide range of designs and capacities. The steam-grid types are designed to introduce steam into the airstream of the system. With a constant steam supply pressure (not to exceed 12 psig), this type responds quickly to system demand. A humidifier should be mounted downstream from the heating coil. However, the open grid may be installed upstream of the heating coil or may have eliminators downstream to separate out water droplets that may form under some operating conditions. The enclosed grid prevents free moisture from being released into the conditioned air; it can be installed at any location where the air can absorb the vapor. The cup or pot type steam humidifier is generally

ESTIMATING CHART 7-13 Centrifugal-type Power Roof Ventilators. (Prices do not include motors. For motor prices see Estimating Charts 4-6 and 4-7.)

1. Material costs

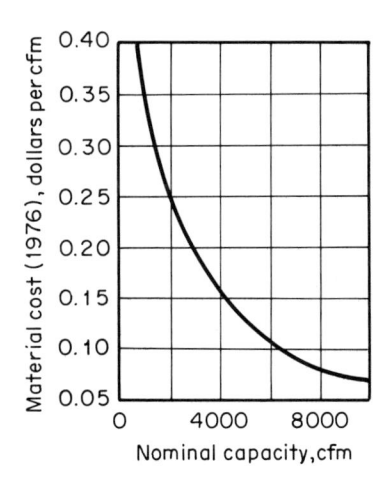

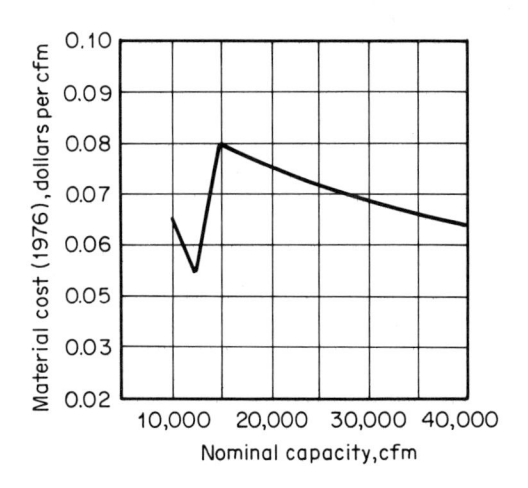

2. Labor man-hours

Nominal capacity, cfm	Wheel diam., in	Man-hours
500	9	3
750	10	3
1,500	12	3
2,500	16	4
5,000	18	6
8,000	20	6
10,000	24	8
15,000	30	8
20,000	33	12
25,000	36	12
30,000	40	16
40,000	44	16

attached to the underside of a duct. Steam is introduced tangentially to the inner periphery of the cup by one or more steam inlets, depending on the capacity of the unit.

Jacketed dry steam humidifier. This type is similar to the steam-grid type, but it uses a steam jacketed manifold and condensate separator to keep condensate from being introduced into the airstream. Units consist of a steam jacketed manifold and condensate separator with steam valve and trap (see Fig. 7-8). Refer to the Estimating Chart 7-14 for estimating data.

Sound-Reduction Units (Traps)

The primary function of a sound-reduction unit is to reduce the noise of fans and pressure-reducing valves. Sound-reduction units may be referred to as *sound traps* or *sound attenuators*. They can be selected to meet the sound-reduction requirements of an air-handling system. Sound traps are available in modular form and can be arranged to suit the available space. The outer casing of a sound attenuator is normally constructed of 22-gauge galvanized steel. Baffles are made of 24-gauge perforated galvanized steel filled with odorless, incombustible, moisture-proof, inorganic acoustical material. These units are available in rectangular (see Fig. 7-9) or round

ESTIMATING CHART 7-14 Jacketed Dry-Steam Humidifiers (Air-operated Humidifiers for Air-handling Systems). (Prices include operator, humidifier, manifold, steam trap, and supply-line strainer.)

1. Material costs

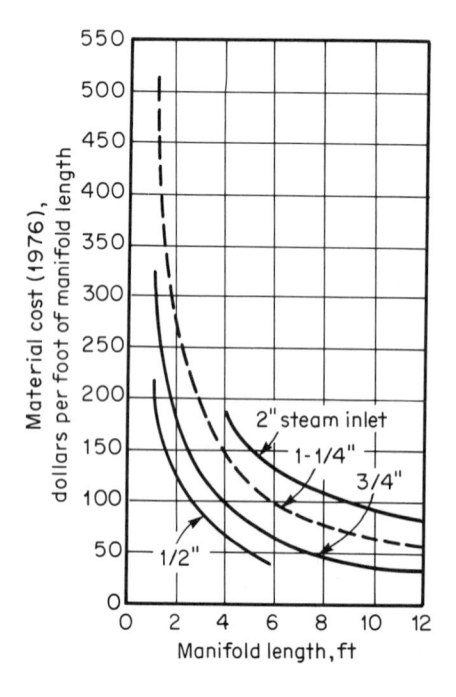

2. Labor man-hours

Steam inlet, in	Man-hours
1/2	3
3/4	4
1-1/4	5
2	6

Note: For automatic humidity controller add $125.00 per humidifier

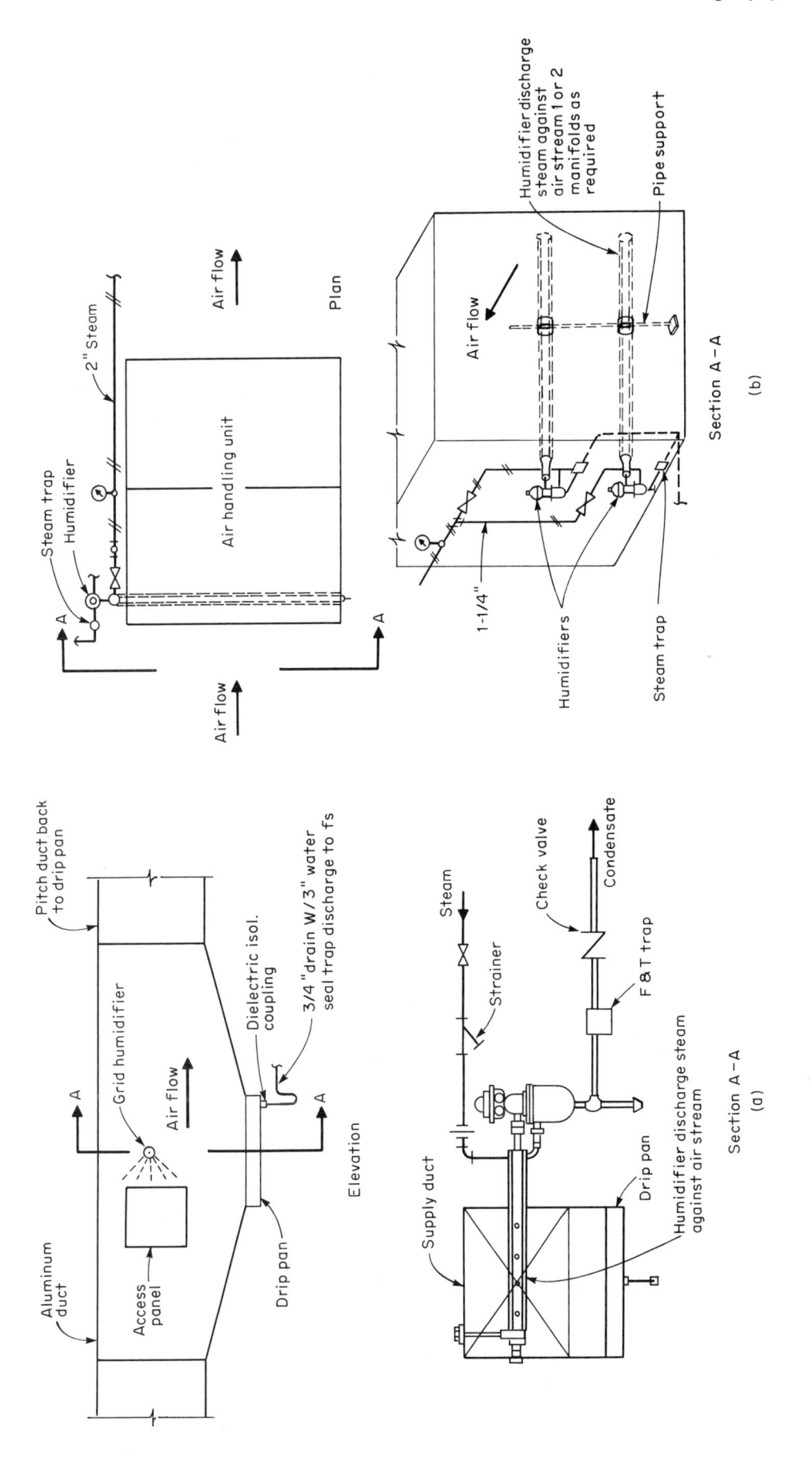

FIGURE 7-8 Steam humidifiers. (a) Duct humidifier; (b) Air-handling-unit humidifier.

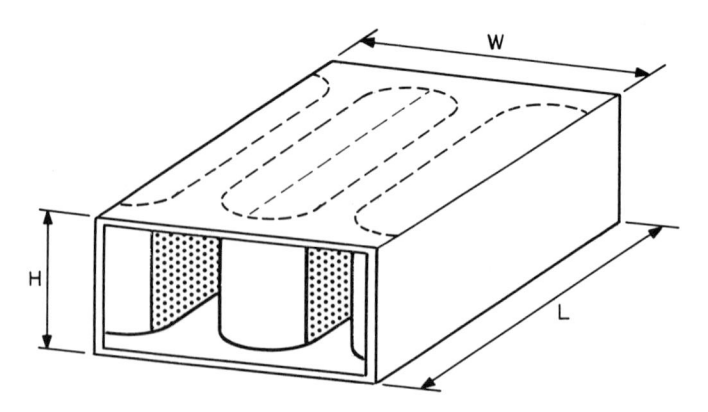

FIGURE 7-9 Rectangular sound attenuator.

ESTIMATING CHART 7-15 Sound-Reduction Units.

1. Material costs

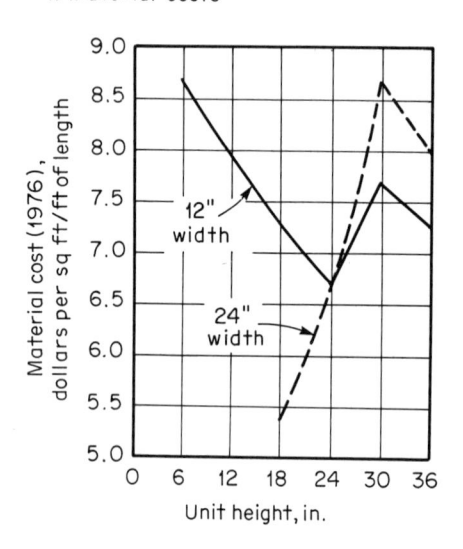

Examples:
 Find out the material cost for a sound trap.
 12"(W) X 24"(H) X 3'(L)

Solution:
 12" X 24" = 6 ft² per ft
 From 12" width chart material:
 cost per ft² per ft = $ 6.70
 ∴ Material cost for this unit
 = 6 ft² per ft X $ 6.70 X 3 ft
 = $ 120.00

Note:
 1. For sizes not indicated use multi-modular of 12" or 24" width.

2. Labor man-hours for units 3 ft in length

Width X height, in.	ft² per ft	Man-hours
12 X 6	3	0.75
12 X 12	4	1.00
12 X 18	5	1.00
12 X 24	6	1.25
12 X 30	7	1.50
12 X 36	8	2.00
24 X 18	7	1.50
24 X 24	8	1.75
24 X 30	9	2.00
24 X 36	10	2.50

Notes:
 1. For units with 5 ft length add 65% to labor man-hours.
 2. For units with 7 ft length add 135% to labor man-hours.

Air-Handling Equipment | 133

configuration. Self-noise generated by the sound-reduction unit at operating face velocity should be at least 5 dB below the permissible output value. This assures that sound energy leaving the sound-reduction unit does not increase the final design level. Refer to Estimating Chart 7-15 for estimating data.

DAMPERS Dampers are used in air-handling systems to ensure proper operation of the various air systems. They may be required by local ordinances (codes). Manual splitter dampers, manual or motorized volume dampers, and self-operating fire dampers are commonly used in HVAC systems.

Manual splitter dampers. Splitter dampers may be required on all supply-duct branches. The minimum dimension of the blade must be equal to the smaller neck size it controls. The blades are normally constructed of 16-gauge steel. The manual splitter damper is equipped with a locking and indicating quadrant mounted on the exterior of the duct. The damper is connected to the

ESTIMATING CHART 7-16   Volume and Fire Dampers.

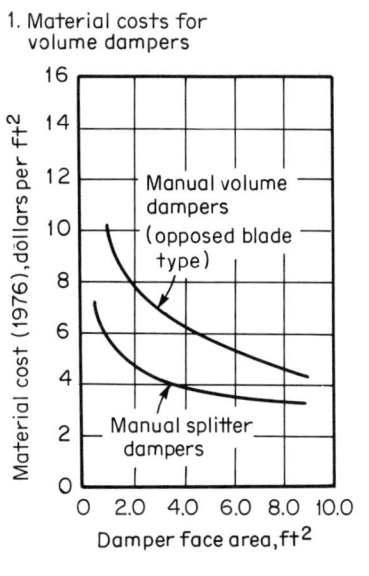

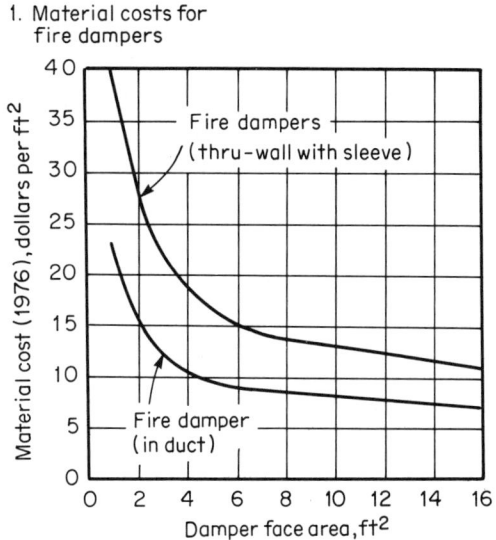

1. Material costs for volume dampers

1. Material costs for fire dampers

2. Labor man-hours for volume dampers

Up to 2 ft²	1/2 man-hour
2 to 4 ft²	3/4 man-hour
Over 4 ft²	1 man-hour

2. Labor man-hours for fire dampers

Face area, ft²	Man-hours, fire damper in duct	Man-hours, fire damper thru-wall
Up to 2	0.75	1.00
2 to 4	1.00	1.50
4 to 6	1.25	2.00
6 to 10	1.75	2.50
10 to 16	2.00	3.00

quadrant with a brass adjusting rod. Refer to Estimating Chart 7-16 for estimating data.

Volume dampers. The manual volume damper is normally mounted in a 16-gauge steel frame which is secured to the duct with bolts. It is equipped with the same components as the splitter damper. Return-air and exhaust-air dampers may be of the single-blade type or the multiblade straight type, depending upon the size. Supply-air dampers may also be of the single-blade type or the multiblade opposed type, depending upon the duct size.

A volume damper may be operated by a motor, in which case it is referred to as a *motorized* volume damper. These dampers are normally used in automatic control systems. Refer to Estimating Chart 7-16 for estimating data.

Fire dampers. Fire dampers should be provided where ducts penetrate firewall, fire-rated partitions, or floors. They are constructed in accordance with the latest requirements of the National Fire Protection Agency (NFPA). Fire dampers may be of the single-blade type for ducts up to 12 in, and the multiblade type for duct sizes over 12 in. The damper blades are mounted on steel rods set in bronze oilite bearings. All blades are constructed of 10-gauge black iron with formed edges. They should be weighted to ensure proper closing. The blade holding frame may be constructed of 10-gauge black iron. The dampers are installed so that the blades close in the direction of the airflow. Fire dampers may also be of the *shutter-type* design. Refer to Estimating Chart 7-16 for fire-damper estimating data.

8

PIPING AND ACCESSORIES

GENERAL A heating and/or cooling medium is conveyed from the point of generation to the distribution units through systems of pipes. The conveyed medium may be a liquid or a gas. This chapter discusses the estimating of the piping and accessories (fittings, valves, hangers, supports, thermometers, and gauges) commonly used for heating, refrigeration, and air conditioning.

PIPING SYSTEMS

A distribution piping system is an arrangement of one or more pipes. The proper piping arrangement is usually selected to meet the needs of the system. Piping systems may be classified as follows:

1. *One-pipe system.* A piping circuit in which the supply and return of the heating or cooling medium are carried in the same main.[1]

2. *Two-pipe system.* A piping arrangement consisting of one supply pipe and one return pipe.

3. *Three-pipe system.* A piping arrangement for a closed water system in which a cold-water supply, a hot-water supply, and a common return are carried in an individual main.

4. *Four-pipe system.* A piping arrangement for a closed water system in which a cold-water supply, a cold-water return, a hot-water supply, and a hot-water return are carried in an individual main.

---

[1] A *main* is a pipe (or duct) distributing to or collecting from various branches. A *branch* is a length of pipe (or duct) consisting of more than one piece of pipe (or duct) and several fittings.

5. *Reverse return system.* A piping arrangement in which the heating or cooling medium from several terminal heating or cooling units is returned along paths arranged so that the length of the supply and return piping is the same for all units. Whenever the terminal units have the same pressure drop through them, this system is recommended.

6. *Direct return system.* A piping arrangement in which the circulated heating or cooling medium is returned to the heat- or cooling-generation equipment by the shortest direct path, resulting in considerable piping savings. Whenever the terminal units have different pressure drops or require balancing valves, the direct return system is recommended.

PIPING MATERIALS

The materials most commonly used in piping systems are:

1. Steel, black and galvanized
2. Wrought iron, black and galvanized
3. Copper, soft and hard

In addition, alloy-steel pipe and stainless-steel pipe may be used for high-temperature service and for highly corrosive fluids. Thermoplastic pipes are now used in the construction industry; polyvinyl chloride (PVC) pipe is suitable up to 150°F and may be used for condensate drain lines and water chemical-treatment piping.

Steel Pipe

Steel pipe, either seamless or welded (butt or electric resistance), may be black or galvanized (zinc-coated). Steel pipe is produced in the following three basic grades:

- Standard weight (S)—schedule 40
- Extra strong (X)—schedule 80
- Double extra strong (XX)—schedule 160

Standard-weight pipe is generally furnished with plain or threaded ends in random lengths of 16 to 22 ft. Extra-strong pipe is generally furnished with plain ends in random lengths of 12 to 22 ft. Steel pipe is generally available in ⅛- to 24-in nominal pipe sizes; however, butt weld is not available above 4-in; lap-weld, electric-resistance-weld, and seamless pipe smaller than 2-in is identified as pressure tubing. Steel pipe joints may be threaded, flanged, or welded. Galvanized-steel pipe usually requires screw fittings.

Wrought-iron pipe sizes are approximately the same as steel-pipe sizes, but the walls are slightly thicker. Wrought-iron pipe resists many corrosive solutions.

Copper and Red-Brass Pipe

Copper and red-brass pipe have always been used in heating, ventilating, refrigeration, and water-supply installations; they are especially useful where resistance to corrosion is required.

Copper pipe is almost pure copper. It is available in ⅛- to 10-in nominal pipe sizes, in standard and extra-heavy weights. Copper pipe is generally furnished in lengths of 12 ft. Lengths of 16 and 20 ft are also available, along with special lengths of up to 30 ft in smaller sizes. The maximum length for larger-diameter pipe is 15 ft. Working pressures normally are not more than 700 psi. Copper pipe joints may be butt-welded, flanged, threaded, or brazed.

Red brass pipe is an alloy of 85% copper and 15% zinc, available in ⅛- to 12-in nominal pipe size and 12- to 16-ft lengths. Working pressures normally range from 8000 psi at 150°F to 3000 psi at 400°F.

In larger sizes, red-brass pipe joints may be flange with lap, socket-weld, or screw-flange. Smaller-size pipes usually have screw or silver-brazed fittings.

Copper Tubing

In addition to copper pipe, several varieties of copper tubing are in use; joints may be flared, compression couplings, or soldered. Copper tubing is almost pure copper metal, seamless, and manufactured with ⅛- to 12-in diameter. Working pressures range from 75 to 500 psi, depending on wall thickness. Maximum temperatures range from 100 to 400°F, again depending on wall thickness.

Copper tubing is classified according to wall thickness as follows:

Type K (heavy wall): available in hard-temper drawn in 20-ft lengths, and soft temper annealed in 60-ft lengths.

Type L (medium wall): available in hard-temper drawn in 20-ft lengths, and soft-temper annealed in 100- and 60-ft lengths.

Type M (light wall): available in hard-temper only, in 20-ft lengths.

Type DWV (light wall): for drainage, waste, and vent service. Available in hard-temper only, in 20-ft lengths.

Hard-temper tubes are used with brazed and solder fittings. Soft-temper tubes are used with flared, compression-coupling, soldered, or brazed fittings.

Polyvinyl Chloride (PVC) Pipe

PVC pipe is rigid pipe, resistant to chemicals and corrosion. There are two types, schedule 40 and schedule 80, available in ⅛- to 8-in nominal size. The maximum working pressure varies with the temperature. The temperature limitation is 140 to 150°F.

Schedule 40 PVC pipe cannot be threaded; it is used with socket fittings only, and joints are solvent-welded. Schedule 80 pipe can be threaded and used with either socket or threaded fittings. PVC joints may also be made with fusion-weld and heat seal.

APPLICATION OF PIPE AND TUBE TO HVAC

Table 8-1 indicates the materials commonly recommended for use in various piping systems. The piping material selected should be checked for design temperature and pressure ratings, and should conform to American Society for Testing and Materials (ASTM) requirements.

Piping Specifications and Standards

Piping specifications should include:

1. Type of materials
2. Applicable standards

TABLE 8-1 Materials Commonly Recommended for Various HVAC Piping Systems

System	Recommended Materials
Chilled water Condenser water Hot water Condensate return Refrigerant	Types K and L hard copper, black steel, wrought iron

Steam supply
  Below 125 psi
 	
  125–250 psi
	Type L hard copper, black steel
	
	Type K hard copper, black steel

Fuel oil and gas lines	Black steel
Condensate drain and make-up water	Hard copper, galvanized steel, PVC

TABLE 8-2 Pipe Schedule

System	Extent	Material	Schedule	Connection
H.P. steam	All	Black steel	80	Welded
H.P. return	2½ in and below	Black steel	80	Screwed
H.P. return	2½ in and up	Black steel	80	Welded
L.P. steam and return	2½ in and below	Yoloy* or black steel	40	Screwed
L.P. steam and return	3 in and up	Yoloy* or black steel	40	Welded
Condenser water	2 in and below	Black steel	40	Threaded and coupled
Condenser water	2½ in and up	Black steel	40	Welded
Chilled water	2 in and below	Black steel	40	Threaded and coupled
Chilled water	2½ in and up	Black steel	40	Welded
Hot-water heating, reheat, and preheat	2 in and below	Black steel	40	Threaded and coupled
Hot-water heating, reheat, and preheat	2½ in and up	Black steel	40	Welded
Indirect drains	All	Galvanized steel	40	Threaded and coupled
City water	All	Type L hard copper	40	Soldered
Vent and relief	All	Black steel	40	Welded
Fuel-oil supply and return piping	All	Black steel	80	Welded
Fuel-oil fill and vent piping	All	Black steel	80	Threaded and coupled
Refrigerant piping	All	ACR copper	80	Soldered
Emergency-generator exhaust pipe	All	Black steel	40	Welded

*Yoloy is a trade name for a nickel-copper-alloy steel pipe material.

3. Weights or wall thickness (i.e., schedule 40, type K, etc.)

4. Type of connections

Normally, the job specifications indicate the various piping materials to be furnished by the mechanical contractor. Tables 8-2 to 8-4 are typical pipe, fitting, and valve schedules, taken from actual job specifications. They are given for reference only.

TABLE 8-3 Fitting Schedule

System	Extent	Material	Schedule	Connection
H.P. steam	All	Steel	80	Welded
H.P. return	All	Steel	80	Welded
L.P. return	2½ in and below	Cast iron	40	Screwed
L.P. return	3 in and up	Steel	40	Welded
Condenser water	2 in and below	Cast iron	40	Screwed
Condenser water	2½ in and up	Steel	40	Welded
Chilled water	2 in and below	Cast iron	40	Screwed
Chilled water	2½ in and up	Steel	40	Welded
Hot-water heating, preheat, and reheat	2 in and below	Cast iron	40	Screwed
Hot-water heating, preheat, and reheat	2½ in and up	Steel	40	Welded
Indirect drains	All	Galvanized malleable	40	Recessed drainage
City water	All	Wrought copper	40	Soldered
Vent and relief	All	Steel	40	Welded
Fuel-oil supply and return piping	All	Cast iron	80	Welded
Fuel-oil fill and vent piping	All	Steel	80	Screwed
Refrigerant piping	All	Wrought copper	80	Soldered

Selections of pipe materials must conform to American Society for Testing and Materials (ASTM) requirements (i.e., schedule 40 or 80 for steel pipe, and type K or L for copper tubing). All fittings and valves must also conform to the latest standards of the American National Standards Institute (ANSI) code. However, consideration should also be given to local code requirements.

PIPE FITTINGS

Pipe fittings are made in various forms, including elbows, tees, couplings, reducers, and unions; they may be screwed, flanged, or welded. The materials generally used are steel, cast iron, malleable iron (heat-treated cast iron), copper, brass, stainless steel, bronze, and plastic (see Table 8-3). The type of fitting which may be used depends upon the pressure and fluid characteristics. As a general rule, in order to facilitate erection and servicing, unions, flanged fittings, and flanges must be installed where equipment and piping accessories will be disassembled quite often.

Fittings for copper tubing are available in soldered, flared, and compression types. The flared and soldered types are used in both large and small sizes, while the compression-type fitting is limited to small tube sizes. Flared tube fittings are widely used in refrigeration systems. Fittings for copper pipe are available in screwed and soldered types. Fittings for steel pipe are available in screwed, socket, and flanged types.

METHODS OF JOINING PIPE

The methods most commonly used for joining pipe are the following:

Threading   The number of threads per inch varies with the pipe size. Thread formation and length should be in accordance with ANSI standards. Right-hand threads are commonly used. All threaded pipe joints should be made up

TABLE 8-4   Valve Schedule

System	Extent	Material	Connection	Class, lb
H.P. steam and return	2 in and below	All bronze	Screwed	300
H.P. steam and return	2½ in and up	C.I. and bronze	Flanged	300
L.P. steam and return	2 in and below	All bronze	Screwed	125
L.P. steam and return	2½ in and up	C.I. and bronze	Flanged	125
Condenser water	2 in and below	All bronze	Screwed	150
Condenser water	2½ in and up	C.I. and bronze	Flanged	150
Chilled water	2 in and below	All bronze	Screwed	150
Chilled water	2½ in and up	C.I. and bronze	Flanged	150
Hot water	2 in and below	All bronze	Screwed	150
Hot water	2½ in and up	C.I. and bronze	Flanged	150
Indirect drains	All	All bronze	Screwed	150
City water	All	All bronze	Screwed	150
Vent and relief	All	All bronze	Screwed	125
Fuel-oil-tank piping	2 in and below	All bronze	Screwed	250
Fuel-oil-tank piping	2½ in and up	C.I. and bronze	Flanged	250

with a lubricant or sealant suitable for the service for which the pipe is to be used.

Threads for fittings are the same, except that it is standard practice to furnish straight tapped couplings for schedule-40 pipe, 2 in NPS and smaller. Screwed joints are limited by the ANSI standard as indicated in Table 8-5. For screwed brass or copper refrigerant pipe, extra-heavy fittings are required.

Soldering and Brazing

Sweat joints are divided into two categories: soldered and brazed.

Soldered Joints

A soldered joint is a gastight joint obtained by joining metal parts with metallic mixtures or alloys which melt at temperatures ranging from 400 to 1000°F. In general, for all services with operating temperatures up to 250°F with low or moderate pressures, a good grade of 50-50 tin-lead solder is most commonly used. If higher strengths are required at temperatures up to 250°F, a 95-5 tin-antimony solder should be used. The use of silver solder (either 35 or 45% alloy) is recommended for copper pipelines carrying refrigerants.

The types of flux which may be used with 50-50 and 95-5 solder are available in paste form. They consist of a petrolatum base impregnated with zinc and ammonium chlorides. The flux removes residues of oxide and serves to protect the surface from oxidation during the heating process, float out the remaining oxides ahead of the molten solder, and promote the wetting action of the solder. A joint has to be cleaned before soldering.

Brazed Joints

A brazed joint is a gastight joint obtained by joining metal parts with alloys which melt at temperatures ranging from 1100 to 1500°F. Brazing materials may be classified as alloys containing silver and copper-phosphorus alloys. The alloys containing silver are normally used for joining copper to bronze or steel; the copper-phosphorus alloys are used for joining copper to copper. Brazed joints are normally used in refrigerant systems.

TABLE 8-5 ANSI Standard for Screwed Joints

Application	Pressure, psig	Nominal Pipe Size, in
Refrigerant	Up to 250	Up to 3
Refrigerant	Over 250	Up to 1¼
Water and brine		Up to 4

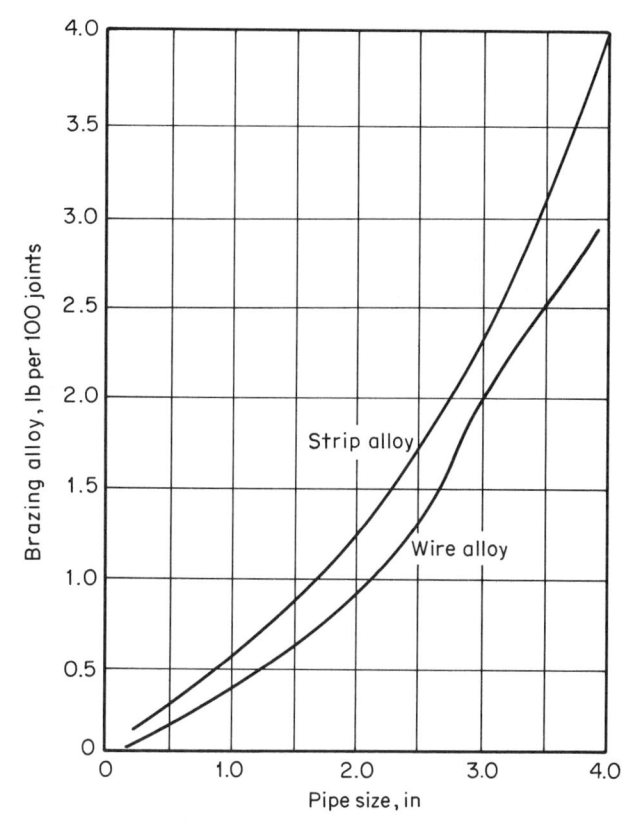

FIGURE 8-1 Consumption of brazing alloys.

Estimating Data for Brazing and Soldering

Figures 8-1 and 8-2 indicate the consumption of brazing alloys and solders in pounds per 100 joints.

Solvent - Cementing

A solvent-cementing joint is a gastight joint obtained by joining plastic parts (PVC) with solvent cement (lubricant) at normal temperatures. Solvent-cementing provides high joint strength and maintains the chemical resistance of the base material. Schedule 80 PVC pipe may be threaded, but schedule 40 PVC pipe must be solvent-cemented. Solvent must be applied to both pipe and coupling. Table 8-6 may be used to estimate the consumption of solvent.

Welding

Fusion welding, commonly used in the erection of steel or wrought-iron piping, is defined as the process of joining metal parts in the molten state. Oxyacetylene, electric arc, and gas-shielded welding is commonly used. The standard rules for the welding of pipe joints are contained in the ANSI standard code for pressure piping; included are welding procedures, qualifi-

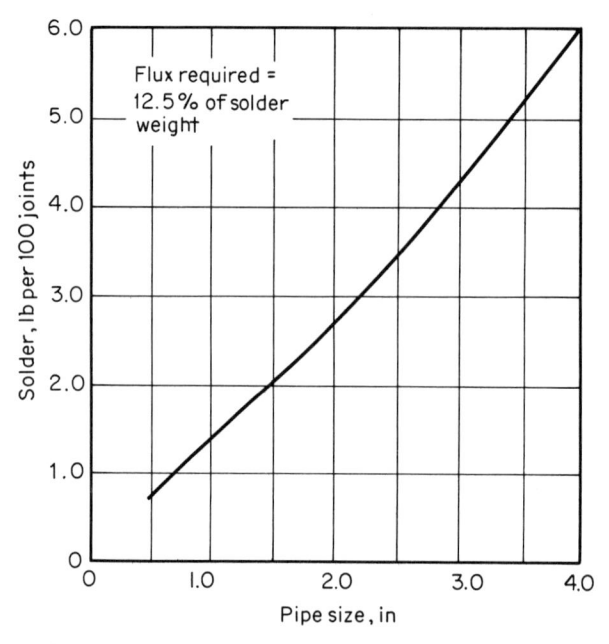

FIGURE 8-2 Consumption of solders.

cations of welders, and testing. In addition, local safety codes may govern the installation of welded piping.

The standard types of steel-welding fittings are

- Butt-welding fittings
- Socket-welding fittings (up to 3 in)

TABLE 8-6 Consumption of Solvent Cement Required to Join PVC Piping*

Pipe Size, in	Number of Joints per Gallon	Feet of Pipe per Gallon of Solvent	
		40-ft Standard Lengths	20-ft Standard Lengths
½	3860	154,400	77,200
¾	2325	93,000	46,500
1	1545	61,800	30,900
1¼	1365	54,600	27,300
1½	1000	40,000	20,000
2	840	33,600	16,800
2½	665	26,600	13,300
3	575	23,000	11,500
4	440	17,600	8,800
5	360	14,400	7,200
6	295	11,800	5,900

*Values apply to socket coupling assemblies. If threaded coupling assemblies are used, divide the solvent quantities by 2.

Flanged fittings also may be welded in ½- to 24-in sizes.

The welding thickness and number of passes can be determined from welding standards. Diameter-inches, i.e., the nominal pipe size multiplied by the number of passes, are the base unit for estimating welding materials and labor. The average labor rate is 1.5 diameter inch/man-hour; this rate is based on normal productivity and normal working conditions. A correction factor may be used to adjust the average labor rate for job difficulty.

Flanged Joints

Flanges may be classified according to the face configuration as

1. Plain-face flanges, the most commonly used
2. Insert-face flanges, used in high-pressure applications

Slip-on, welding-neck, lap-joint, threaded, blind, and reducing flanges are presently used and available in standard nominal sizes.

In order to obtain a tight flanged joint, a gasket (coated with the recommended lubricant) must be inserted between the contact faces of the flanges; then the faces are drawn up and tightened with bolts.

Victaulic Joints

The victaulic grooved piping method is a reliable and economical system for joining pipe. Square grooved-end pipe, fittings, or valves must be used with victaulic couplings. A victaulic coupling consists of a seal gasket, housing clamps, and two track-type oval neck bolts. Victaulic couplings are used to join all standard-weight and lightwall steel pipe up to 2500 psi working pressure. They provide expansion, flexibility, and vibration reduction.

UNDERGROUND PIPING SYSTEMS

The basic function of an underground piping system is to convey a given quantity of fluid from one point to another without leakage of fluid or undue loss of energy. Underground piping systems are normally used whenever a district central heating and cooling plant is used to serve institutional, residential, or industrial facilities. Distribution piping systems may be located in buildings, passageways at grade containing pipe chase, tunnels, or galleries below grade. Underground piping systems may require a wide variety of components, such as manholes, expansion joints, pipe anchors, and valves. There are various types of systems, the most commonly used being underground conduits. This type of system consists of a carrier pipe or pipes surrounded by thermal insulation and encased in an exterior conduit. Metal (steel or cast iron), fiberglass-reinforced plastic (FRP), plastic (PVC), and asbestos-cement conduits are used. The carrier pipe is nearly always steel, although copper is sometimes used. When steel conduits are used, coating and cathodic protection should be applied to protect the conduit from corrosion. Fiberglass and plastic conduits should only be used for chilled-water systems.

Conduits are generally furnished in 20- to 40-ft lengths, which are assembled in the field. All field joints must be watertight. Watertightness is normally tested by applying an air pressure of from 15 to 20 psig to the interior of the conduit after welding and before backfilling. (For excavation and backfilling, see Chap. 17.) Carbon-steel pipe, coated and wrapped, may also be used in underground distribution systems.

HANGERS AND SUPPORTS

The basic function of a hanger is to support a length of pipe. Hangers and supports are also used to prevent excessive stresses, excessive vibration, and possible resonance with imposed vibrations.

Pipe hangers, supports, guides, and sway braces should be fabricated in full conformance with the ANSI code for pressure piping. However, the design of the pipe-support system is left to the preference of the designer. This design involves the following:

- Determination of hanger and support locations
- Selection of the appropriate hanger or support
- Arrangement of guides, anchors, sway bracings, or vibration dampeners
- Calculation of hanger and support loads

The spacing of hangers or supports depends on the pipe weight and size, the locations of valves and heavy fittings, and the superstructure available for support. The recommended maximum spacing of hangers and supports and minimum hanger rod size are indicated in Table 8-7. Hangers should be located at each valve or other concentrated load, and near each change of direction.

TABLE 8-7   Maximum Spacing between Pipe Supports and Minimum Hanger Rod Size

Nominal pipe size, in	1	1½	2	2½	3	3½	4	5	6	8	10	12	14	16	18	20	24
Maximum span, ft	7	9	10	11	12	13	14	16	17	19	22	23	25	27	28	30	32
Minimum rod diameter, in	⅜	⅜	⅜	½	½	½	⅝	⅝	¾	⅞	⅞	⅞	1	1	1¼	1¼	1¼

TABLE 8-8 Functions of Flow-Control Valves

Function	Valve Type
Stopping flow	Ball Butterfly Globe Gate Plug
Throttling Flow	Ball Butterfly Globe Plug
Preventing backflow	Check

TABLE 8-9 Functions of Pressure-Control Valves

Function	Valve Type
Reducing pressure	Pressure-reducing valve
Maintaining a definite pressure	Pressure regulators
Relieving pressure	Relief valve

Anchors and guides are used to guide expansion or to fix the pipe. They serve the purpose of preventing excessive strains in the piping itself by ensuring the direction of thermal movement. A sway braced is used in all flexibly supported lines to prevent undue or unwanted movement or vibration. Hangers and supports are available in various types. There is a support for almost every conceivable requirement (Fig. 8-3).

VALVES[2]

Valves are used in a piping system to control fluid flow and system pressure. A properly selected valve must give the required performance with long life and minimal maintenance. The valves commonly used in piping systems may be classified as

1. Flow-control valves, including ball, butterfly, check, gate, globe, and plug cock valves

2. Pressure-control valves, including pressure regulators and relief valves

Each valve has a definite purpose in the control of fluids in the system. Tables 8-8 and 8-9 indicate the functions of various types of valves.

Valves are made of bronze, iron, and steel (see Table 8-4). A valve consists of a body, bonnet, stem, and disk. There are several designs for each of these components, as follows.

Bonnet and Body Construction

1. *Threaded bonnets.* Recommended for low-pressure service; not recommended for frequent dismantling and reassembly of valve

[2]Carrier Air Conditioning Co., *Handbook of Air Conditioning System Design*, McGraw-Hill, New York, 1966, has been used as a reference with permission.

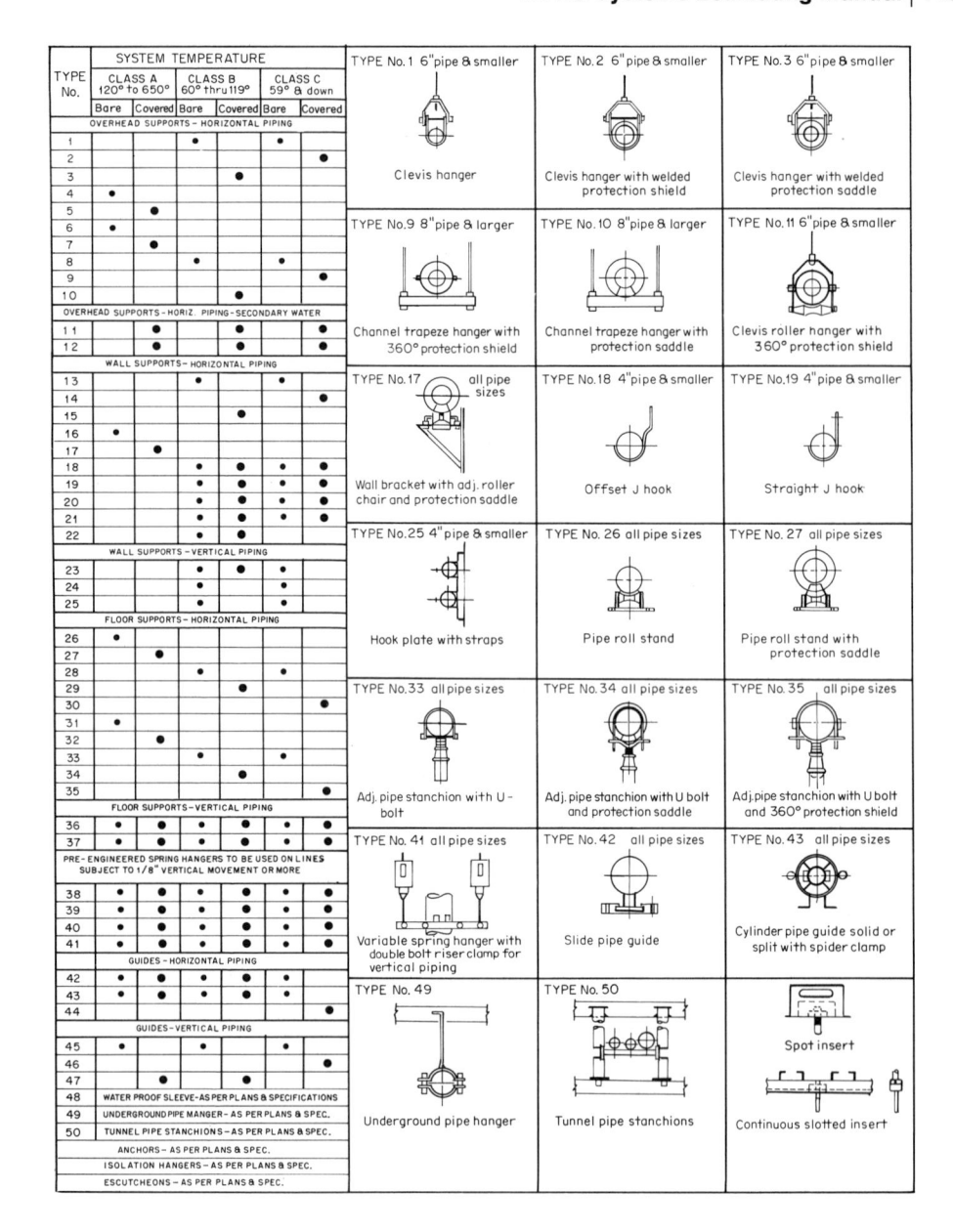

FIGURE 8-3 Types of hangers and supports.

2. *Union bonnets*. Normally made in sizes up to 2 in; easy to dismantle and reassemble; make tight joints

3. *Bolted bonnets*. Recommended for high-pressure service; mainly used in large valves, but also available in small valves; easy to dismantle and reassemble

4. *Welded bonnets*. Used in small steel valves only; recommended for high-

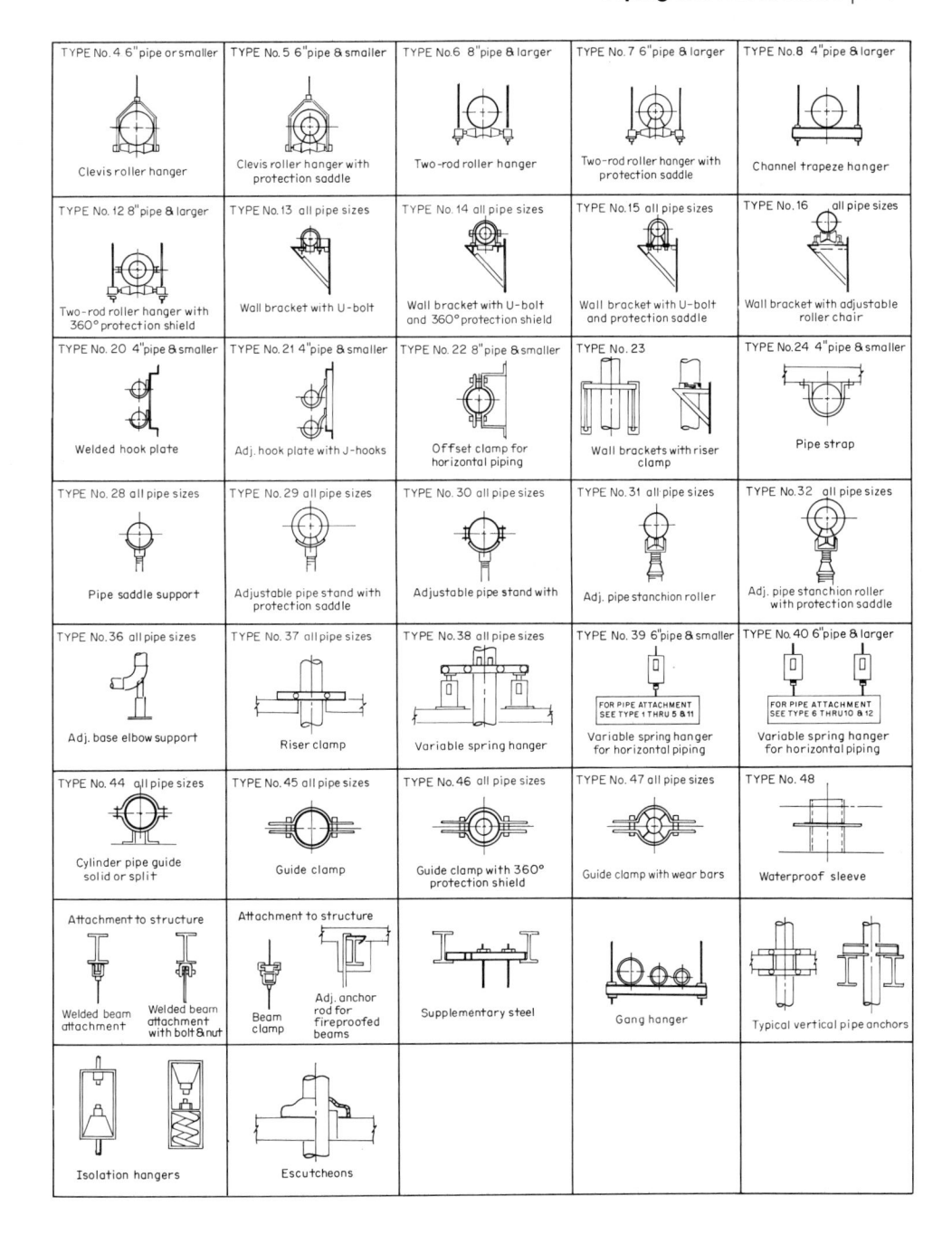

TYPE No. 4 6"pipe or smaller	TYPE No. 5 6"pipe & smaller	TYPE No. 6 8"pipe & larger	TYPE No. 7 6"pipe & larger	TYPE No. 8 4"pipe & larger
Clevis roller hanger	Clevis roller hanger with protection saddle	Two-rod roller hanger	Two-rod roller hanger with protection saddle	Channel trapeze hanger
TYPE No. 12 8"pipe & larger	TYPE No. 13 all pipe sizes	TYPE No. 14 all pipe sizes	TYPE No. 15 all pipe sizes	TYPE No. 16 all pipe sizes
Two-rod roller hanger with 360° protection shield	Wall bracket with U-bolt	Wall bracket with U-bolt and 360° protection shield	Wall bracket with U-bolt and protection saddle	Wall bracket with adjustable roller chair
TYPE No. 20 4"pipe & smaller	TYPE No. 21 4"pipe & smaller	TYPE No. 22 8"pipe & smaller	TYPE No. 23	TYPE No. 24 4"pipe & smaller
Welded hook plate	Adj. hook plate with J-hooks	Offset clamp for horizontal piping	Wall brackets with riser clamp	Pipe strap
TYPE No. 28 all pipe sizes	TYPE No. 29 all pipe sizes	TYPE No. 30 all pipe sizes	TYPE No. 31 all pipe sizes	TYPE No. 32 all pipe sizes
Pipe saddle support	Adjustable pipe stand with protection saddle	Adjustable pipe stand with	Adj. pipe stanchion roller	Adj. pipe stanchion roller with protection saddle
TYPE No. 36 all pipe sizes	TYPE No. 37 all pipe sizes	TYPE No. 38 all pipe sizes	TYPE No. 39 6"pipe & smaller	TYPE No. 40 6"pipe & larger
Adj. base elbow support	Riser clamp	Variable spring hanger	Variable spring hanger for horizontal piping	Variable spring hanger for horizontal piping
TYPE No. 44 all pipe sizes	TYPE No. 45 all pipe sizes	TYPE No. 46 all pipe sizes	TYPE No. 47 all pipe sizes	TYPE No. 48
Cylinder pipe guide solid or split	Guide clamp	Guide clamp with 360° protection shield	Guide clamp with wear bars	Waterproof sleeve
Welded beam attachment / Welded beam attachment with bolt & nut (Attachment to structure)	Beam clamp / Adj. anchor rod for fireproofed beams (Attachment to structure)	Supplementary steel	Gang hanger	Typical vertical pipe anchors
Isolation hangers	Escutcheons			

FIGURE 8-3 Types of hangers and supports. *(Continued)*

pressure, high-temperature steam service; difficult to dismantle and reassemble

5. *Pressure-seal bonnets.* Recommended for high-temperature steam

Valve-Stem Operation

1. *Rising stem with outside screw.* Recommended for steam and high-temperature, high-pressure water service

2. *Rising stem with inside screw.* Recommended in smaller valves

3. *Nonrising stem with outside screw.* Generally used in gate valves; not desirable for use with fluids that cause corrosion

4. *Sliding stem (quick opening).* Useful where quick opening and closing are desirable

Pipe Ends and Valve Connections

1. *Screwed ends.* Suitable for all pressures; require unions to facilitate dismantling; recommended for small valves

2. *Welded ends.* Used mainly for high-temperature and high-pressure service and for tight joints; available in butt and socket-weld ends

3. *Flanged ends.* Generally used in large valves; easy to dismantle and reassemble

4. *Soldered ends.* Generally used in low-temperature service with copper pipe

5. *Brazed ends.* Generally used in low-pressure service; made of brass

6. *Flared ends.* Commonly used in metal and plastic tubing; Available in sizes up to 2 in

Disk Construction

1. *Gate valve*
   - Solid-wedge disk
   - Split-wedge disk
   - Flexible-wedge disk
   - Double-disk parallel-seat

2. *Globe, angle, and Y valves*
   - Plug disk
   - Narrow seat (conventional disk)
   - Composition disk
   - Needle valve (expansion valve)

3. *Refrigerant valves.* Back-seating globe valves of either packed or diaphragm packless type

4. *Check valves.* Composition disk of either swing or lift type

5. *Relief valves.* Usually loaded by spring weight or electric device

EXPANSION JOINTS

Expansion joints are used in heating systems to allow for pipe expansion. There are three types of expansion joints involving temperatures of up to 400°F. These are:

1. *Packless* expansion joints include bellows expansion joints, rubber, corrugated copper, and other metals.

2. *Slip-type packed* expansion joints include single and double type. These joints allow for expansion by sliding a female member over a male member. The packing is used to obtain a tight joint.

3. *Flexible-ball* pipe joints are commonly used in mains and risers for high-pressure service. They are used with screwed and welded fittings.

Expansion U bends are also used to allow for pipe expansion (cold-spring method). They include the regular U bend, offset U bend, and U bend with two or four fittings.

STRAINERS

Strainers are normally installed in the pipeline ahead of control valves, pumps, steam traps, and other equipment that should be protected against the damage caused by dirt and pipe scale. The various types of strainers are:

1. *Y-type* strainers are made of cast-iron or semisteel bodies with a stainless-steel perforated screen. The perforations depend on the degree of protection required. Where Y strainers are used on 2-in and larger piping or on water-pump suction lines, the strainer may be equipped with a drain valve.

2. Simplex basket-type strainers consist of cast-iron bodies and removable tops with stainless-steel baskets.

Basket-type strainers are also available with duplex basket and are commonly used in fuel-oil piping systems. Strainers are available with screwed ends and flanged ends (2 in and larger).

STEAM TRAPS

A steam trap is used in a steam system to hold steam in a heating unit or piping system and allow condensate and air to pass through. The steam remains trapped until it gives up its latent heat and changes to condensate. The size of a steam trap depends on the amount of condensate (pounds per hour), pressure differential between inlet and discharge at the trap, and the safety factor used to select the trap. The types of steam traps commonly used in steam systems are indicated in Table 8-10.

TABLE 8-10 Types and Applications of Steam Traps

Trap Type	Discharge	Applications
Float	Continuous	Draining condensate from steam headers and steam heateing coils; a low-pressure trap should be equipped with a thermostatic air vent
Thermostatic	Intermittent	Draining condensate from radiators, convectors, unit heaters, steam coils; recommended for 15 psi maximum
Float and thermostatic	Intermittent	Draining condensate from blast heaters, steam coils, unit heaters; recommended for large volumes of condensate and 15 psi maximum
Upright bucket	Intermittent	Draining condensate and air from blast coils, steam mains, unit heaters; at least 1 psi pressure drop is required; recommended for systems that have pulsating pressures
Inverted bucket	Intermittent	Draining condensate and air from blast coils, unit heaters, steam coils; recommended for discharging abnormal amounts of air
Flash	Intermittent	Recommended for pressure drops of 5 psi or more
Impulse (thermo)	Intermittent for normal loads and continuous at heavy loads	Recommended for pressure drops of at least 25% of the inlet pressure
Lifting bucket	Intermittent	Recommended for steam heating systems up to 150 psig

AIR VENTS   Air vents should be installed at the high points of all closed water systems that cannot vent back to an open expansion tank. Systems using a closed expansion tank require vents at all high points. A runoff drain should be provided at each vent to carry possible leakage to a suitable drain line. Air vents are available in automatic and manual types.

THERMOMETERS AND GAUGES

Thermometers and gauges are used in a system to determine the water temperature or pressure. Water thermometers are usually selected for a range of 30 to 200°F. They should be equipped with wells. Pressure gauges are selected so that the normal reading of the gauge is near the midpoint of the pressure scale. Thermometers and gauges are normally installed on the supply and/or return lines of pumps, chillers, heat exchangers, air-handling units, and in other locations as indicated on the drawings and specifications.

PIPE SIZES

Pipe sizes are determined according to flow rates, piping friction losses, and fluid velocities. However, in the preliminary-design stage, pipe sizes are normally not shown on the job drawings. In order to obtain an accurate estimate for piping systems, the estimator can quickly rough out pipe sizes. Figure 8-4 shows a B & G System Syzer. This device was engineered by the Bell and Gosset Company to simplify the designing of hydronic heating and cooling systems. It contains all the data necessary for determining flow rates, pipe sizes, and system pressure drops. The System Syzer can be used by the estimator to determine pipe sizes. It can be ordered from the local representative of the Bell & Gosset Company. It is not distributed from the home office.

ESTIMATING

Takeoff Procedure

The takeoff procedure is discussed in general terms in Chap. 1.

The following steps will be of help in obtaining an accurate quantity takeoff:

1. Prepare the required takeoff sheets. Typical takeoff sheets are shown in Figs. 8-5 through 8-8.

2. Read the specifications carefully; then list the types of materials and joints specified for each system.

3. Use one takeoff sheet for each type of material which requires a particular type of joint. List on this sheet all quantity takeoffs for the various piping systems which require this material and joint and are covered with the same insulating material.

4. Takeoff each system, as indicated on the floor plans, section details, and flow diagrams. The takeoff must include pipe, fittings, valves, joints, hangers, and all piping accessories.

5. Measure the length in feet. Count the fittings, valves, joints, hangers, etc.

6. List in one column of the takeoff sheet all quantities which may have the same size.

HVAC Systems Estimating Manual | 154

FIGURE 8-4 B & G System Syzer. *(Used by permission of Bell & Gosset Company, ITT Fluid Handling Division.)*

Piping system _____						Project Name _____					
Material _____						Job No. _____					
Pipe sizes, inches											
Feet	No. of joints	Feet	No. of joints	Feet	No. of joints	Feet	No. of joints	Feet	No. of joints	Feet	No. of joints
Total											

P-8P

FIGURE 8-5 Pipe takeoff sheet.

FIGURE 8-4 B & G System Syzer. *(Continued)*

FIGURE 8-6 Fitting takeoff sheet.

Material ____________________ Project Name ____________________

Class ____________________ Job No. ____________________

Type of valve and device	Value and device sizes, inches							

P-8V

FIGURE 8-7 Valves and devices takeoff sheet.

7. After completing the takeoff, add the quantities of each size, and transpose the sum to the piping estimating sheet (see sheet P-8E in Fig. 8-9).
8. Use the labor figures shown in this section and manufacturers' price lists to find and note the unit price and man-hour figure for each size. Calculate the total material cost and labor man-hour figure.

Note: To figure solder and brazing alloys, use the information given in Figs. 8-1 and 8-2.

Project Name ____________________

Job No. ____________________

Type of hangers	Pipe sizes, inches							

P-8H

FIGURE 8-8 Hanger and support takeoff sheet.

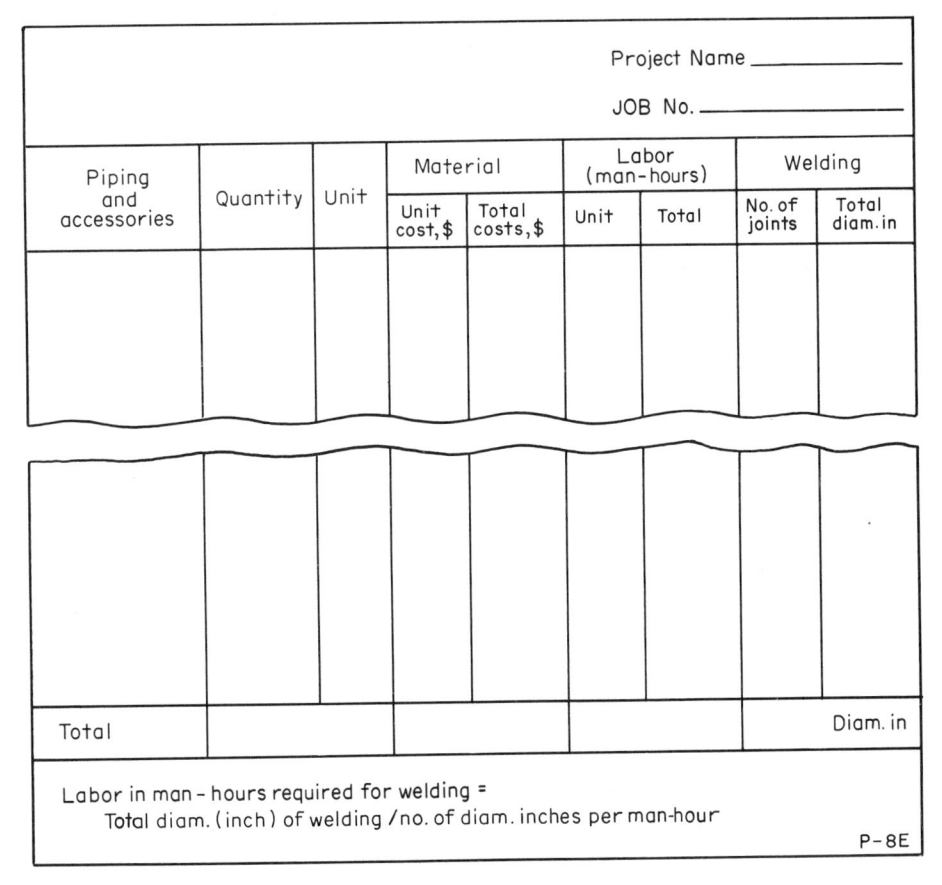

[Figure: Piping and accessories estimating sheet form with columns for Piping and accessories, Quantity, Unit, Material (Unit cost $, Total costs $), Labor man-hours (Unit, Total), Welding (No. of joints, Total diam. in). Bottom row labeled Total. Formula at bottom: Labor in man-hours required for welding = Total diam. (inch) of welding / no. of diam. inches per man-hour. Form ID: P-8E]

FIGURE 8-9 Piping and accessories estimating sheet.

Piping and Accessories Takeoff and Estimating Sheets

A well-designed takeoff sheet reduces the work load and the chance of error. Piping and accessories takeoff sheets are classified in Table 8-11.

Sheet P-8E (see Fig. 8-9) is an estimating sheet which includes columns for quantity (transposed from takeoff sheets), material cost, labor man-hours, number of joints, and welding. This sheet may be used to calculate the material cost and labor man-hours for each classification (i.e., pipe, fittings, valves, and hangers); these subtotals may then be transposed to the job estimating sheets given in Chap. 1 (see Fig. 1-4).

TABLE 8-11 Piping and Accessories Takeoff Sheets

Classification	Takeoff Sheet Number*
Pipe	P-8P
Fittings	P-8F
Valves and devices	P-8V
Hangers	P-8H

*See Figs. 8-5 through 8-8.

ESTIMATING CHART 8-1 Labor Man-hours for Threaded Pipes.

Nominal Pipe Size, in	Black and Galv. Steel Pipe T & C Schedule 40	Black and Galv. Steel Pipe T & C Schedule 80	Copper and Red Brass Pipe T & C Standard Weight	Copper and Red Brass Pipe T & C Extra Heavy	PVC Pipe T & C Schedule 80	Additional for Field-threaded Steel and Brass
1/8	.06	.06	.06	.06		.02
1/4	.07	.07	.07	.07	.06	.02
3/8	.08	.08	.08	.08	.06	.03
1/2	.09	.09	.09	.09	.07	.03
3/4	.10	.10	.10	.10	.08	.03
1	.11	.11	.11	.11	.09	.03
1 1/4	.14	.15	.14	.15	.11	.03
1 1/2	.16	.17	.16	.17	.14	.04
2	.20	.22	.20	.22	.19	.05
2 1/2	.26	.29	.26	.29	.23	.05
3	.31	.34	.31	.34	.26	.12
3 1/2	.33	.37	.33	.37		.14
4	.35	.39	.35	.39	.29	.15
5	.44	.49	.44	.49		.19
6	.50	.55	.50	.55	.40	.22

ESTIMATING CHART 8-2 Labor Man-hours for Pipes with Various Types of Joints.

Nominal Pipe Size, in	Threadless (TP) Brass Pipe with Silver Brazed Joint	Copper Tubing with 50-50 Soldered Joint				PVC Pipe with Solvent-cemented Socket Joint	
		Type K	Type L	Type M	Type DWV	Schedule 40	Schedule 80
1/4	.10	.09	.09	.09			.07
3/8	.11	.10	.10	.10			.07
1/2	.14	.13	.12	.11		.08	.08
3/4	.17	.16	.15	.14		.09	.09
1	.21	.20	.19	.18		.11	.11
1 1/4	.27	.26	.25	.24	.23	.14	.14
1 1/2	.30	.28	.27	.26	.25	.17	.17
2	.35	.33	.32	.30	.30	.21	.22
2 1/2	.44	.42	.40	.38		.26	.27
3	.50	.47	.45	.43	.42	.30	.32
3 1/2	.58	.55	.53	.50			
4	.65	.62	.59	.56	.55	.35	.37
5	.78	.74	.71	.67	.66		
6	.96	.91	.87	.83	.81	.52	.55

Notes:

1. For 1 1/4-in and smaller types K and L soft copper tubing with flared joints, use the labor figures given above.
2. For types K, L, and M with 95-5 solder joints, add 10%.
3. For types K, L, and M with silver-solder joints, add 15%.
4. For ACR tube joined with 95-5 solder or silver solder, use the labor figures for type K with 95-5 solder or silver solder.
5. For PVC pipe with fusion-welded or heat-sealed joints, deduct 20% from schedule 40 and 15% from schedule 80.

ESTIMATING CHART 8-3 Labor Man-hours for Black-Steel-Pipe Welded Joints*,†

Nom. Pipe Size, in	Schedule 40	Schedule 80	Nom. Pipe Size, in	Schedule 40	Schedule 80
¾	.05	.06	6	.22	.24
1	.06	.07	8	.29	.32
1¼	.07	.08	10	.35	.38
1½	.08	.09	12	.43	.47
2	.09	.10	14	.50	.55
2½	.10	.11	16	.57	.62
3	.11	.12	18	.65	.71
4	.14	.16	20	.72	.79
5	.16	.18	24	.80	.88

*These labor units do not include the labor for making welding joints. Add labor for welding according to the information given in the welding section.

†*Notes:* For underground steel conduits, add 20% to the labor figures given above for setting the conduit in place and leveling it. Also add the following labor costs for making the welded joint:

1. Labor for joining a carrier pipe or pipes.

2. Labor for joining a conduit. For 10-gauge steel conduit, use 3 diameter-in/man-hour as an average labor figure for welding.

Conduits are normally furnished in heavy 40-ft lengths. They are moved and put in place by a rigging crew with rigging equipment. The costs involved in this process must be added to the total cost of the conduit system.

ESTIMATING CHART 8-4 Labor Man-hours for Hangers and Supports

	Rod Size, in								
	3/8	1/2	5/8	3/4	7/8	1	1¼	1¼	1¼
	Pipe Size, in								
Classification	½–2	2½–3½	4–5	6	8–12	14–16	18	20	24
Riser clamp	.20	.25	.32	.32	.35	.40	.45	.45	.50
Clevis hanger	.80	.85	.89	.94	.99	1.08	1.20	1.30	1.45
Clevis roller hanger	.85	.91	.96	1.00	1.11	1.20	1.32	1.45	1.60
Two-rod roller hanger	1.62	1.69	1.79	1.87	197	2.16	2.38	2.61	2.88
Channel trapeze hanger	1.46	1.54	1.62	1.69	1.78	1.96	2.15	2.37	2.60
Wall bracket with U bolt	.78	.80	.85	.89	.90	1.05	1.15	1.32	1.60
Additional for saddle	.50	.50	.50	.50	.50	.85	.85	.85	.85
Additional for shield	.18	.18	.18	.18	.18	.25	.25	.25	.25
Additional for spring type (per spring)	.75	.75	.75	.75	.75	.75	1.80	1.80	1.80
Slide pipe guide	1.00	1.00	1.00	1.00	1.15	1.30	1.60	1.60	1.80
Cylinder pipe guide	2.00	2.30	2.30	3.40	3.60	4.30	4.30	4.70	4.70
Anchors (wide flange and plates)	3.10	3.90	4.10	4.25	4.70	5.50	6.00	6.10	8.00
Sheet-metal sleeve (12-in cuts)	.30	.30	.30	.30	.40	.50	.50	.60	.60
Steel sleeve (12-in cuts)	.50	.50	.50	.50	.75	.95	1.35	1.80	2.25

ESTIMATING CHART 8-5 Labor Man-hours for Brass and Bronze Valves

Valve Size, in	Flanged Valves Gate and Globe (150 lb)	Threaded Valves				Sweat Connection Valves		95-5 Soldered Connection Valves	
		Angle, Ball Gate, Globe, and Lubr. Plug Cock (150–200 lb)	Diaphragm (150 lb), Gate, Globe (300 lb)	O.S. & Y Gate (125 lb)	Check, Non-Lubr. Plug Cock (125–300 lb)	Gate and Globe (125 lb)	Check Non-Lubr. Plug Cock (125 lb)	Globe Refrigeration (500 lb)	Check Refrigeration (500 lb)
½		.41	.41		.40	.33	.32		
¾		.51	.54		.50	.40	.39	.42	.41
1	.64	.56	.60		.54	.42	.41	.46	.45
1¼	.68	.74	.77		.70	.53	.52	.58	.57
1½	.76	.89	.95		.84	.62	.59	.68	.65
2	.95	1.10	1.20	1.26	1.03	.73	.70	.80	.76
2½	1.10	1.47	1.60	1.61	1.36	.96	.90	1.06	1.00
3	1.26	1.66	1.80	1.88	1.56	1.10	1.05	1.21	1.15

ESTIMATING CHART 8-6 Labor Man-hours for Ball, Butterfly, and Diaphragm Valves

Valve Type	Class, lb	½	¾	1	1¼	1½	2	2½	3	4	5	6	8	10	12
Ball valve															
Socket weld	150	.75	.90	1.15	1.30	1.60	2.00								
Flanged	150			.75	.80	.88	.94	1.04	1.18	2.20	2.30	2.40			
C.I. threaded		.40	.50	.54		.86	1.00		1.65						
Flanged-wafer butterfly valve															
Wrench operated	150						.40	.42	.50	.64	.79	.84	1.16	1.60	2.25
Gear operated							.65	.70	.80	.90	1.05	1.10	1.60	2.10	2.75
Diaphragm valve															
Flanged	150	.44	.60	.67	.68	.76	.92	1.06	1.16	2.00	2.40	2.80	4.00	5.50	6.30
PVC socket weld		.30	.38	.40	.50	.59	.70								

ESTIMATING CHART 8-7 Labor Man-hours for Iron-Body, Cast-Iron, and Cast-Steel Valves

Valve Type	Class, lb	2	2½	3	4	5	6	8	10	12	14	16	18	20	24
Flanged-Iron-Body Valves															
Angle gate and globe	125	1.45	1.76	2.06	2.56	3.28	4.16	5.00	7.00	7.75	10.20	11.50	12.30	13.82	16.12
	250	1.90	2.30	2.68	3.35	4.28	5.43	6.55							
Gate and globe OS & Y	125	1.45	1.78	2.10	2.60	3.44	4.36	5.15	7.25	8.00	10.80	11.75	12.35	14.00	16.20
	250	1.92	2.32	2.74	3.37	4.40	5.50	6.70	9.35	10.78	12.05	12.10	14.85	15.86	18.50
Swing check	125	1.29	1.52	1.74	2.15	2.56	3.75	4.75	7.00	7.75	10.20	11.50	12.30		
	250	1.66	1.95	2.23	2.76	3.28	4.80	5.91							
Lift check	125			1.37	1.62	1.90	2.63	3.20	4.82	6.00	8.30	9.87	10.30	11.81	13.50
	250	1.50	1.71	1.92	2.26	2.65	3.68	4.41	6.72	8.10					
Cast-Iron Valves															
Gate (threaded)	200	1.30		2.30	3.20		6.0								
Gate OS & Y (threaded)	200	1.33		2.36	3.30										
Gate (flanged)	150 200	1.26		2.37	2.80		4.27	5.30	6.80	8.00	10.60	12.00	13.40	14.00	17.00
Gate OS & Y (flanged)	200	1.28		2.41	2.90		4.32	5.50	7.10	8.36	11.10	12.50	13.70	14.30	18.00
Cast-Steel Valves															
Gate (welded end)	150	3.50	4.25	4.90	6.42	7.75	9.30	11.80	14.90	18.40	21.50	24.00	26.40	29.20	34.60
	300	4.20	5.15	5.90	7.70	9.30	10.85	14.50	18.50	25.00	30.00	34.00	42.00	50.00	70.00
Gate OS & Y (flanged)	300	2.25	2.62	3.06	3.96	4.62	5.96	7.36	10.12	11.35	12.65	13.00	16.20	18.10	21.75
Angle and globe (welded end)	150	3.50	4.25	4.90	6.50	7.95	9.60	12.00							
	300	4.00	5.00	5.50	7.50	9.00	10.40	14.00	18.80	25.00					
Globe OS & Y (flanged)	150	1.28	1.42	1.75	3.00	4.11	4.47	5.57							
	300	1.96	2.44	2.78	3.61	4.41	5.68	7.04	10.10	11.31					

ESTIMATING CHART 8-8 Labor Man-hours for Plug Valves

Valve Type	Class, lb	1½	2	2½	3	4	5	6	8	10	12	14	16	18	20	24
3-way flanged	125	1.26	1.50	1.85	2.50	3.25	4.00	5.00	7.00	10.35	10.98		15.00			
4-way flanged			2.20	2.50	3.00	5.50	10.00	12.00	12.50	25.00	27.00		35.00			
Wrench-operated non-lubr., flanged	150		1.25	1.30	1.50	2.40		4.00	5.25	6.90	7.50	9.10	9.15	12.00		
Wrench-operated lubr. flanged		85	1.50	1.85	2.20	2.50	3.00	4.50	5.50	7.50	8.60	11.95	12.00	13.65	14.55	16.96
Gear-operated lubr. flanged							3.50	5.05	6.20	8.40	9.60	13.40	13.45	15.30	16.30	19.00

ESTIMATING CHART 8-9 Labor Man-hours for Pressure Regulating Valves (PRV)*

Valve Type	Maximum Pressure, lb	½	¾	1	1¼	1½	2	2½	3	4	5	6	8	10	12
Steam PRV															
Threaded	250	1.40	1.70	2.00	2.30	2.60	3.00								
Flanged	15							2.65	3.00	4.80	5.90	6.95	8.00	11.15	13.95
Flanged	125							3.00	3.50	5.90	7.15	7.95	9.35	14.00	15.40

*For temperature and pressure-temperature regulating valves, refer to Estimating Charts 11-4 and 11-5.

ESTIMATING CHART 8-10 Labor Man-hours for Relief Valves*

Valve Type	Pressure, lb	1½	2	2½	3	4	6
C.I. flanged relief valve	250	1.85	2.20	3.10	3.92	5.18	6.00

*For ½-in threaded relief valve, estimate 0.75 man-hour.

ESTIMATING CHART 8-11 Labor Man-hours for Expansion Joints

Type	Class, lb	1/2	3/4	1	1 1/4	1 1/2	2	2 1/2	3	4	5	6	8	10	12	14	16
Straight Swivel Joints																	
C.S. & mall. screwed	Std.	.88	1.08	1.24	1.37	1.46	1.73	2.56	2.95	3.80		7.52	10.07				
S.S. screwed	Std.	.97	1.23	1.39	1.57	1.64	1.98	2.92	3.40	4.40							
Alum. screwed	Std.		1.14	1.30	1.46	1.51	1.78	2.67	3.10	4.00							
Flanged end	150					1.46	1.58	1.60	2.65			3.48	4.25	6.45	7.50		
Straight Ball Joints																	
Welded end	Std.						4.83		6.80	9.22		9.76	13.00	16.29	19.60	22.92	26.62
Welded end	XH VY						5.56		8.00	10.47		10.97	14.61	18.29	22.00	25.00	29.00
Flanged end	150											4.95	6.79	9.15	10.15	12.63	15.07
Flanged end	300											6.74	8.85	11.28	13.70	16.50	18.00
Slip-type Joints																	
S.S. and bronze	150		1.30	1.45	1.65	1.75	2.05	3.00	3.50								
Single joint	125				1.70	1.78	1.93	2.00	2.31	3.76	4.45	4.67	6.34	8.55	9.07	11.24	13.06
Double joint	125				1.95	1.99	2.17	2.29	2.76	4.45	5.30	5.75	7.35	9.72	10.60	13.32	15.23
Single Joint	250				1.90	2.07	3.05	3.28	3.86	4.49	5.03	6.65	8.35	10.97	12.89	15.47	16.27
Double joint	250					2.25	2.41	3.39	3.68	4.48	5.42	6.20	7.86	9.42	12.54	14.96	17.07	17.27
Corrugated Joints																	
One corrugation flanged end	125								2.50	3.90	4.42	4.67	5.82	8.30	9.30	10.97	12.00
One corrugation flanged end	250								3.80	4.40	4.75	6.80	8.37	11.30	13.17	15.39	15.80
One corrugation welded end	125								6.28	7.85	9.07	10.62	13.54	16.52	20.10	23.13	26.60
One corrugation welded end	250								6.30	7.91	9.20	10.80	13.90	17.60	20.30	23.50	27.00
Multi-corrugation: add value for each additional corrugation									.10	.15	.20	.30	.35	.40	.50	.55	.60

ESTIMATING CHART 8-12 Labor Man-hours for Strainers

Strainer Type	Class, lb	½	¾	1	1¼	1½	2	2½	3	4	5	6	8	10	12
Iron body, screwed (Y)	250	.76	.96	1.03	1.10	1.22	1.62	2.16	2.60						
Iron body, flanged (Y)	125						1.89	2.21	2.52	3.21	4.13	5.60	6.84		
Steel screwed simplex basket	125	.90	1.05	1.20	1.30	1.45	1.70	2.60	3.05	4.32	6.40	7.40			
C.I. flanged simplex basket	125		1.12	1.16	1.21	1.37	1.50	2.01	2.06	3.28	4.67	5.60	6.02	8.11	9.50
C.I. flanged duplex basket	125			1.75	1.94	2.02	2.40	3.70	3.80	5.90	7.85	8.90	9.20	10.60	11.30

ESTIMATING CHART 8-13 Labor Man-hours for Steam Traps

Trap Type	½	¾	1	1¼	1½	2
Float	.80	.95	1.60	1.70	2.00	2.15
Thermostatic	.60	.85	1.10			
Float-thermostatic	.80	1.05	1.25	1.40	1.60	2.00
Bucket	.83	1.10	1.30	1.65	1.85	2.30

ESTIMATING CHART 8-14 Labor Man-hours for Air Vents

Type	Man-hours
½-in Hydroscopic	0.30
½-in Float operated	0.65
½-in Float operated with 10-ft capillary	1.25

ESTIMATING CHART 8-15 Labor Man-hours for Thermometers

Type	Man-hours
Direct reading with well	1.00
Remote with 10-ft capillary	1.75
Thermometer well only	0.50

ESTIMATING CHART 8-16 Labor Man-hours for Gauges

Type	Man-hour
Gauge cock and siphon	1.00
Gauge cock and snuber	1.00
Gauge cock only	0.40

Material Costs

Material prices for piping and accessories can best be obtained from manufacturers' price lists. These prices may vary, depending upon price trends, volume of order, discount rate, and availability of materials. For these reasons, material prices for piping and accessories are not included in this chapter.

The total material cost for piping and accessories must include the subtotal material costs for pipe, fittings, valves, hangers, and miscellaneous devices, as well as brazing alloys, solder, and welding materials.

Labor Man-hours

The labor figures indicated in Estimating Charts 8-1 through 8-3 are in units of man-hours per foot of pipe. The labor figures include handling and erecting the pipe, based on normal working conditions. They also include the labor required for making one joint every 20 ft of pipe and installing one fitting every 10 ft of pipe. These figures are applicable to threaded, unthreaded, brazed, and soldered joints; they do not, however, include the labor required for making welded joints. The welding man-hours must be added to the labor figures (see the section on welding in this chapter).

The labor figures for hangers, supports, and other devices include the labor required for handling and installing one unit of each (see Estimating Charts 8-4 through 8-16).

9

SHEET-METAL WORK (DUCTWORK)

GENERAL The duct, plenum, apparatus casing, and acoustic liner are the various elements requiring sheet-metal work. The function of ductwork is to supply the conditioned air to the space served, as well as to remove the exhaust and return air from that space. It is important to construct ducts that are sufficiently airtight; will not vibrate or breathe when the airstream varies in pressure; and will ensure an even flow of air without undue pressure loss. The major component of duct cost is the construction labor. Historical cost data indicate that the cost of ductwork and related insulation may be in the range of 20 to 40% of the total cost of a heating, ventilating, and air-conditioning system. Duct construction must conform to the latest standards of the American Society of Heating, Refrigerating and Air Conditioning Engineers (ASHRAE); the Sheet Metal and Air Conditioning Contractors National Association (SMACNA); the National Fire Protection Association (NFPA); and local codes.

MATERIALS The materials commonly used in duct construction are:

1. *Galvanized steel sheets.* Galvanized steel is the most commonly used material for duct construction, owing to its low cost. This material has a galvanized coating of 1¼ oz (total for both sides) per square foot of sheet. A galvanized steel duct can be painted when used on the outside of a building to increase its corrosion resistance.

2. *Black-steel sheets (hot or cold rolled).* Black steel is normally used for ducts and hoods conveying high-temperature air or gases in such applications as boiler breeching, kitchen exhaust ducts, and hoods. This material can also be used for belt guards and fire dampers. When it is used for drain pans, supplementary waterproof mastic coatings must be provided.

3. *Copper and stainless-steel sheets.* These materials have maximum resistance to moisture. They can be used for exposed ducts, shower and

swimming-pool exhaust ducts where extreme humidity occurs, and fume hoods and fume exhaust ducts when other metals are not satisfactory. Stainless steel is used more than copper, owing to its higher rigidity and strength and lower cost.

4. *Aluminum sheets.* Aluminum sheet is lighter than other materials and has a superior resistance to corrosion. Aluminum ducts are suitable for interior use where extreme humidity occurs, and for exterior use without being painted. Aluminum ducts are recommended for low-pressure systems only.

5. *Asbestos.* Asbestos is used for ducts not exposed to moisture and for fume exhaust ducts.

6. *Transite or asbestos board.* These are used for fume exhaust systems where extremely high temperatures occur and for underground air ducts.

7. *Plastic (PVC).* This is used for corrosive-fume exhaust systems.

8. *Fibrous glass.* A fibrous-glass duct is a composite of rigid fiberglass board, a special factory-applied jacket, and the field-applied joining system. A fibrous-glass duct is made in molded round sections or in rigid board form for fabrication into rectangular shapes. It is commonly used in low-velocity applications where insulation or acoustic lining is required.

9. *Flexible ducts.* Flexible ducts can be made of aluminum factory-insulated hose. They can be furnished in 10-ft lengths. Flexible ducts are used for connectors to air-terminal boxes, diffusers, and light troffers. Connections can be clamped with hose clamps on the ends and sealed with cemented sealer to obtain airtight connections.

PRESSURE AND VELOCITY CLASSIFICATIONS

Duct construction is normally classified in terms of operating pressure and velocity. Table 9-1 indicates the classifications which are commonly used in air-distribution systems. However, ducts for medium- and high-pressure service are used in modern buildings to meet space limitations when large air quantities are required.

TABLE 9-1 Pressure and Velocity Classifications for Ductwork

System	Velocity, ft/min	Static Pressure, in of water
Low pressure, low velocity	Up to 2000	Up to 2
Medium pressure, high velocity	Over 2000	2–6
High pressure, high velocity	Over 2000	6–10

TABLE 9-2 Weights and Thicknesses of Galvanized Steel, Black-Steel, and Stainless-Steel Sheets

U.S. Gauge	Galvanized Steel Weight, lb/ft²	Galvanized Steel Thickness, in	Black Steel Weight, lb/ft²	Black Steel Thickness, in	Stainless Steel Weight, lb/ft²	Stainless Steel Thickness, in
26	0.906	0.0217	0.75	0.0179	0.787	0.0187
24	1.156	0.0276	1.00	0.0239	1.05	0.025
22	1.406	0.0336	1.25	0.0299	1.313	0.0312
20	1.656	0.0396	1.50	0.0359	1.575	0.0375
18	2.156	0.0516	2.00	0.0478	2.10	0.05
16	2.656	0.0635	2.50	0.0598	2.625	0.0625
14	3.281	0.0785	3.125	0.0747	3.281	0.0781
12	4.531	0.1084	4.375	0.1046	4.593	0.1093
10	5.781	0.1382	5.625	0.1345	5.906	0.1406

WEIGHTS AND THICKNESSES OF METAL SHEETS

The standard weights and thicknesses of galvanized steel, black-steel, and stainless-steel sheets are given in Table 9-2. For aluminum alloy and copper, refer to Table 9-3.

DUCT TYPES Ducts can be made in round, rectangular, or flat oval shapes. Round ducts are less expensive to fabricate and to install than the rectangular and oval types. Flat oval ducts combine the advantages of round and rectangular ducts, have considerably less flat surface that is susceptible to vibration, and

TABLE 9-3 Weights and Thicknesses of Aluminum and Copper Sheets

B & S Gauges	Aluminum Sheets (3003) Weight, lb/ft²	Aluminum Sheets (3003) Thickness, in	Copper Sheets Weight, oz/ft²	Copper Sheets Thickness, in
26	0.226	0.016		
24	0.285	0.02	16	0.0216
22	0.365	0.025	20	0.027
20	0.456	0.032	24	0.0323
18	0.57	0.04	30	0.04
16	0.713	0.05	38	0.051
14	0.898	0.063	48	0.0647
12	1.00	0.071		
10	1.13	0.08		

require less reinforcement than corresponding sizes of rectangular duct. They are furnished in varying aspect ratios.

Round and flat oval ducts are recommended for medium- and high-velocity systems. The rectangular low-velocity duct is less expensive than the rectangular high-velocity duct, owing to the use of simple joining methods.

RECOMMENDED CONSTRUCTIONS AND GAUGES FOR VARIOUS TYPES OF DUCTS

The recommended constructions, gauges, connections, and reinforcing for various types of ducts are described in Tables 9-4 through 9-10. These tables, taken from actual job specifications, are given for general reference. Duct construction must conform to voluntary industry standards and local codes. Further construction details can be found in the applicable SMACNA duct-construction standards.

TABLE 9-4 Recommended Construction for Galvanized Steel Rectangular Ducts—Low Pressure, S.P. up to 2 in, Velocity up to 2000 ft/min*

Steel U.S. Standard Gauge†	Maximum Side, in	Type of Transverse Joint Connections	Bracing
26	Up to 12	S, drive, pocket, or bar slips	None
24	13–18	S, drive, pocket, or bar slips	None
24	19–30	Hemmed S, pocket, or bar slips	1 × 1 × ⅛ in angles, 5 ft C-C
22	31–42	Hemmed S, pocket, or bar slips	1 × 1 × ⅛ in angles, 5 ft C-C
22	43–54	1½-in angle connections, bar, or pocket slips	1½ × 1½ × ⅛ in angles, 5 ft C-C
20	55–60	1½-in angle connections, bar, or pocket slips	1½ × 1½ × ⅛ in angles, 5 ft C-C
20	61–84	1½-in angle connections, angle, or bar slips	1½ × 1½ × ⅛ in angles, 2.5 ft C-C
18	85–96	1½-in angle connections, angle, or pocket slips	1½ × 1½ × 3/16 in angles, 2.5 ft C-C
18	Over 96	2-in angle connections, angle, or pocket slips	2 × 2 × ¼ in angles, 2.5 ft C-C

*Constructions in this table are among those recommended in SMACNA's *Low Velocity Duct Construction Standards,* 4th ed., 1969.
†Aluminum ductwork must be one gauge size heavier than that specified for galvanized ductwork.

TABLE 9-5 Recommended Construction for Galvanized Steel Round Ducts—Low Pressure, S.P. up to 2 in, Velocity up to 2000 ft/min

U.S. Gauge	26	24	22	20	18	
Duct diameter, in	Up to 14	15–26	27–36	37–50	51–60	61–84

Galvanized Steel Ductwork

The recommended construction for galvanized steel ductwork is described in Tables 9-4 through 9-9. Note that the duct-dimensions columns in the tables for rectangular ducts refer to the maximum dimension, i.e., the top, bottom, or side of a rectangular duct. This maximum side determines the recommended metal gauge for all four sides of the duct. For example, for 36 × 12 in duct, use 36 in as the maximum side to determine, from the applicable table, the recommended metal gauge for the duct. Note, however, that rectangular ducts are reinforced on the basis of the width of each side, separately evaluated. The maximum side does not set the reinforcement for the shorter sides.

Black-Steel Ductwork

The recommended construction for black-steel ductwork is the same as for galvanized steel ductwork, except that the duct should be riveted or continuously welded. The recommended gauge for a black steel duct of a given cross-sectional area is as shown on page 173.

TABLE 9-6 Recommended Construction for Galvanized Steel Rectangular Ducts—Medium Pressure, S.P. from 2 to 6 in*

Steel U.S. Standard Gauge	Maximum Side, in	Type of Transverse Joint Connections	Bracing
24	0–12	1¼ × 1¼ × ⅛ in galvanized matched angle frames	None
24	13–18	1¼ × 1¼ × ⅛ in galvanized matched angle frames	1 × 1 × 16 ga. angles, 48 in C-C
22	19–36	1¼ × 1¼ × ⅛ in galvanized matched angle frames	1 × 1 × ⅛ in angles, 40 in C-C
22	37–48	1¼ × 1¼ × ⅛ in galvanized matched angle frames	1½ × 1½ × ⅛ in angles, 30 in C-C
20	49–60	1½ × 1½ × ⅛ in galvanized matched angle frames	2 × 2 × ⅛ in angles, 24 in C-C
20	61–72	2 × 2 × 3/16 in galvanized matched angle frames	2½ × 2½ × 3/16 in angles, 24 in C-C

*Constructions in this table are among those recommended in SMACNA's *High Velocity Duct Construction Standards,* 2d ed., 1969.

TABLE 9-7 Recommended Construction for Galvanized Steel Rectangular Ducts—High Pressure, S.P. 6 in and over*

Steel U.S. Standard Gauge	Maximum Side, in	Type of Transverse Joint Connections	Bracing
22	0–12	1¼ × 1¼ × ⅛ in galvanized matched angle frames	None
22	13–24	1¼ × 1¼ × ⅛ in galvanized matched angle frames	1 × 1 × ⅛ in angles, 48 in C-C
22	25–36	1¼ × 1¼ × ⅛ in galvanized matched angle frames	1¼ × 1¼ × ⅛ in angles, 32 in C-C; or 1½ × 1½ × ⅛ in angles, 40 in C-C
22	37–48	1½ × 1½ × ⅛ in galvanized matched angle frames	2 × 2 × ⅛ in angles, 30 in C-C
20	49–60	2 × 2 × ⅛ in galvanized matched angle frames	2 × 2 × 3/16 in angles, 24 in C-C
20	61–72	2 × 2 × 3/16 in galvanized matched angle	2½ × 2½ × 3/16 in angles, 24 in C-C
18	73–84	1¼ × 1¼ × ⅛ in galvanized matched angle frames with tie rod in center	1½ × 1½ × ⅛ in angles, 24 in C-C with tie rod in center

*Constructions in this table are among those recommended in SMACNA's *High Velocity Duct Construction Standards*, 2d ed., 1969.

TABLE 9-8 Recommended Construction for Galvanized Steel Round Ducts—Medium and High Pressures

Type	Duct Diameter, in						
	Up to 8	9–14	15–16	27–36	37–50	51–60	61–84
	U.S. Gauge						
Spiral lock seam duct	26	26	26	22	20	18	
Longitudinal seam duct	24	24	22	20	20	18	16
Welded fittings	22	20	20	20	18	18	18

TABLE 9-9 Recommended Construction for Galvanized Steel Flat Oval Ducts—Medium and High Pressures

Type	Dimension of Major Axis, in					
	Up to 24	25–36	37–48	49–50	51–70	Over 71
	U.S. Gauge					
Spiral lock seam duct	24	22	22	20	20	18
Longitudinal seam duct	20	20	18	18	16	16
Welded fittings	20	20	18	18	16	16

1. For 154 in² or less use No. 16 USSG

2. For 155 to 200 in² use No. 14 USSG

3. For 201 to 254 in² use No. 12 USSG

4. For 225 in² or larger use No. 10 USSG

Field joints are matched flanges with asbestos jackets.

Rectangular Fibrous-Glass Ducts (1-in-thick Board)

Fibrous-glass boards, formerly identified as standard-duty, heavy-duty, and extra-heavy-duty, are now identified by ratings of 475E1, 800E1, and 1400E1, respectively. The reinforcement given in Table 9-10 is designed for installation on the side of a duct of the dimension listed. The reinforcement can be made of a 3-in-width tee or channel. The height, gauge, and spacing of tee- and channel-shaped reinforcements are indicated in Table 9-10.

Transverse and longitudinal joints must be sealed with a continuous closure system, and reinforcement must be positioned at the joints. Further construction details can be found in the third edition (1972) of the SMACNA fibrous-glass-duct construction standards.

ACOUSTICALLY LINED DUCTWORK

Metal ducts are frequently installed with an acoustic lining to reduce airborne noises in the duct. The acoustic liner also functions as thermal insulation where the duct wall temperature exceeds the dew point of the air in the duct. In this case, the metal duct wall may serve as a vapor barrier. Most duct liners consist of a flexible or semirigid fibrous-glass blanket with

TABLE 9-10 Recommended Reinforcement for 1-in-Thick-Board Rectangular Fibrous-Glass Ducts (2000 ft/min Maximum Velocity)*

Board Rating	S.P., in	Up to 15	16–18	19–24	25–30	31–36	37–42	43–48	49–60	61–72	73–84	85–96
475 El Board:	Up to ½	None							1—22 @ 24			1¼—22 @ 24
Reinforcement	½–1	None			1—22 @ 24					1—18 @ 24	1¼—18 @ 24	
height, in—gauge @ cc, in	1–2	None	1—22 @ 24		1—22 @ 16					1—18 @ 16	1¼—18 @ 16	1½—18 @ 16
800 El Board:	Up to ½	None					1—22 @ 48		1½—22 @ 48	1—22 @ 24		
Reinforcement	½–1	None			1—22 @ 48		1—22 @ 24			1¼—22 @ 24		
height, in—gauge @ cc, in	1–2	None			1—22 @ 24		1¼—22 @ 24			1—18 @ 16	1¼—18 @ 16	1½—18 @ 16
1400 El Board:	Up to ½	None						1—22 @ 48		1—18 @ 48	1¼—18 @ 48	1½—18 @ 48
Reinforcement	½–1	None				1—22 @ 48	1¼—22 @ 48	1—22 @ 24		1—18 @ 24	1¼—18 @ 24	
height, in—gauge @ cc, in	1–2	None			1—22 @ 24		1¼—22 @ 24	1¼—18 @ 24	1½—18 @ 24	1¾—18 @ 24		

*Reinforcements in this table are among those recommended in SMACNA's *Fibrous Glass Duct Construction Standards*, 3d ed., 1972.

special surface treatment. Rigid, coated fibrous-glass board is sometimes used to line air-handling-unit casings and plenums. The surfaces to be lined are covered with a fire-retardant adhesive. The liners are secured by mechanical fasteners. The spacing of fasteners and edge treatment of the liner vary according to the velocity classification. Prefabricated round and flat duct and fittings are available with acoustic liner. Acoustic lining must have a maximum flame-spread rating and a maximum smoke-developed rating in conformance with NFPA and ASTM standards.

The dimensions of lined ducts, as shown on job drawings, are the inside dimensions of the duct after the liner has been installed. The metal duct should be larger than the size shown on the drawing; twice the thickness of the liner should be added to each side of a rectangular duct and to the diameter of a round duct. For example, suppose a 1-in-thick duct liner is installed in a rectangular metal duct shown on job drawing as a 36 × 12 in duct. The actual size of the metal duct should be 38 × 14 in. If a round duct is shown on the drawing as having a 22-in diameter, then 24-in-diameter round metal duct should be installed. The acoustic lining is normally installed by the sheet-metal contractor.

APPARATUS CASINGS AND PLENUMS

Casings and plenums are normally made of galvanized steel, aluminum sheet, and shapes. Casings and plenums may consist of a single metal wall with field-applied acoustic lining or thermal insulation. A double-wall panel with an internal fill of thermal insulation or acoustic liner may also be used. The casings of built-up air-handling units are normally made of double-wall panels. Panel manufacturers normally prepare detailed design of the double-wall casings on a custom basis.

ACCESS DOORS

Access doors are normally installed in the casings, plenums, ductwork, and boiler breeching, as well as at fire dampers, to provide easy access for servicing and inspection. Access doors may be made of double-wall panel for air-conditioning installations, or single metal walls for nonconditioned installations. Access doors are provided with heavy angle-iron frames and fitted with hinges and fasteners. The number of hinges and fasteners varies according to the access-door height. Access doors are available commercially.

BELT GUARDS

Belt guards are normally made of black-iron sheets, galvanized steel sheets, or flattened expanded metal with angle-iron frames. An opening must be provided on the front side to permit instrument readings of shaft speeds. The guard may be supported on brackets extending from the fan and motor, or provided with an angle iron (on the bottom edge) that can be fastened to the floor. Belt guards are designed and arranged so as to be readily removable.

HANGERS FOR DUCTS

The duct hanging system consists of three elements. These are the upper attachment to overhead structural features, the hanger itself, and the lower attachment to the duct. Ductwork must not be hung from overhead piping or ceiling hanger irons. Duct hangers are made of strap irons, rods, angle irons, or a combination (as in trapeze hangers). Rectangular ducts are normally supported by two metal strap hangers screwed to the duct sides. Round ducts are normally supported either by one strap hanger or by two strap or rod hangers bolted to the duct bands.

Trapeze-type hangers are recommended for large ducts. They are made of shelf angles which may be attached to the supporting rods, straps, or angles by welding, bolting, or pushnuts.

All types of hangers are fastened to the supporting structural member by clamps, anchor bolts, metal screws, bolting, or nailing. The recommended maximum spacing between duct hangers is 10 ft for round ducts with diameters up to 84 in and rectangular ducts with largest dimension up to 60 in. For rectangular ducts with the largest dimension over 61 in, 8-ft spacing may be recommended. The hanger size depends on the weight and size of the duct.

DUCT SIZES Duct sizes are determined according to airflow rates, duct friction losses, and air velocities. However, in the preliminary design stages, duct sizes are normally not shown on job drawings. In order to obtain an accurate estimate for a ductwork system, the estimator can quickly rough out duct sizes.

Figure 9-1 shows a Trane Ductulator, a device engineered by the Trane Company to simplify the designing of ductwork systems. It contains all the data necessary for determining the proper sizes of ducts and checking existing systems. The ductulator can be used by the estimator to determine duct sizes. It can be ordered from Trane Company representatives.

SURFACE AREAS OF DUCTS

Since the unit of weight of all metal sheets is pounds per square foot of surface area, it is necessary to calculate the surface area of all ductwork to

HVAC Systems Estimating Manual | 176

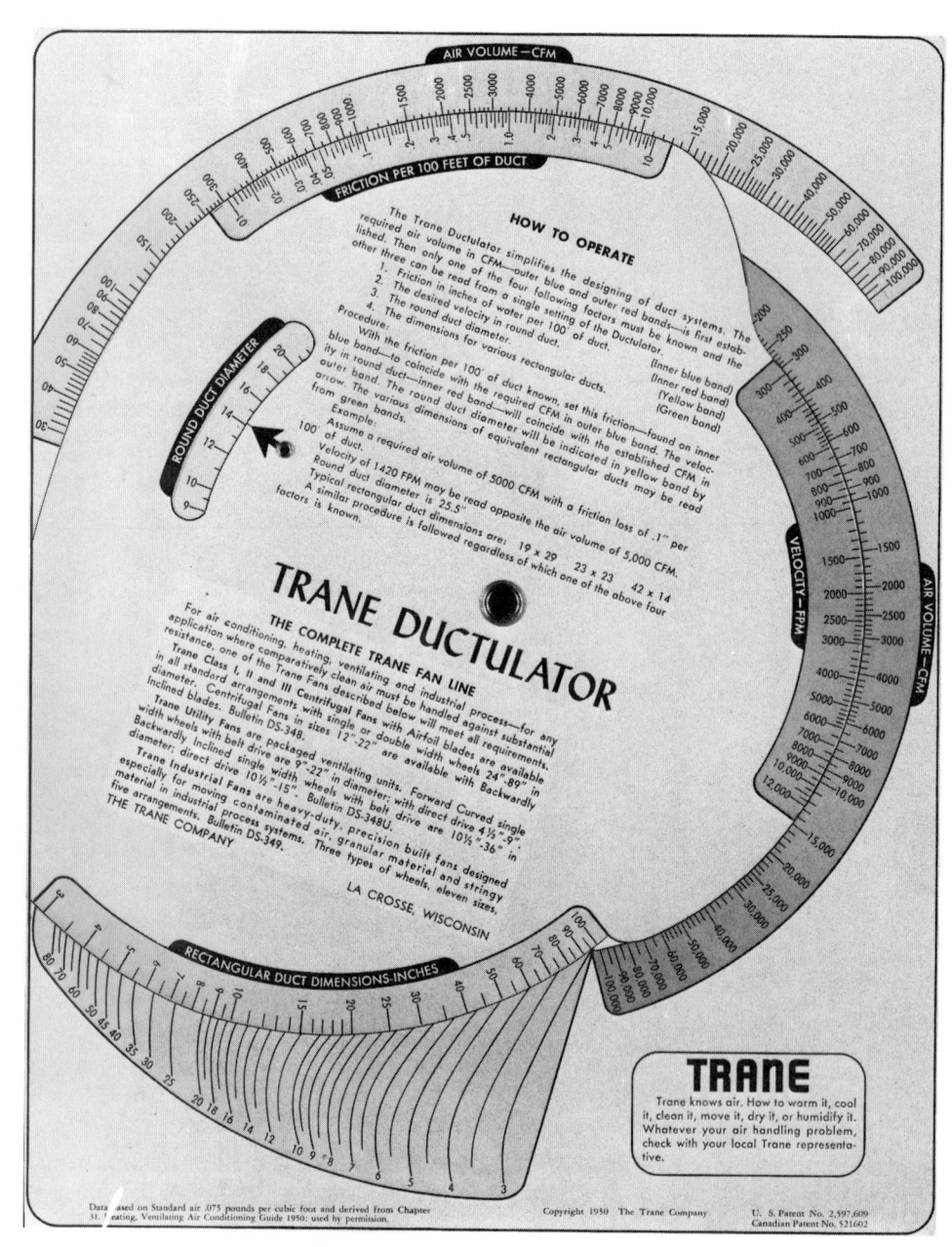

FIGURE 9-1 **Trane Ductulator** (*Copyright © The Trane Company* 1950. *Used by permission*).

be used. The following formulas can be used to calculate the surface area in square feet per foot of duct length:

1. *For round duct:* $\quad S = D \times 0.262 \quad$ ft²/ft $\hfill (9\text{-}1)$

2. *For rectangular duct:* $\quad S = \dfrac{W + H}{6} \quad$ ft²/ft $\hfill (9\text{-}2)$

where S = surface area, ft²/ft (of duct length)
D = duct diameter, in
W = rectangular-duct width, in
H = rectangular-duct depth, in

Tables 9-11 and 9-12 list the surface areas, in square feet per foot, for various sizes of rectangular and round ducts. As an example, for 36 × 12 in duct, width + depth = 48 in. Table 9-11 shows that the area of this duct is 8 ft²/ft.

DUCTWORK TAKEOFF RULES[1]

The rules shown in Fig. 9-2 are recommended for ductwork takeoff. These rules are among those recommended by the Sheet Metal Contractors Association of New York City, Inc. (recently known as the Sheet Metal Industry Promotion Fund of New York City). Consideration should also be given to prevailing local trade practices.

ESTIMATING OF DUCTWORK

Takeoff Procedure

The takeoff procedure is discussed in general in Chap. 1. In addition, the following steps may be used to obtain an accurate quantity takeoff for ductwork.

1. Prepare the required takeoff sheet. A typical ductwork takeoff sheet (DT-9) is shown in Fig. 9-3.

2. Read the specifications carefully; then list the types of materials, recommended construction, and gauges specified for each system.

3. Include all the takeoff quantities, for the various ductwork systems, which may have the same material, construction, acoustic lining, and/or insulation on the same takeoff sheet(s).

4. Take off each system as indicated on the floor plans, sections, and details. Ducts, elbows, transformations, etc., are measured in feet. Casings, plenums, and access doors are listed by their physical dimensions.

5. To obtain an accurate quantity takeoff, use the ductwork takeoff rules discussed in the previous section (see Fig. 9-2).

6. After completing the takeoff, list the surface area, in square feet per foot, for each duct size in the proper column (see Fig. 9-3). Surface areas can be obtained from Tables 9-11 and 9-12.

7. Calculate the total surface area for each duct size (total surface area equals duct length in feet times area in square feet per foot). Then

---

[1]*Sheet Metal Industry Uniform Trade Practice, Revised Hand Manual*, Sheet Metal Industry Promotion Fund of New York City, New York, May 1969, has been used as a reference, with permission.

TABLE 9-11 Surface Areas of Rectangular Ducts

Width + Depth, in	Area, ft²/ft	Width + Depth, in	Area, ft²/ft	Width + Depth, in	Area, ft²/ft
10	1.67	47	7.83	84	14.00
11	1.83	48	8.00	85	14.17
12	2.00	49	8.17	86	14.34
13	2.17	50	8.34	87	14.50
14	2.34	51	8.50	88	14.67
15	2.50	52	8.67	89	14.83
16	2.67	53	8.83	90	15.00
17	2.83	54	9.00	91	15.17
18	3.00	55	9.17	92	15.34
19	3.17	56	9.34	93	15.50
20	3.34	57	9.50	94	15.67
21	3.50	58	9.67	95	15.83
22	3.67	59	9.83	96	16.00
23	3.83	60	10.00	97	16.17
24	4.00	61	10.17	98	16.34
25	4.17	62	10.34	99	16.50
26	4.34	63	10.50	100	16.67
27	4.50	64	10.67	101	16.83
28	4.67	65	10.83	102	17.00
29	4.83	66	11.00	103	17.17
30	5.00	67	11.17	104	17.34
31	5.17	68	11.34	105	17.50
32	5.34	69	11.50	106	17.67
33	5.50	70	11.67	107	17.83
34	5.67	71	11.83	108	18.00
35	5.83	72	12.00	109	18.17
36	6.00	73	12.17	110	18.34
37	6.17	74	12.34	111	18.50
38	6.34	75	12.50	112	18.67
39	6.50	76	12.67	113	18.83
40	6.67	77	12.83	114	19.00
41	6.83	78	13.00	115	19.17
42	7.00	79	13.17	116	19.34
43	7.17	80	13.34	117	19.50
44	7.34	81	13.50	118	19.67
45	7.50	82	13.67	119	19.83
46	7.67	83	13.83	120	20.00

TABLE 9-12 Surface Areas of Round Ducts

Diameter, in	Area, ft²/ft	Diameter, in	Area, ft²/ft	Diameter, in	Area, ft²/ft
4	1.05	16	4.19	28	7.33
5	1.31	17	4.45	29	7.59
6	1.57	18	4.72	30	7.85
7	1.83	19	4.98	31	8.11
8	2.09	20	5.24	32	8.38
9	2.36	21	5.50	33	8.65
10	2.62	22	5.76	34	8.91
11	2.88	23	6.02	35	9.17
12	3.15	24	6.28	36	9.43
13	3.40	25	6.54	37	9.69
14	3.67	26	6.80	38	9.95
15	3.93	27	7.07	39	10.21
				40	10.47

include the areas of all ducts which are made of the same metal gauge in the column for that metal gauge in sheet DT-9 (Fig. 9-3).

8. Add the areas in each metal-gauge column and transpose the sum to ductwork summary sheet DS-9 (see Fig. 9-4). Individual sheets must be prepared for all systems with the same material, construction, acoustic lining, and/or insulation.

9. Calculate the total weight of metal ducts (see the next section). Metal ducts are normally estimated on an installed per-pound basis.

10. Calculate the surface areas of casings and plenums. Single-metal-wall plenums and casings are normally estimated on an installed per-pound basis. Double-wall panels are estimated on a per-square-foot basis.

11. Fibrous-glass ducts are normally estimated on a per-square-foot basis.

12. Use the labor rates shown in Estimating Chart 9-2 and the material prices in manufacturers' price lists. Use sheet DS-9 (Fig. 9-4) to calculate the total material cost and labor man-hours. Then transpose the material cost and labor man-hours to the job estimating sheet given in Chap. 1 (refer to Fig. 1-4).

On many jobs, ductwork is furnished and installed by the sheet-metal subcontractor. If so, the sheet-metal contractor provides the prime mechanical contractor with the total cost of sheet-metal work, including the applicable markup. The prime contractor may add a percentage markup to the subcontractor's cost to cover miscellaneous costs such as overhead, supervision, and profit. The amount of this markup varies,

Rule 1. Elbow lengths, square throat, square heel

Exterior dimensions of elbow establish its length.

All dimensions are to be aggregated to next foot or half-foot.

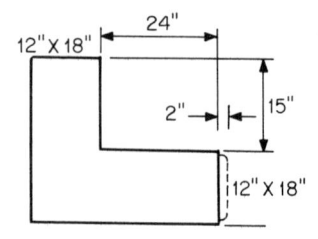

Example elbow, to be computed:

2'0" = 2'0"
+ 12" = 1'0"
+ 12" = 1'0"
+ 15" = 1'3"
= 5'3"
= 5'6"

Rule 2. Elbows, radius throat, radius heel

Use throat measurement plus cheek size in both directions. Then add for total length.
Throat measurement 12" plus cheek 18" = 30"
Throat measurement 24" plus cheek 18" = 42"
Total 72" = 6' length

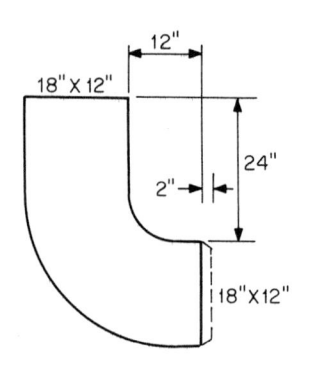

Rule 3. Duct lengths: duct perimeters

(a) Minimum perimeters shall be 4 ft. (Any ducts whose perimeters measure less than 4 ft shall be considered as having a perimeter of 4 ft because below this perimeter labor remains constant.

(b) Lengths shall be computed to next foot or half-foot.

(c) 24 ga. is minimum gauge used for weight calculations.

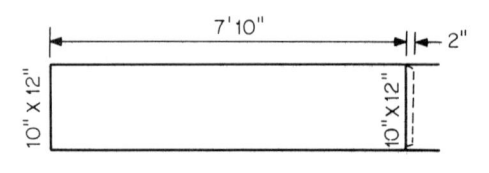

Example:
(a) Duct shown as actual perimeter of 44" to be computed at 48" = 4'.
(b) Length 7'10" to be computed at 8'.
Result: 8' – 12" X 12" Duct.

Rule 4. Sets or raises

(a) The length of a duct set, extending in any direction, shall be considered on the extension plus the set.

(b) The aggregate length shall be computed to next foot or half-foot.

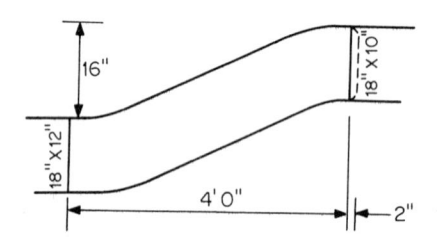

Example: Set to be computed
4'0" + 16" =
4'0" + 1'4" =
5'4" = 5'6" of 18" X 12" duct

FIGURE 9-2 Ductwork takeoff rules.

depending upon the size of the job. However, in average jobs, the contractor may add 5 to 7.5% of the subcontractor's cost.

Weights of Metal Ducts

After the surface area for each metal gauge is transposed to ductwork takeoff summary sheet DS-9 (Fig. 9-4), a conversion to weight may be

Rule 5. Transformations

In computing the square footage of a transformation, reducing duct from one size to another, the computation shall be made on the basis of the larger dimensions.

Example 1:

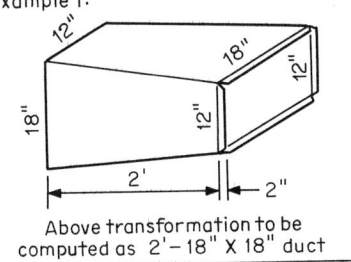

Above transformation to be computed as 2'−18" X 18" duct

Example 2.

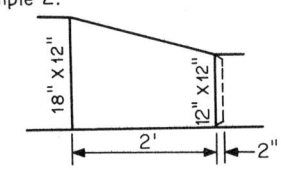

Above transformation to be computed as 2'−18" X 12" duct

Rule 6. Collars carrying registers, grilles and diffusers are to be figured a minimum of 1' in length.

Example:
    Collar length 10" = 1'
    Collar length 3" = 1'

Rule 7. Square foot−duct

(a) Compute perimeter and adjust to next half-foot.
(b) Compute length and adjut to next half- foot.
(c) Calculate footage.

Example: Duct 14" X 12" X 3'10" long

(a) Perimeter 52"= 4'4" = 4'6"
(b) Length 3'10" = 4'
(c) Computation 4' X 4 1/2' =18 ft²

Rule 8. Square foot
    Fire dampers
    Louvers
    Multiblade dampers
    Ducturns

(a) Adjust each dimension to next half-foot.
(b) Multiply for area and adjust to next square foot.
(c) Minimum of 3 ft² for any sundry items.

Examples: Fire damper 40" X 32"
    = 3' 4" X 2' 8"
    = 3 1/2' X 3'
    = 10 1/2 ft²
    = 11 ft²

Use minimum of 12" for any side in computation which is under 12".

Example:
    36" X 7" = 36" X 12" = 3 ft²

FIGURE 9-2 Ductwork takeoff rules. *(Continued)*

ESTIMATING CHART 9-1 Material Prices for Ductwork

Material	Price*
1. Galvanized steel sheet	$0.18/lb
2. Black-iron sheet	0.16/lb
3. Stainless-steel sheet (304)	1.50/lb
4. Aluminum sheet (3003-H14)	1.00/lb
5. Copper sheet	2.25/lb
6. 1-in-thick board, rectangular fibrous-glass duct	1.10/ft²
7. 1-in acoustic lining	0.15/ft²

*1976 average prices (for reference only).

made. To estimate the total weight of the ductwork, list the weight per square foot for each metal gauge in the proper column of sheet DS-9. [The weights and thicknesses of various metal sheets were discussed early in this

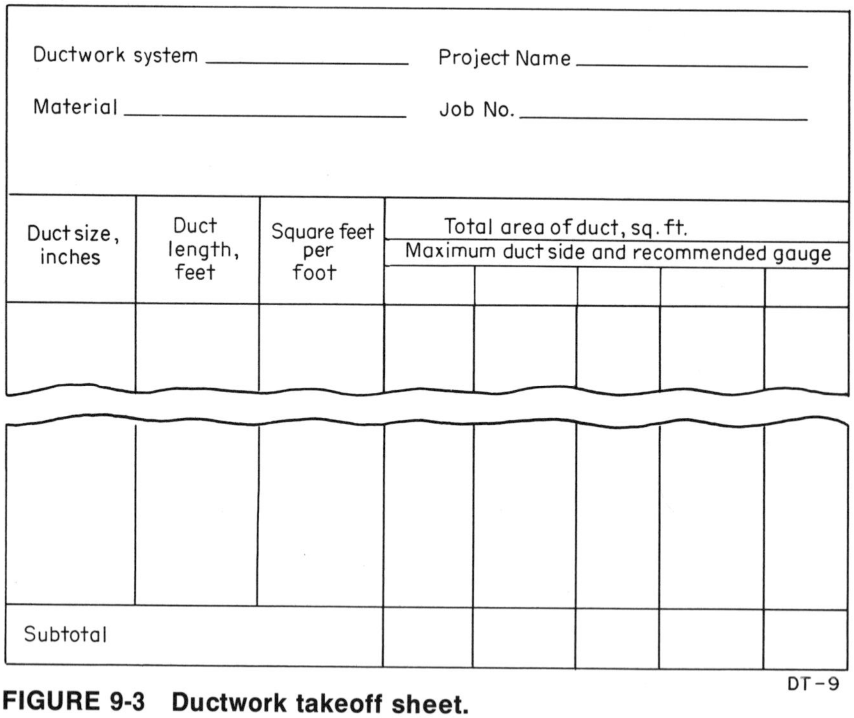

FIGURE 9-3 Ductwork takeoff sheet.

chapter (refer to Tables 9-2 and 9-3).] The weight per square foot multiplied by the number of square feet gives the weight in pounds. To adjust for joints, angles, rivets, hardware, and hangers, add the percentages given in Table 9-13 to the computed weights.

The percentages in Table 9-13 are based on normal ductwork installation. They may be adjusted according to the simplicity or complexity of a particular ductwork design, as well as the number of fittings and joints. The sum of the weights for each metal gauge equals the total weight of ductwork in pounds.

TABLE 9-13 Recommended Adjustment Percentages for Ductwork

Gauge	Recommended Adjustment, %		
	Low and Medium Pressure	High Pressure	Flanged Connections
Up to 22	19	24	29
20	20	25	30
18	23	28	33
16	20	25	30
14	18	23	28
12	17	22	27
10	15	20	25

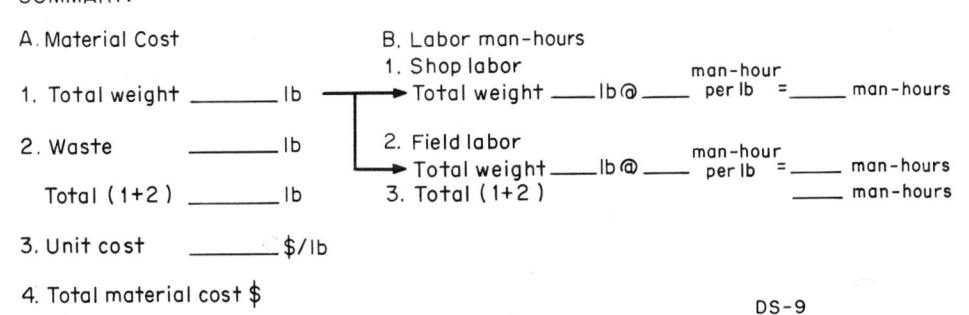

FIGURE 9-4 Ductwork takeoff summary sheet.

To estimate the labor man-hours, multiply the total weight computed above by the man-hours-per-pound figure indicated in sheet DS-9 (Fig. 9-4). To adjust for material wasted during the fabrication of the ductwork, add 15% to the total weight computed above. Then multiply by the material unit cost, in dollars per pound, to obtain the total material cost for the ductwork, as indicated in sheet DS-9.

Material Costs Material prices for ductwork can be best obtained from manufacturers' price lists. Their prices may vary, depending upon price trends, volume of order, discount rate, and availability of materials. The material prices in Estimating Chart 9-1 are intended for reference only; they are based on 1976 average costs.

Labor Man-hours

Metal ductwork is traditionally estimated on an installed per-pound basis. Highly mechanized techniques are now widely used for the fabrication of straight sections of sheet-metal and fibrous-glass ducts. The mechanization of straight-duct production has reduced its fabrication cost. The gains in the productivity of straight ducts have been offset by the essentially constant productivity and rising wages involved in fitting fabrication. For this

ESTIMATING CHART 9-2   Labor Man-hours for Ductwork

Material and System	Unit	Shop	Field	Total
1. *Galvanized and black iron conventional-construction rectangular ductwork*				
Low pressure	lb	.05	.07	.12
Medium pressure	lb	.05	.08	.13
High pressure	lb	.08	.11	.19
2. *Galvanized and black iron flanged-construction rectangular ductwork*				
Low pressure	lb	.05	.08	.13
Medium pressure	lb	.06	.08	.14
High pressure	lb	.08	.12	.20
3. *Galvanized and black iron round ductwork*				
Low-pressure construction	lb	.04	.06	.10
Medium-pressure construction	lb	.04	.07	.11
High-pressure construction	lb	.06	.10	.16
4. *Aluminum ductwork*				
Conventional construction	lb	.14	.21	.35
Flanged construction	lb	.16	.26	.40
5. *Copper ductwork*				
Conventional construction	lb	.07	.11	.18
Flanged construction	lb	.09	.14	.23
6. *Stainless steel ductwork*				
Conventional construction	lb	.07	.11	.18
Flanged construction	lb	.09	.14	.23
7. *Rectangular fibrous glass ductwork* (1-in thick board)	ft²		.14	.14
8. *Acoustic lining*	ft²	.03		.03
9. *Flexible ductwork* Spin-in fitting and hose clamps are included	ft		.09	.09

reason, some contractors may estimate the labor of fabrication and installation of ductwork on a per-fitting and per-joint basis, and on a per-pound basis for straight ducts, especially when the weight of the fittings and branches accounts for more than half the weight of the straight duct.

The labor figures (in man-hours per pound) indicated in Estimating Chart 9-2 are based on average job conditions. Specific job conditions may require that the figures be modified. The estimator should judge whether or not to adjust these figures according to the following considerations:

1. Larger jobs result in lower unit labor figures, and vice versa.

2. A greater number of fittings and joints increases the labor cost.

3. Ceiling height, number of floors, special structural features, and other conditions may also affect the erection cost.

Ductwork labor costs are normally divided into *shop* labor costs and *field* labor costs. Shop labor costs include fabrication detailing, fabrication, and overhead costs. Field labor costs include handling, erection, job-site hauling, scaffolding, and overhead costs.

The shop and field labor figures indicated in Estimating Chart 9-2 are in man-hours per unit of quantity (i.e., pound, square foot, or foot), and are based on an average job size and an assumption of normal productivity. The shop labor figures include ductwork fabrication. The field labor figures include handling, erecting, and joining of ducts. The estimator should analyze the job conditions carefully to determine the factors that will affect the rate of production, and use the applicable correction factors to adjust the labor figures in Estimating Chart 9-2.

After the labor costs for ductwork fabrication and erection are computed, all other costs which may be encountered and all overhead costs should be added (see Chap. 1).

10
THERMAL INSULATION

GENERAL Thermal insulation is used in heating, air-conditioning, and refrigeration systems to insulate piping, ducts, vessels, and equipment in order to conserve energy, prevent surface condensation, control heat input to the contained fluid, or provide personal protection. Thermal insulation can be used to control heat flow in temperature ranges from absolute zero through 3000°F and higher, when the proper material and thickness are applied.

MATERIALS Insulating materials are commercially available in various types, including cellular glass (foam glass), fiberglass, calcium silicate, mineral slag wool, expanded polyurethane, and foamed rubber. These materials are furnished in a wide variety of forms, such as loose fill, cement, flexible, semirigid, and rigid, in standard shapes, lengths, and thicknesses.

The ability of an insulating material to retard the flow of heat is given by its thermal conductivity, or conductance value. A low thermal conductivity or conductance value distinguishes thermal insulation. Table 10-1 indicates the physical and thermal properties of various types of insulating materials. In general, an increase in mean temperature increases the thermal conductivity (K factor) for the same density.

The proper selection of thermal insulation is normally governed by the temperature of surfaces to be insulated. Surface temperatures may be either higher or lower than ambient temperature.

For *temperatures below ambient (cold services)*, insulation of a suitable type and thickness is finished with a vapor barrier and jacket to prevent condensation, as well as to protect the insulation from weather and mechanical damage. Fiberglass, either board, blanket, or molded pipe covering, is most commonly used in HVAC systems for cold-service applications. Insulation may be furnished with a factory-applied vapor barrier and jacket.

For *temperatures above ambient (hot services)*, insulation of a suitable type and thickness is covered with a jacketing as required to provide protection

TABLE 10-1 Physical and Thermal Properties of Insulating Materials

Insulating Material	Density, lb/ft^3	K Factor @ Mean Temperature
1. Cellular glass	9	0.38 @ 100°F
2. Fiberglass	3	0.25 @ 50°F
3. Expanded polyurethane	2	0.15 @ 75°F
4. Calcium silicate	13	0.55 @ 500°F
5. Mineral wool, premoulded	10	0.49 @ 500°F
6. Mineral-wool blanket	8	0.49 @ 500°F

against mechanical damage and weather. Insulation may be furnished with a factory-applied jacket. The insulating materials which are commonly used in HVAC systems for hot-service applications are:

1. Calcium silicate, either block or preformed pipe covering

2. Fiberglass, either block, board, or preformed pipe covering

Accessory materials for thermal insulation normally include mechanical and adhesive fasteners, exterior and interior finishes, vapor-barrier coatings, jackets and weather coatings, sealants, lagging adhesives, membranes, and flashing compounds.

THICKNESS OF INSULATION

Tables 10-2 to 10-4 are typical piping, equipment, and duct insulation schedules. They are given for reference only, and were taken from actual job specifications. They indicate the thicknesses and densities of insulation for piping, equipment, and ducts. The indicated service and ambient temperatures are intended to show the minimum and maximum conditions occurring in the various systems, and are to be used as a guide by the insulation manufacturer. The indicated densities and thicknesses are based on fiberglass as a standard and serve as a guide to the intended amount of protection.

Recently, the selection of the proper insulation thicknesses has been governed by economic considerations, owing to the continuous increase in fuel costs and shortages of fuel. The economic insulation thickness in applications such as refineries and power and process plants may range up to 9 in.

GENERAL REQUIREMENTS FOR HVAC-SYSTEM INSULATION[1]

The general requirements for piping, equipment, and ductwork insulation are discussed in this section.

Piping Insulation

Small pipes are insulated with cylindrical half sections of insulation furnished with factory-applied jackets. Large pipes may be insulated with curved or flat segmental or cylindrical half, third, or quarter sections of insulation. Fittings and valves are insulated with preformed fitting insulation, fabricated fitting insulation, individual pieces cut from the surplus of sectional pipe insulation, or insulating cement. Fitting insulation should be consistent with pipe insulation.

Insulation with certain types of factory-applied jacketing may be secured on small piping by cementing the overlapping jacket. On large piping, supplemental wiring or banding may be required. Insulation on large piping requiring separate jacketing is wired or banded, depending on type. Insulation with factory-applied metal jacketing is secured according to the specific design of the jacket and its joint closure.

The thermal-insulation finish for piping operating at *temperatures above ambient (hot service)* is governed by location. The simplest finish for indoor pipe insulation, if concealed, is a laminate of kraft paper, fiberglass reinforcing, and a vapor-seal membrane. If exposed, the finish can be either the above or a presized glass cloth. Such finishes are designed to meet fire-safety requirements. For maximum fire safety, unusual exposure conditions, or appearance, factory- or field-applied metal jackets may be used. Insulation, factory-applied prefinished jackets, and finishes must have a maximum flame-spread rating and maximum smoke-developed rating in accordance with NFPA and ASTM standards.

Outdoor pipe-insulation finish must protect the insulation from weather, chemical exposure, and mechanical abuse while having a reasonable appearance. The jacketing may be asphalt, organic paper, asbestos paper, plastic film, aluminum foil, medium-gauge aluminum, galvanized steel, or stainless steel. Fittings may be finished with asphalt, resin-base mastics, preformed aluminum covers, or factory-combined insulation and jacketng.

Vapor-seal finishes for piping operating at *temperatures below ambient (cold service)* are generally designed to meet requirements of operating temperature, fire safety, and appearance. Indoor pipe insulation may be vapor-sealed, with jacketing commonly consisting of laminates of paper, aluminum foil, plastic film, and fiberglass reinforcing in various combinations.

[1] The *ASHRAE Handbook of Fundamentals*, 1972 edition, has been used as a reference, with the permission of ASHRAE.

TABLE 10-2 Typical Piping Insulation Schedule

System	Extent, in	Temperatures, °F Service	Temperatures, °F Ambient	Thickness, in	Density, lb/ft³
H.P. steam and return piping	½–4 5–12 14 and up	250–350 250–350 250–350	40–100 40–100 40–100	1½ 2 2½	4 4 4
L.P. steam and return piping	½–4 5–12 14 and up	212–250 212–250 212–250	40–100 40–100 40–100	¾ 1 1½	4 4 4
Hot-water piping to panel system	1–3 4–12 4–6	100–200 100–200 100–150	50–80 50–80 50–80	½ 1 ¾	4 4 4
Chilled-water piping to panel system	1–3 4–12 14 and up 4–6	40–80 40–80 40–80 60–70	60–100 60–100 60–100 60–100	½ 1 1½ ¾	4 4 4 4
Condenser-water piping	Inside of building	85–105	69–90	None	None
Condenser-water piping	Outside of building	85–105	10–100	2	4
Radiant-panel runouts	All	60–120	60–100	½	Foamed plastic
Cold-water piping	Inside of building	50–70	40–100	¾	4
Cold-water piping	Outside of building	50–70	10–100	2	4
Cold-water piping	Outside building below ground		40–60	Two layers	Plastic pipe wrap
Refrigeration piping	Section Liquid Hot gas	−20–35 −20–135 90–125	−10–100 −10–100 50–100	1½ 1½ 1	6 6 6
Emergency-generator exhaust	All	800–1200	50–100	2 (rigid)	6

Outdoor pipe insulation may be vapor-sealed in the same manner as indoor piping, with an added weather-protection jacketing applied without damage to the vapor seal and sealed to keep out water. Fitting insulation is usually vapor-sealed by field application of a vapor-seal adhesive or vapor-seal tape.

Piping subject to unusual freezing conditions requires heat-tracing along the side of the pipe. Electric resistance-heating cable is commonly used in pipe heat-tracing.

The piping insulation presently in use is available in standard forms in accordance with industry practice, as shown on page 191.

TABLE 10-3 Typical Equipment Insulation Schedule

System	Extent, in	Temperatures, °F Service	Ambient	Thickness, in	Density, lb/ft³
Hot-water heat exchangers	All	212–300	50–100	1	6
Refrigeration machines	Chilled-water boxes, refrigerant piping, and pumps	40–60	50–100	1½	6
Chilled-water pumps	All	40–60	50–100	1½	6
Expansion tanks	All	100–200	50–100	1 (rigid)	6

½–22 in	2-segments (half sections)
24–32 in	4-quadrants (quarter sections)
33–36 in	Beveled and scored block
Above 36 in	Flat block

Equipment Insulation

Equipment, tanks, and vessels are normally insulated with flat blocks, beveled lags, curved segments, blanket forms, or sprayed asbestos insula-

TABLE 10-4 Typical Duct Insulation Schedule

System	Extent, in	Temperatures, °F Service	Ambient	Thickness, in	Density, lb/ft³
Supply and return air-conditioning ductwork	Outside of mechanical equipment rooms	55–80	50–90	1 (flexible)	¾
Supply and return air-conditioning ductwork	Inside of mechanical equipment rooms	55–80	50–90	1 (rigid)	6
Fresh-air and spill-air ductwork	All	−10–100	50–100	1 (rigid)	6
Boiler breeching	All	100–400	50–100	2 (rigid)	6
Kitchen-range-hood exhaust ductwork	All	100–200	50–100	1½ (rigid)	6

tion. On small-diameter cylindrical vessels, the insulation may be secured by banding around the circumference. On larger cylindrical vessels, banding may be supplemented by angle-iron support ledges that support the insulation against slippage. When the diameter exceeds 10 to 15 ft, slotted angle iron may be run lengthwise along the cylinder at intervals, to secure the banding and to avoid excessive lengths of banding. On large flat and cylindrical surfaces, banding or wiring may be supplemented by fastening to various types of welded studs at short intervals.

The thermal-insulation finish for equipment operating at *temperatures above ambient (hot service)* are governed by location. On smaller indoor equipment, the insulation is commonly finished with a coat of hard-finish cement and hexagonal mesh, properly secured. On smaller outdoor equipment, the finish is the same as for smaller indoor equipment, with the addition of a coat of weather-resistant mastic. Suitable sheet-metal jacketing is required for larger indoor or outdoor equipment.

Equipment operating at *temperatures below ambient (cold service)* requires thermal-insulation finish to provide a degree of vapor sealing in accordance with operating temperature, to avoid the entry of moisture from the surrounding air, since this may increase the thermal conductivity of the insulation. For ordinary low temperatures, the insulation may be finished with hexagonal mesh properly secured, a coat of hard cement, and several coats of suitable paint. Where a greater degree of vapor sealing is required, two coats of vapor-seal mastic, reinforced with open-mesh glass fabric, may be used. For an even greater degree of vapor sealing, the insulation may be finished with two coats of asphalt vapor-seal mastic reinforced with asphalt-saturated and perforated asbestos felt at the edges (lapped and cemented). Where this type of vapor sealing is required indoors, acceptable appearance and protection from mechanical damage may require the addition of hexagonal mesh, hard-finish cement, and finish painting.

For outdoor equipment, the insulation may be finished with heavier or additional coats of vapor-seal mastic, reinforced with open-mesh glass fabric, to provide adequate protection against mechanical damage and weather. For appearance, the mastic may be finish-painted or covered with a suitable sheet-metal jacket.

Duct Insulation

Thermal insulation is normally used in ductwork systems to provide reductions in equipment size and operating cost, prevent condensation on low-temperature ducts, control temperature changes in long ducts, and control airborne noise (for interior duct lining, see Chap. 9).

In general, thermal insulation is especially important in areas with high dry-bulb and dew-point temperatures. All ducts exposed to outdoor conditions, as well as cooling ducts passing through unconditioned spaces, should be insulated, but the individual circumstances will determine the need for insulation on cooling ducts within conditioned spaces. Sheet-metal ducts are normally insulated with rigid and semirigid fiberglass boards or the flexible-blanket type. The insulation for exterior surface application may have factory-applied vapor barriers or facings or field-applied vapor barriers. Exterior duct insulation may be secured onto ducts with adhesive and supplemental preattached pins and clips, with wiring, or with banding. Indoor heating ducts need not be insulated in most cases. Vapor barriers are not required on the exterior insulation of ducts used for heating. Cooling ducts in any unconditioned space should be insulated and provided with a vapor barrier to prevent condensation. Joints and laps in the vapor barrier must be sealed with vapor-barrier compound, impermeable tape, or strips of vapor-barrier material and adhesive.

ESTIMATING OF THERMAL INSULATION

The takeoff procedure is discussed in general in Chap. 1; piping and ductwork takeoff procedures are also discussed in Chaps. 8 and 9, respectively. In addition, the following steps may be used to obtain an accurate quantity takeoff for thermal insulation:

1. Read the specifications carefully; then make a list of the type of materials specified for ductwork, equipment, and piping.

2. Obtain, from the ductwork and piping takeoff sheets, all the quantities which have to be insulated (see Chaps. 8 and 9).

3. List all the quantities which may be covered with the same insulating material in an individual group.

4. In order to estimate the quantity of insulation for the fittings and valves used in the piping systems, obtain their quantities from the valves and fittings takeoff sheets (see Chap. 8). Convert the quantities of each type and size to equivalent lengths of pipe in feet by using the multipliers given in Table 10-5.

TABLE 10-5 Fitting and Valve Multipliers for Equivalent Length of Pipe

Item	Equivalent Length of Pipe, ft
Screwed fitting	2
Flanged valve	7
Socket-weld valve	4
Elbow (3 in and larger)	5
Flanged cross	12
Flanged pair	4

ESTIMATING CHART 10-1 Labor Man-hours for Fiberglass Insulation with Vapor Barrier (Cold Service)

1. Piping insulation

Pipe size, in	Productive Labor, man-hour/ft — Insulation thickness, in				
	1	1 1/2	2	2 1/2	3
Thru 1 1/2	0.078	0.078	0.11	0.11	0.125
2 – 2 1/2	0.082	0.082	0.091	0.13	0.143
3	0.082	0.082	0.091	0.13	0.143
4	0.085	0.085	0.095	0.133	0.148
5	0.089	0.089	0.105	0.143	0.16
6	0.095	0.095	0.11	0.11	0.182
8	0.11	0.11	0.13	0.13	0.20
10	0.11	0.11	0.13	0.13	0.20
12	0.118	0.118	0.143	0.143	0.222
14	0.118	0.118	0.143	0.143	0.222
16	0.125	0.125	0.148	0.148	0.235
18	0.133	0.133	0.154	0.154	0.25
20	0.154	0.154	0.182	0.182	0.286
24	0.182	0.182	0.222	0.222	0.364

Single layer ← → Multiple layer

2. Duct and equipment rigid insulation

Insulation thickness, in	Productive Labor, man-hour/ft^2
1	0.08
1 1/2	0.08
2	0.08
2 1/2	0.09
3	0.09

Single layer ↑ ↓ Multiple layer

For hex. mesh add .010 MH/ft^2
For cementing add .032 MH/ft^2
For canvas add .032 MH/ft^2
For painting add .018 MH/ft^2

3. Duct blanket insulation
   Thru 2" thick – 0.05 man-hour/ft^2

Note: Jackets are factory applied

5. Equipment and breaching insulation is estimated based on the basis of surface area. Calculate these areas, using the physical dimensions of the various types of equipment which require insulation.

6. Thermal insulation is traditionally estimated on an installed per-square-foot basis for ductwork and equipment, and a per-foot basis for piping and fittings.

7. Normally, piping and equipment insulation quantity is not adjusted for waste, because the surplus is used for fittings and for filling between segments. However, a waste factor of 3 to 5% is needed on purchased material, to cover breakage in shipping and field handling.

8. A waste factor of 10 to 15% may be needed on flexible insulation material purchased, for overlaps and waste.

9. To cover the cost of insulation accessories (cement, fasteners, bands, etc.), an allowance of 10 to 15% may be added to the cost of the

Thermal Insulation | 195

insulating material. The labor cost for installing these accessories is included in the labor man-hours for insulation.

10. If field-applied metal jacketing is required, such jacketing is normally handled by the sheet-metal contractor.

11. Use the labor rates shown in Estimating Charts 10-1 through 10-4 and manufacturers' price lists to calculate the total material cost and labor man-hours.

12. Transpose the material cost and labor man-hours to the job estimating sheets given in Chap. 1 (refer to Fig. 1-4).

On many jobs, thermal insulation is normally furnished and installed by the insulation subcontractor. If so, the insulation contractor provides the prime mechanical contractor with the total cost of insulation work, including the applicable markup. The prime contractor may add a percentage markup to the subcontractor's cost to cover miscellaneous costs such as overhead,

ESTIMATING CHART 10-2 Labor Man-hours for Fiberglass Insulation without Vapor Barrier (Hot Service).

1. Piping insulation

Pipe size in	Productive Labor, man-hour/ft — Insulation thickness, in				
	1	1 1/2	2	2 1/2	3
Thru 1 1/2	0.067	0.067	0.095	0.095	0.105
2 — 2 1/2	0.07	0.07	0.077	0.11	0.12
3	0.07	0.07	0.077	0.11	0.12
4	0.073	0.073	0.08	0.114	0.125
5	0.077	0.077	0.083	0.12	0.133
6	0.08	0.08	0.089	0.089	0.148
8	0.095	0.095	0.105	0.105	0.174
10	0.095	0.095	0.105	0.105	0.174
12	0.10	0.10	0.11	0.11	0.18
14	0.10	0.10	0.11	0.11	0.18
16	0.105	0.105	0.118	0.118	0.20
18	0.114	0.114	0.125	0.125	0.21
20	0.133	0.133	0.148	0.148	0.25
24	0.16	0.16	0.18	0.18	0.31

Single layer ← | → Multiple layer

2. Duct and equipment rigid insulation

Insulation thickness, in	Productive Labor, man-hour/ft²
1	0.067
1 1/2	0.067
2	0.067
2 1/2	0.072
3	0.072

Single layer ↑ / ↓ Multiple layer

For hex. mesh add .010 MH/ft²
For cementing add .032 MH/ft²
For canvas   add .032 MH/ft²
For painting add .018 MH/ft²

3. Duct blanket insulation
   Thru 2" thick-0.04 man-hour / ft²

Note: Jackets are factory applied

supervision, and profit. The amount of this markup varies, depending upon the size of the job. However, in average jobs, the prime contractor may add 5 to 7.5% of the subcontractor's cost.

Material Costs

Material prices for insulating materials can be obtained from manufacturers' price lists. The prices may vary with price trends, order volume, discount rate, and availability of materials. For this reason, material prices are not included in this chapter.

Labor Man-hours

Labor man-hours for thermal insulation may be obtained from Estimating Charts 10-1 through 10-4. The labor figures include handling, erecting, securing, and applying the necessary accessories for 1 ft of pipe insulation

ESTIMATING CHART 10-3 Labor Man-hours for Calcium Silicate Insulation (Hot Service up to 1200°F).

1. Piping insulation

Pipe size, in	Productive Labor, man-hour/ft				
	Insulation thickness, in				
	1	1 1/2	2	2 1/2	3
Thru 1 1/2	0.089	0.10	0.11	0.125	0.143
2 — 2 1/2	0.095	0.105	0.12	0.133	0.148
3	0.10	0.11	0.125	0.143	0.16
4	0.105	0.118	0.129	0.148	0.167
5	0.11	0.125	0.143	0.16	0.182
6	0.114	0.125	0.143	0.16	0.182
8	0.125	0.154	0.154	0.182	0.19
10		0.167	0.167	0.19	0.20
12		0.182	0.182	0.20	0.222
14		0.19	0.19	0.222	0.25
16		0.20	0.20	0.25	0.267
18		0.222	0.222	0.267	0.286
20		0.25	0.25	0.286	0.333
24		0.286	0.286	0.333	1.00

→ Beveled and scored block

2. Duct and equipment block insulation

Insulation thickness, in	Productive Labor, man-hour/ft^2
1	0.09
1 1/2	0.09
2	0.09
2 1/2	0.095
3	0.095

For hex. mesh add .010 MH/ft^2
For cementing add .032 MH/ft^2
For canvas   add .032 MH/ft^2

Note: Jackets are factory applied on sectional insulation.
      Jackets on segmental insulation are precut and field applied.

ESTIMATING CHART 10-4 Labor Man-hours for Insulation Weatherproofing (Metal Jacketing).

1. Pipe with 0.016 in. thick aluminum jacketing

Insulation OD, in.	2.88	3.50	4.00	4.50	5.00	5.56	6.63	7.63	8.63	9.63
*Productive labor, man-hour/ft	0.024	0.025	0.027	0.031	0.034	0.038	0.045	0.051	0.058	0.065

Insulation OD, in.	10.75	11.75	12.75	14.00	15.00	16.00	17.00	18.00	19.00	20.00
*Productive labor, man-hour/ft	0.072	0.079	0.086	0.095	0.10	0.108	0.115	0.112	0.13	0.135

Insulation OD, in.	21.00	22.00	23.00	24.00	25.00	26.00	27.00	28.00	29.00	30.00
*Productive labor, man-hour/ft	0.142	0.15	0.156	0.162	0.17	0.176	0.182	0.19	0.197	0.20

Add 10% to above labor units for 0.01 in. thick stainless steel, 0.01 in. thick vinyl-coated galvanized steel, or 1/32 in. thick plastic jacketing.

2. Duct and equipment

Jacket type	*MH/ft²
0.020" thick — Aluminum 0.024" thick — Aluminum 0.026" USS ga. — Alum. steel 26 USS ga. — Galv. iron 24 USS ga. — Alum. steel 24 USS ga. — Galv. steel	0.026
0.010" thick — St. steel 0.010" thick — Vinyl coated galv. steel 1/32" thick — Plastic	0.029

*Metal jacketing is installed by sheet metal workers.

and 1 ft² of duct and equipment insulation, based on an average job size and an assumption of normal working conditions. The estimator should analyze the job conditions carefully to determine the factors that will affect the rate of production, and use the applicable correction factors to adjust the labor figures in Estimating Charts 10-1 through 10-4.

11

AUTOMATIC CONTROL SYSTEMS

GENERAL Heating and air-conditioning systems are designed to produce the maximum space loads, but they must function at partial loads when the internal loads and exterior conditions change. Automatic controls are, therefore, required on these systems to vary their output when loads are below maximum. An automatic control system must also produce satisfactory conditions in different parts of the space, as they are affected unequally by changes of load. Zoning is therefore required. Zones may be major portions of a building or individual rooms, according to the desired degree of sophistication, initial costs, and operating costs. Variations in the cost of the controls are most dramatic in multizone, individual-room, peripheral, and year-round air-conditioning applications. However, the degree of sophistication of an automatic control system is governed by the nature and type of building, the size of the heating and/or cooling system, and the desired result, including the conservation of energy.

FEEDBACK CONTROL SYSTEMS

The feedback control system normally functions according to the theory of closed-loop control. A schematic diagram for a closed-loop (feedback control) system is shown in Fig. 11-1. The various components of the automatic control system are also indicated in the diagram. In the closed control loop shown in Fig. 11-1, a controller (4) measures a controlled variable (2) by means of a sensing element (3) and compares this measurement with the desired set point (1). The transducer element of the controller responds to any variation between the set point and the controlled variable, and regulates the pressure of the compressed air (pneumatic system) or the electric current (electric or electronic system) applied to a controlled device (6). The controlled device reacts to changes of pressure or current and regulates the flow of a controlled medium (7). The result of the action of the controlled device is then transmitted back to the controller for another measurement, to check if the controlled device responded properly to the

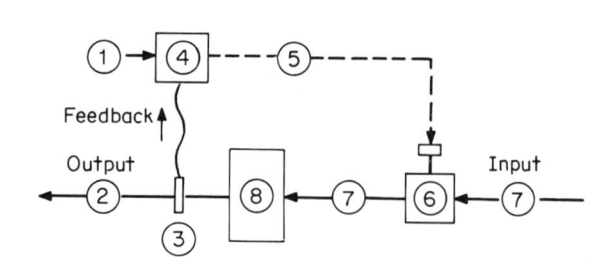

FIGURE 11-1 Schematic closed loop (feedback control system). (1) Set point (desired temperature, humidity, or pressure); (2) controlled variable (temperature, humidity, or pressure); (3) controller sensing element; (4) controller (thermostat, humidifier, or pressure controller); (5) air or electrical current; (6) controlled device (valve, damper, motor, or electric relay); (7) controlled medium (air, electricity, gas, oil, steam, or water); (8) apparatus (coil, duct, fan, pump, etc.) in individual room or space.

controller signal. This completes the cycle of the closed-loop or feedback control system. Although most control systems form a closed loop, an open loop may be used in a control system to vary the flow of heat to a building when its load is affected by changes in outdoor temperature. This type of system has no feedback. An outdoor thermostat is normally used in an open-loop system, and it responds only to changes in outdoor temperature.

DEFINITIONS OF CONTROL TERMS[1]

In order to understand the specifications and flow diagrams for an automatic control system, it is very important to know the definitions of the terms that are most common in automatic controls. Controller, controlled device, controlled variable, and controlled medium are defined in Fig. 11-1. Their functions are also discussed in the section on feedback control systems. In addition, other related terms may be defined as follows:

1. *Set point.* The set point represents the desired value of the controlled variable, the value at which the controller is set.

2. *Actual control point.* The control point is the actual value of the controlled variable, measured by the controller. It varies according to the demand on the system and variations in the interior loads and exterior conditions. The difference between the set point and the actual control point is known as the *offset*.

3. *Overall control.* The overall control is the necessary control on the system as a whole. It responds to the major factors affecting the system load.

4. *Zone control.* The zone control is the necessary control on a number of rooms or areas having similar orientation or occupancy. It is

[1]The *ASHRAE Handbook—1973 Systems Volume* has been used as a reference, with the permission of ASHRAE.

employed to control the heating or cooling effect within the space served by the system.

5. *Individual room control.* The individual room control is the necessary control required to maintain the desired temperature within a room, regardless of orientation or occupancy.

6. *Two-position control.* A two-position control is a type of control in which the controlled device can be reset only to a maximum or minimum position, or can be either ON or OFF. An example of a two-position control is a thermostat which opens and closes a valve or starts and stops a burner.

7. *Controller differential.* Controller differential is the difference between the maximum and minimum settings at which the controller operates from one position to another. Controller differential usually applies to two-position and floating controls.

8. *Floating-action control.* A floating-action control is a type of control similar to a two-position control, except that the controlled device can be positioned at a constant rate toward its open or closed position. Generally, there is a neutral zone between the two positions; it allows the controlled device to stop at any position whenever the controlled variable is within the controller differential. When the controlled variable gets outside this differential, the controller moves the controlled device in the proper direction. A static-pressure controller positioning a damper to maintain static pressure is an example.

9. *Proportional control.* A proportional control is a type of control in which the controlled device is positioned in proportion to a system requirement. It responds immediately to slight changes in the controlled variable. It does not run through its complete stroke, as is the case with a floating control. An example of a proportional control is a thermostat in a fan discharge duct, actuating an automatic valve in a heating- or cooling-medium supply to a coil, in order to regulate the air temperature leaving a coil.

10. *Throttling range.* The throttling range is the total amount of change in the controlled variable required for the controller to move the controlled device through its complete stroke, from one extreme to the other. It is often adjustable to meet system requirements.

11. *Proportional–automatic-reset control.* This type of control combines floating control with proportional control to achieve the stability of proportional control with a relatively wide throttling range, and the invariable control point of floating control. The system functions in the same way as proportional control. The automatic reset shifts the control point back to the set point whenever any offset occurs. In this system, the proportional control is usually repeated by the reset control. The number of repeats per minute is known as the *reset rate*. The

reset rate in most controllers is adjustable and must be carefully matched to the system characteristics. Because automatic-reset control is necessarily slow-acting, it should be used only when load changes are of reasonably long duration and when the maximum offset resulting from proportional control alone is outside acceptable limits.

TYPES OF CONTROL SYSTEMS

Five types of control systems are presently used in heating, ventilating, and air-conditioning systems. These systems are classified according to the primary source of energy, as follows:

1. *Pneumatic control system.* A pneumatic control system usually utilizes compressed air at a pressure of 15 to 25 psig as the source of energy. The compressed air is usually supplied to the controller, which in turn regulates the pressure supplied to the controlled device.

2. *Electric control system.* An electric control system utilizes electrical energy of either low or line voltage as the energy source. The electrical energy supplied to the controlled device is regulated by the controller, either directly or through relays or pneumatic-electric transducers.

3. *Electronic-electric control system.* An electronic-electric control system also utilizes electrical energy, but it employs an electronic amplifier to increase the minute voltage variations of the measuring element to the values required for the operation of standard electrically controlled devices.

4. *Electronic-pneumatic control system.* An electronic-pneumatic control system utilizes compressed air for the operation of the controlled device by converting the output of an electronic amplifier to suitable air-pressure changes by means of an electronic-pneumatic transducer. It is also possible to have a system that converts a pneumatic signal to an electrical output with a pneumatic-electric transducer.

5. *Self-contained control system.* A self-contained (also called *mechanical-action*) control system combines the controller and the controlled device in one unit and transmits the effect of temperature changes on the sensing element directly to the operator device of the valve or damper, without the use of any auxiliary source of power. The thermostatic water-mixing valve is an example.

AUTOMATIC-CONTROL-SYSTEM COMPONENTS[2]

An automatic control system usually consists of controllers, controlled devices, and auxiliary control equipment.

---

[2]The *ASHRAE Handbook—1973 Systems Volume* has been used as a reference, with the permission of ASHRAE.

Controllers

An automatic controller consists of a sensing element and a transducer.

Sensing Elements

The sensing element measures changes in the controlled variable and produces a proportional effect on the transducer. This effect may be a change in force, position, or electrical resistance, depending on the type of sensing element used. Sensing elements can be divided into four general classes, depending on the controlled variable being measured.

Temperature-sensing elements are normally classified as follows:

1. *Position-output type.* The element may be a bimetal strip, or a rod and tube of dissimilar metals.

2. *Forced-output type.* The element may be sealed below, with or without a remote bulb.

3. *Electrical-resistance type.* The element is made of wire with electrical resistance.

Humidity-sensing elements usually consist of hygroscopic organic materials (human hair, wood, paper, or animal membrane) or an electrical resistance.

Pressure-sensing elements can be subdivided into two classes, depending upon pressure range.

1. For pressures or vacuums measured in pounds per square inch or inches of mercury, the element is usually a bellows, diaphragm, or Bourdon tube.

2. For low ranges of pressure or vacuums measured in inches of water, such as the static pressure in an air duct, the element may be an inverted bell immersed in oil, a large slack diaphragm, or a large, flexible, metal bellows. Pitot tubes may be used to measure flow, velocity, liquid level, and static pressure.

Water-flow-sensing elements include orifice plates, pitot tubes, venturis, flow nozzles, turbine meters, magnetic flowmeters, and vortex-shedding flowmeters.

There are also sensing elements for other purposes, such as flame detection and measuring smoke density. They are often necessary for the complete control of a HVAC system.

Transducers

The transducer converts the effect produced by the sensing element to an effect suitable for operation of the controlled device. In pneumatic systems, a mechanical transducer regulates the pressure of compressed air supplied to the controlled device; in electric and electronic systems, the transducer regulates the flow of electrical energy. Transducers can be divided into four types, according to the type of energy (pneumatic or electrical) or the type of control action produced.

Electric transducers are normally combined with sensing elements having a force or position output to obtain two-position, floating, or proportional control action.

1. For two-position control, the transducer may be a simple electric contact which starts a pump or actuates a damper operator. Single-pole-double-throw (SPDT) switching circuits are used to control a three-wire unidirectional motor operator.

2. For floating control, the transducer output is a SPDT switching circuit with a neutral zone where no contact is made. This type is used to control a reversible motor operator.

3. For proportional control, two types of transducers systems are in common use:

    a. The sensing element moves a wiper arm across a potentiometer to vary the voltage applied to a balancing relay. The relay controls the operation of a reversible motor operator and is rebalanced by a variable-voltage feedback signal from a second potentiometer in the motor operator.

    b. The sensing element positions a floating-type switching circuit with a neutral zone, which controls the operation of a reversible motor operator. The control is rebalanced by a small solenoid, energized by a variable-voltage feedback signal from a potentiometer in the motor operator.

Electronic transducers use an electronic amplifier to detect the small voltage changes generated by the sensing elements. The amplifier output actuates an electric relay or relays. This type of electronic transducer may be employed to obtain two-position, floating, or proportional control action, depending upon the amplifier-and-relay arrangement. For proportional control, relays control the operation of a reversible motor operator, and rebalance is obtained by a variable voltage feedback signal impressed on the amplifier from a potentiometer in the motor operator.

Electronic-pneumatic transducers use the output of an electronic amplifier to actuate an electropneumatic relay, resulting in a variable air pressure applied to the controlled device. The control action (also called control *mode*) may be two-position or proportional, depending upon the amplifier and relay arrangement.

Pneumatic transducers are normally combined with sensing elements having a force or position output to obtain a variable air-pressure output. They are normally employed to obtain proportional control action. Two types are in common use.

1. The nonrelay type uses a restrictor in the air supply and a bleed nozzle. The sensing element positions a flapper which varies the nozzle opening, resulting in a variable air-pressure output applied to the controlled device. This type is limited to applications requiring small volumes of air.

2. In a relay-type pneumatic transducer, the variable force produced by the sensing element (either directly or indirectly through a restrictor, nozzle, and flapper arrangement) actuates a relay of substantial air-handling capacity. The force produced by the output pressure change is rebalanced against the force produced by the sensing element.

The nonrelay and relay types are further classified by construction as direct- or reverse-acting. A direct-acting thermostat increases the output air pressure when the temperature rises; a reverse-acting thermostat increases the output air pressure when the temperature drops.

Indicating and Recording Controllers

Generally, controllers may be of the nonindicating, indicating, or recording type. The nonindicating type is most common in HVAC control; however, when an indication is desired, a separate thermometer, pressure gauge, relative-humidity indicator, etc., is required. In the indicating controller, the indicating pointer may be added to the sensing element or attached to it by a linkage so that the value of the controlled variable is indicated on a suitable scale. A recording controller is similar to an indicating controller except that the pointer is replaced with a recording pen which provides a permanent record on a chart driven by a clock motor.

Types of Controllers

Controllers are mainly thermostats, humidistats, and pressure controllers.

Thermostats. The various types of thermostats in present use may be classified as follows:

1. The *room thermostat* is mounted on a wall and responds to room temperature.

2. The *insertion thermostat* is mounted on a duct, with its sensing element extending into the duct.

3. The *immersion thermostat* is mounted on a pipe or a tank. It has a fluidtight connection to allow the measuring element to extend into the fluid. A separable socket or well is often used with the immersion thermostat, to avoid the need for draining the system when the thermostat must be removed.

4. The *remote-bulb thermostat* is normally used when the point of temperature measurement is at some distance from the desired thermostat location. A thermostat mounted on a local control panel is an example. The remote-bulb element may be of the insertion or the immersion type.

5. The *surface thermostat* is mounted on a pipe or similar surface. It is often used for sensing the surface temperature.

6. The *day-night thermostat* controls at a reduced temperature at night. It may be indexed (to change from day to night operation) individually or in groups from a remote point by manual or time switch. Some electric types have an individual clock and switch built into the thermostat.

7. The *pneumatic day-night thermostat* uses a two-pressure air-supply system. Changing the pressure at a central point from one value to the other actuates switching devices in the thermostat and indexes them from day to night, or vice versa.

8. The *heating-cooling thermostat* is used to actuate a controlled device (damper or valve) that regulates a heating source at some times, and a cooling source at other times. It is often manually indexed in groups by a switch, or automatically indexed by a thermostat sensing the temperature of the controlled medium, the outdoor temperature, or some other suitable variable.

9. The *pneumatic heating-cooling thermostat* uses a two-pressure air-supply system as described for pneumatic day-night thermostats.

10. The *multistage thermostat* is arranged to operate two or more successive steps in sequence.

11. The *submaster thermostat* has its set point raised or lowered over a predetermined range in accordance with variations in output from a master controller. The master controller can be a thermostat, manual switch, pressure controller, or similar device. As an example, a master thermostat measuring outdoor temperature can be used to adjust a submaster thermostat controlling the water temperature in a heating system.

Humidistats.   Humidistats are of either the room or the insertion type. Submaster humidistats of either type are sometimes used with a master thermostat to reduce humidity in cold weather and prevent condensation on windows.

A wet-bulb thermostat is often used for humidity control in conjunction with proper control of the dry-bulb temperature. This type is provided with a wick or other means for keeping the bulb wet; a rapid air motion is also required to ensure a true wet-bulb measurement. The dew-point thermostat is designed to measure the dew point.

Pressure or static-pressure controllers. These are designed for mounting either directly on a pipe, or remotely on a panel or wall.

Controlled Devices

Controlled devices are normally either automatic control valves or automatic control dampers.

Automatic control valves. An automatic valve is normally used in a heating or cooling system to control the flow of steam, water, or some other fluid. It may be positioned by impulses from the controller. An automatic valve may be equipped with a throttling plug or V port of special design to provide the desired flow characteristics. It is commonly constructed with a renewable composition disk, made of a material suitable to the fluid handled by the valve and the operating temperature and pressure. Automatic valves should be properly selected and sized for specific applications.

The various types of automatic valves may be classified as follows:

1. A *single-seated valve* is designed for tight closing.

2. A *pilot-piston valve* utilizes the pressure of the controlled medium as an aid in operating the valve. It is usually single-seated and is used where large forces are required for valve operation.

3. A *double-seated* or *balanced valve* is designed so that the fluid pressure acting against the valve disk is essentially balanced in order to reduce the force required of the operator. It is widely used where the fluid pressure is too high to permit a single-seated valve to close. It cannot be used where tight closing is required.

4. A *three-way mixing valve* has two inlet connections, one outlet connection, and a double-faced disk operating between two seats. It is used to mix, as required, two fluids entering through the inlet connections and leaving through the outlet connection.

5. A *three-way diverting valve* has one inlet connection, two outlet connections, and two separate disks and seats. It is used to divert the flow to either of the outlets or proportion the flow to both outlets.

6. A *butterfly valve* consists of a heavy ring enclosing a disk which rotates on an axis at or near its center.

Automatic valves are normally operated by electric or pneumatic operators. Valve operators are of three general types:

1. A *solenoid* consists of a magnetic coil operating a movable plunger. It is used for two-position control, and it is generally limited to relatively small valve sizes (up to 3 in).

2. An *electric motor* operates the valve stem through a gear train and linkage. Electric-motor operators are classified in three types.

    a. An *unidirectional* motor operator is used for two-position control. The valve opens during one half-revolution of the output shaft and closes during the other half-revolution. Once started, it continues until the half-revolution is completed, regardless of subsequent action by the controller. Limit switches built into the operator stop the motor at the end of each stroke. If the controller has been satisfied during this interval, the operator will continue to the other position.

    b. A *spring return* is used for two-position control. Electrical energy drives the valve to one position and holds it there. When the circuit is broken, the spring returns the valve to its normal position.

    c. A *reversible* motor operator is used for floating and proportional control. The motor can run in either direction and can stop in any position; sometimes it is equipped with a return spring. In proportional-control applications, a potentiometer (for rebalancing the control circuit) is also driven by the motor.

3. A *pneumatic* operator consists of a spring-opposed flexible diaphragm or bellows attached to the valve stem in such a way that an increase in air pressure moves the valve stem and simultaneously compresses the spring. Pneumatic operators are designed primarily for proportional control; however, two-position control may be accomplished by use of a two-position controller or a two-position pneumatic relay, to apply either full pressure or no pressure to the valve operator.

Pneumatic valves and valves with spring-return electric operators can be classified as normally open or normally closed. A normally open valve will assume the open position when all operating forces are removed. A normally closed valve will assume the closed position when all operating forces are removed.

Automatic control dampers. Automatic dampers are normally used to control a flow of air or gas; they function like valves in this respect. The damper types commonly used are:

1. *Single-blade* dampers are generally restricted to small sizes because of the difficulty of securing proper operation with high velocity.

2. *Multiblade* dampers have two or more blades linked together. The blades may be arranged for parallel operation, in which all blades rotate in the same direction, or opposed operation, in which adjacent blades rotate in opposite directions (refer to Chaps. 6 and 7 for differences in airflow characteristics).

3. *Mixing* dampers are composed of two sections interlinked so that one section opens as the other closes (refer to Chap. 7).

Damper operators are similar to valve operators, except that pneumatic damper operators usually have a longer stroke, or the stroke is increased by means of a multiplying lever. Damper operators are mounted on the damper frame and connected by a linkage directly to a damper blade, or are mounted outside the duct and connected to a crank arm attached to a shaft extension of one of the blades. On large dampers, two or more operators may be required; they usually are applied at different points on the damper. Normally open or normally closed operation is obtained by properly mounting the operator and connecting the linkage.

Auxiliary Control Equipment

In addition to controllers and controlled devices, most control systems require auxiliary control equipment to perform various functions. Auxiliary control equipment may be classified as follows:

Auxiliary control equipment for electric control systems includes:

1. *Transformers* to provide current at the required voltage.

2. *Electric relays* for control of electric heaters or to start and stop burners, refrigeration compressors, fans, pumps, or other apparatus for which the electric load is too large to be handled directly by the controller. They are also used in time-delay, circuit-interlocking, safety applications.

3. *Potentiometers* for manual positioning of proportional control devices or for remote set-point adjustment of electronic controllers.

4. *Manual switches* for manual performance of a variety of operations. These may be of the two-position or multiple-position type with a single pole or multiple poles.

5. *Auxiliary switches* on valve and damper operators for providing a selected sequence of operations.

Auxiliary control equipment for pneumatic systems includes:

1. *Air compressors* and accessories, including driers and filters to provide a source of clean, dry air at the required pressure.

2. *Electropneumatic relays*, which are electrically actuated air valves for operating pneumatic equipment in accordance with variations in electrical input to the relay.

3. *Pneumatic electric relays*, which are actuated by air pressure to make or break an electric circuit.

4. *Pneumatic relays*, which are actuated by the pressure from a controller to perform numerous functions. These include:

   a. *Two-position relays*, which permit a controller actuating a proportional device to also actuate one or more two-position devices. They are also used in various automatic switching operations.

   b. *Proportional relays*, which are used to reverse the action of a proportional controller, select the higher or lower of two pressures, average two or more pressures, respond to the difference between two pressures, add or subtract pressures, amplify or retard pressure changes, and perform other similar functions.

5. *Proportional relays*, which are used to assure accurate positioning of a valve or damper operator in response to changes in pressure from a controller. They are affected by the position of the operator and the pressure from the controller; whenever the two are out of balance, these relays will change the pressure applied to the operator until balance is restored.

6. *Switching relays*, which are pneumatically operated air valves for diverting air from one circuit to another or for opening and closing air circuits.

7. *Pneumatic switches*, which are manually operated devices for diverting air from one circuit to another or for opening and closing air circuits. They may be of the two-position or multiple-position type.

8. *Gradual switches*, which are proportional devices for manually varying the air pressure in a circuit.

Auxiliary control devices common to both electric and pneumatic systems include:

1. *Step controllers* for operating a number of electric switches in sequence by means of a proportional electric or pneumatic operator. They may be used for the sequential operation of electric heating elements and other equipment, in response to the demands of a proportional controller.

2. *Power controllers* for controlling the electric power input to resistance-type electric heating elements. They are arranged to regulate the power input to the heater in response to the demands of proportional electronic or pneumatic controllers.

3. *Clocks or timers* for turning apparatus on and off at predetermined times, for switching control systems from day to night operation, and for other time-sequence functions.

CONTROL OF HEATING SYSTEMS[3]

Heating systems require overall control and, usually, zone control. However, an individual room control may also be required.

Overall control in steam systems is effected by the control of steam pressure and temperature in subatmospheric systems, and by on-off control in other systems. Zone control in steam systems is similar to overall control, except that it is accomplished by zones.

Overall control in water systems is most often accomplished by varying the water temperature inversely as the outdoor temperature varies; in very small residential systems, it is accomplished by starting and stopping the pump. Zone control in water systems is accomplished by varying the temperature of the water supplied to the zone by mixing or by throttling the flow of water to the zone. Whenever throttling control is used, it is necessary to connect a relief valve across the main in order to maintain a constant flow through the pumps and the heat-generation units as the various zones throttle.

In indirect warm-air systems, overall control is accomplished by throttling the flow of the heating medium through the coil or by passing it around the coil. Zone control of indirect warm-air systems may be accomplished in any of three ways:

- By separate air-handling units assigned to each zone, with their coils controlled as described for overall control.

- By units with hot and cold decks and dual-duct distributing systems, with local mixing at the zones.

- By induction units located in hung ceilings in connection with high-pressure systems, to mix air from the hung-ceiling space with primary air from the high-pressure duct to the desired delivery temperature. This application requires high illumination level.

Overall control of direct-fired warm-air systems is usually accomplished by stopping and starting the burner; the supply fan is, in turn, started and stopped according to the temperature of the air leaving the furnace. Direct-fired warm-air systems sometimes are controlled at the burner only, with the fan running continuously, especially in larger systems with high-capac-

---

[3]T. Baumeister, *Marks' Standard Handbook for Mechanical Engineers*, 7th ed., McGraw-Hill, New York, 1967, has been used as a reference with permission.

ity outlets. Zone control of direct-fired warm-air units is usually effected by the assignment of separate units.

In most applications, electric heating systems need not be controlled on an overall basis. Zone control of electric heating systems is usually accomplished either by local control of each heat emitter or by control of selected groups of circuits in distribution panels by means of thermostat-actuated contactors.

CONTROL OF COOLING AND DEHUMIDIFYING SYSTEMS[4]

Control of cooling and dehumidifying systems involves overall control as well as zone control. However, in small residential, commercial, and institutional applications, overall control is the only control used; it is accomplished by starting and stopping the refrigeration of package or perimeter units. Humidification control in comfort applications consists in varying the dry-bulb temperature only; it is accomplished by using reheat coils and dryers.

Zone control in cooling and dehumidifying systems is accomplished by varying the input to zone cooling coils, zone reheating coils, or room cooling coils. Bypasses around coils are also used. In dual-duct systems, air is mixed to the desired delivery temperature for each zone. Room control of an air-water induction-type unit is accomplished by varying the input of chilled water to the unit cooling coil, so that the dry-bulb temperature will be affected. In this type of induction system, dehumidification is controlled by the maintenance of a fixed dew point at the primary-air unit. Fan-coil perimeter units usually have three-speed fans for air volume control. Automatic modulating valves can be applied to chilled water as well. A self-contained valve-thermostat may be used for this purpose. Throttling dampers, with or without static-pressure control, are sometimes used for zone control.

During intermediate seasons, outside air can be used as coolant in lieu of refrigeration when the air temperature permits. Control is accomplished by mixing the outside air and return air with modulating dampers. Usually, the outside-air damper is controlled according to the mixed-air temperature, and the return-air damper is controlled according to the static pressure in the mixing boxes. Since a varying input of outside air requires a varying quantity of exhaust air, the action of the outside-air damper is coordinated with an exhaust-air damper to maintain a constant positive pressure in the building. Such a system requires an additional fan in the return-air system.

---

[4]T. Baumeister, *Marks' Standard Handbook for Mechanical Engineers*, 7th ed., McGraw-Hill, New York, 1967, has been used as a reference with permission.

Automatic Control Systems | 213

SEQUENCE OF OPERATION OF AUTOMATIC CONTROL SYSTEMS

Figures 11-2 through 11-4 will facilitate this discussion of the operation of control systems. Figure 11-2 is the automatic-temperature-control diagram for an all-water-type heat-pump system. The sequence of operation of this control system is as follows:

Normal Operation

1. With the system water temperature between 60 and 90°F, neither the steam-to-hot-water convertor, the heat-rejector fans, nor the evaporative cooler will operate. The temperature of the water being supplied to the heat pumps is indicated by a temperature-indicating gauge on the control-panel door.

2. As the system water temperature rises above 80°F, the control energizes the damper motors to open the positive-closure dampers. The evaporative-cooler pump starts at 83°F. At 90°F, the next heat-rejector fan starts, with all modulating dampers open, through the end switches.

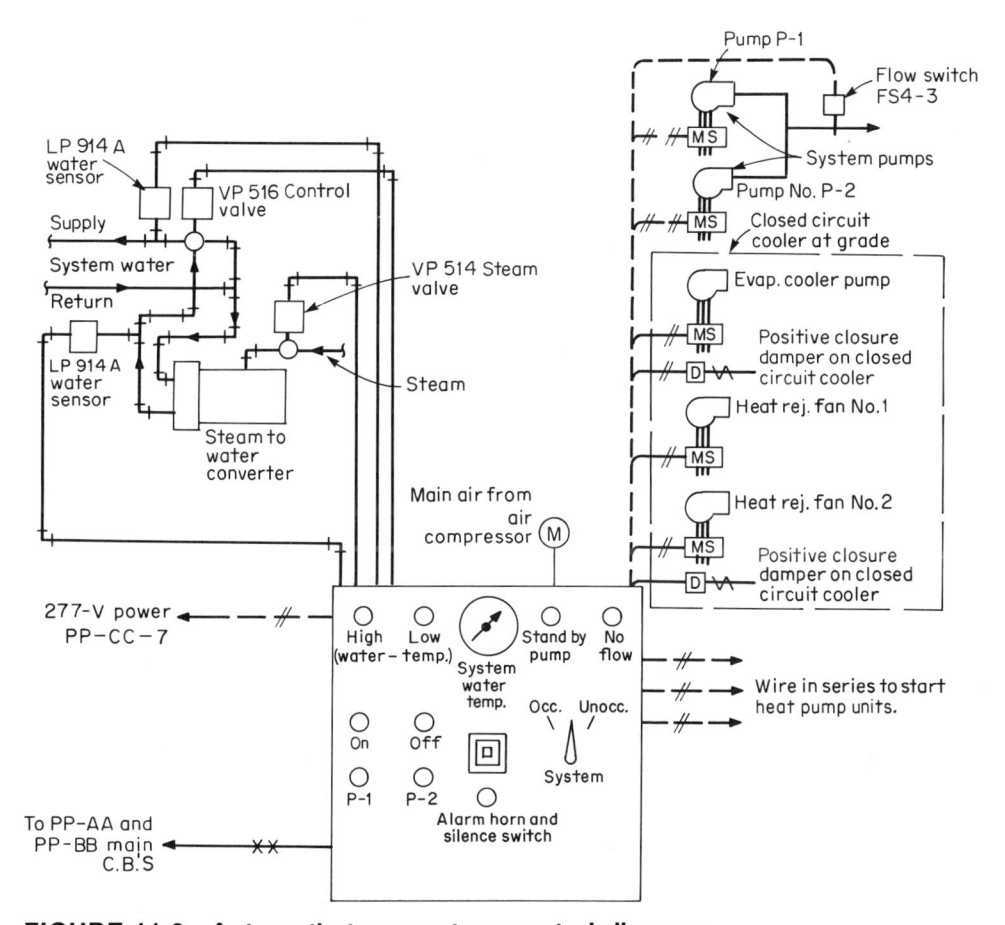

FIGURE 11-2 Automatic-temperature-control diagram.

3. As the system water temperature drops, the reverse sequence occurs: the heat-rejector fans and the evaporative-cooler pump are stopped. At approximately 60°F, the three-way control valve at the steam-water convertor begins modulating open as required to supply supplementary heat to the system. The steam control valve modulates through the action of the immersion-discharge controller to maintain its setting.

4. A pump-selector switch allows the selection of either system condenser water pump for continuous operation, with the other pump on standby for emergency operation. The pumps should be alternated from time to time to equalize operating time and wear.

Safety Controls

1. If the system water temperature rises to 105°F, an alarm sounds and a red light indicates a high water temperature.

2. If the system water temperature drops to 50°F, an alarm sounds and a red light indicates a low water temperature.

3. If the system water flow stops, a flow switch automatically starts the standby pump, and a red light indicates that the standby pump is operating so that the primary pump can be repaired. In addition, an alarm sounds and a red light indicates no flow, for approximately 1 min after the standby pump is energized. If the standby pump fails to restore the system water flow, the alarm and no-flow light remain on.

4. To reset the safety controls for normal operation following the correction of a pump failure, the operator pushes the OFF button and then immediately pushes the ON button.

5. A relay connected to the shunt trips the main breakers in PP-AA, PP-BB (see Fig. 11-3) to stop the heat pumps if the system water becomes too hot or too cold, or if there is no system water flow.

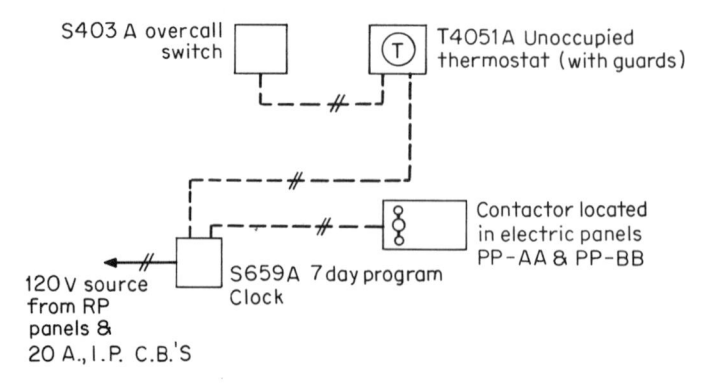

FIGURE 11-3 Heat pump unit control.

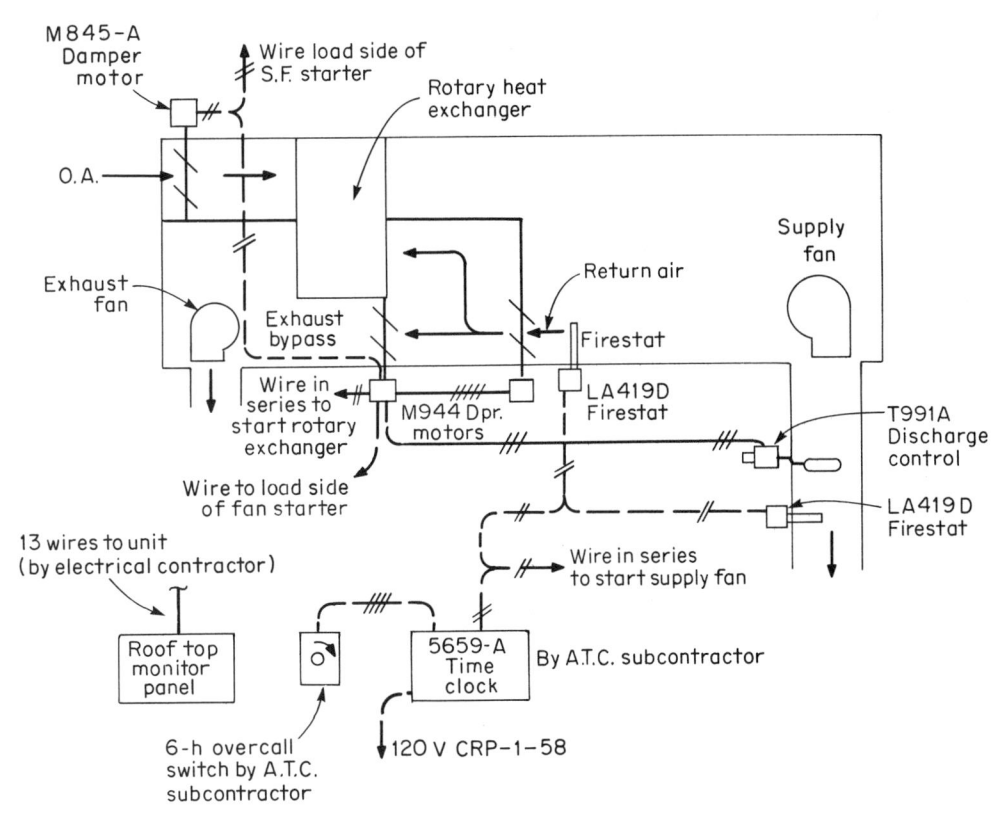

FIGURE 11-4 Rooftop heat-recovery-unit control.

Other Controls   Figure 11-3 shows the heat-pump-unit control. The heat pump has a control box containing a manual-changeover electric-type low-voltage occupied-room thermostat and a five-button HEAT-OFF-COOLING-HIGH-LOW switch. The thermostat remote bulb is furnished with a factory-mounted-and-wired heat anticipator to operate on the heating cycle.

The heat pump is programmed from *occupied* to *unoccupied* by a seven-day time clock in conjunction with the zone contactor. The zone override switch allows the operator to return the zone to *occupied* operation during an *unoccupied* heating cycle. The zone occupied thermostat cycles its contactor to provide a reduced space temperature during the *unoccupied* heating cycle. The heat pump remains off during the *unoccupied* cooling cycle.

Figure 11-4 indicates the rooftop heat-recovery-unit (100% outside air) control. The controls are furnished by the unit manufacturer and are factory installed and factory wired. The controls include damper operators, thermostats, and all control specialties and adjuncts as indicated. When the outside air temperature is below 60°F as determined by an outdoor thermostat, the supply-air thermostat controls the motorized exhaust-air bypass damper to bypass exhaust air around the rotary heat exchanger to maintain the thermostat set point.

When the outside air temperature is above 78°F as determined by the outdoor thermostat, the supply-air thermostat is inactive, the motorized exhaust-air bypass damper is closed, and the rotary heat exchanger rotates continuously. When the outside temperature is between 60 and 78°F, the unit delivers 100% outdoor air, with the rotary heat exchanger off and the bypass damper full open. The system is programmed from its seven-day program clock. When it is indexed to *occupied*, the supply and exhaust fans run continuously. During the clock's *unoccupied* cycle, the supply and exhaust fans are off and the outside-air damper is closed. The seven-day program clock is furnished by the temperature-control subcontractor.

CENTRAL CONTROL PANELS

The concept of centralized control is almost universally applied to facilities being built today. In its simplest form, it might be merely a START-STOP switch controlling a remote fan or pump, and located at the most convenient control point. In its most complex form, it may be a computer-operated system collecting, processing, and displaying data and performing numerous functions automatically for a number of buildings. Centralized control is usually used in a facility to reduce operating cost, maintain a given level of performance, and provide occupant safety.

The application of centralized control requires special consideration during the early stages of building design. Items to be considered include the locations of equipment necessary for centralized control, wiring and piping runs to interconnect the systems, electric power requirements, owner requirements, and manpower requirements.

ESTIMATING The automatic temperature-control system for a project is generally furnished and installed, as shown on the drawings and specifications, by a temperature-control subcontractor. Since the automatic temperature-control system is composed of controllers, controlled devices, and auxiliary control equipment, including related items such as air piping and/or temperature-control wiring, the automatic-control specification will detail the responsibilities of the individual contractors.

The estimator must carefully read the specification to determine whether the automatic control valves and dampers are furnished and installed by the temperature-control subcontractor, or installed by the mechanical contractor or the sheet-metal contractor under the supervision of the control contractor. In addition, the estimator must determine whether the control contractor or the electrical contractor must provide (furnish and install) the control wiring. The estimator must also determine whether or not the equipment manufacturers will furnish packaged units with the basic controls.

Once the mechanical estimator has determined which items are included in the mechanical contract and the sheet-metal contract, the estimating procedure for these items should begin. The automatic-control subcontractor generally estimates the total cost of the system; then the mechanical contractor incorporates these costs, along with the markup, in the job bid estimate.

Since automatic control systems differ according to the various designs of heating and cooling systems, as well as the required degree of control-system sophistication, it is extremely difficult to estimate automatic controls on a rational basis. However, for the purpose of early-stage design estimates, it may be noted that the cost of automatic temperature controls for an average building represents approximately 10% of the mechanical contract, not including provisions for a central control console if required. Also, for the purpose of advanced-stage design estimates, the cost of an automatic temperature-control system may be estimated on a per-control-point basis. As an example, consider an all-water-type fan-coil unit connected to a two-pipe system, with control accomplished by a room thermostat and automatic control valve. Such a fan-coil unit may be said to have two control points (thermostat and automatic control valve). In this way, the control points incorporated in each controlled unit may be counted; the sum of all the control points for the individual units is the total number of control points included in the control system.

The average cost for furnishing and installing a control point, including provisions for auxiliary equipment, air piping, and/or control wiring, is approximately $200.00 per point for pneumatic control systems, and $150.00 per point for electronic-electric systems. The cost of a central control panel is approximately $50.00 to $60.00 per control point. This information should not be used for bidding purposes. Since the automatic control system is generally let as a subcontract, the mechanical estimator rarely needs to produce a complete estimate for this system. In general, the mechanical estimator estimates that portion of the work which may be installed by the mechanical contractor. Because of this, the estimating information which appears in this section is given only for the purpose of estimating the portion of the automatic control system normally installed by the mechanical contractor. It is advisable to consult with the automatic-control subcontractor in order to obtain the actual cost of an automatic

ESTIMATING CHART 11-1   Labor Man-hours for Automatic Control Valves with Operators (125 lb, Bronze-Threaded)

Valve Type	½	¾	1	1¼	1½	2
Two-port valve	1.00	1.20	1.40	1.55	1.75	2.00
Three-port valve	1.25	1.55	1.75	2.00	2.20	2.65

Valve Size, in

ESTIMATING CHART 11-2 Labor Man-hours for Automatic Control Valves with Operators—125 lb, Cast Iron Flanged

Valve Type	Valve Size, in							
	2½	3	4	5	6	8	10	12
Two-port valve	2.45	3.25	4.50	5.55	6.55	7.20	13.50	14.00
Three-port valve	3.40	4.20	6.90	7.50	7.70	8.35	14.90	15.50
Single butterfly valve	2.35	2.75	3.15	3.50	3.90	4.35	6.35	7.00
Double butterfly valve	5.00	6.20	7.90	9.00	9.45	12.00	15.00	16.00

ESTIMATING CHART 11-3
Labor Man-hours for Control Air Compressor (packaged air compressor with receiver, accessories, piping, etc.)

Motor HP	Man-hours
Up to 15	12
Over 15	24

ESTIMATING CHART 11-4 Labor Man-hours for Temperature Regulating Valves*

Valve Type	Class, lb	Valve Size, in					
		½	¾	1	1¼	1½	2
Steam-threaded	250	1.45	1.75	2.00	2.30	2.60	3.15

Valve Type	Class, lb	Valve Size, in							
		2½	3	4	5	6	8	10	12
Steam-flanged	15	2.67	3.00	4.82	5.96	6.95	8.06	11.20	13.96
	125	3.10	3.55	5.95	7.20	7.96	9.36	14.20	15.40
	250	4.40	4.95	6.70	7.45	9.20	11.90	16.20	17.70

ESTIMATING CHART 11-5 Labor Man-hours for Pressure and Temperature Regulating Valves*

Valve Type	Class, lb	Valve Size, in					
		½	¾	1	1¼	1½	2
Steam-threaded	250	1.55	1.80	2.10	2.40	2.70	3.25

Valve Type	Class, lb	Valve Size, in							
		2½	3	4	5	6	8	10	12
Steam-flanged	15	2.76	3.06	4.90	6.05	7.00	8.12	11.21	14.00
	125	3.15	3.60	6.00	7.22	8.00	9.40	14.22	15.41
	250	4.50	5.00	6.75	7.50	9.26	11.92	16.20	17.70

*For pressure regulating valves refer to Estimating Chart 8-9.

ESTIMATING CHART 11-6 Labor Man-hours for Two-Port Bronze Solenoid Valves

Valve Size, in							
½	¾	1	1¼	1½	2	2½	3
1.02	1.22	1.50	1.60	1.85	2.10	3.00	3.90

ESTIMATING CHART 11-7 Labor Man-hours for Magnetic Flowmeters (in-line flanged)

Meter Size, in					
1	2	3	4	6	8
2.75	3.00	4.00	5.75	7.00	8.00

ESTIMATING CHART 11-8 Labor Man-hours for Wafer-type Orifice Flowmeters

Meter Size, in						
1¼	1½	2½	3	4	6	8
0.50	0.55	0.60	0.62	0.65	0.75	0.90

ESTIMATING CHART 11-9 Labor Man-hours for Flow Control Valves

Valve Type	Valve Size, in					
	1½	2	3	4	6	8
Threaded	1.45	1.65	3.00			
Flanged			2.15	2.30	3.25	3.65

ESTIMATING CHART 11-10 Labor Man-hours for Cast-Iron Venturi

Valve Type	Class, lb	Venturi Size, in					
		4	5	6	8	10	12
Flanged	125	2.75	3.50	4.95	6.20	8.00	9.00
Wafer		0.75	0.80	1.10	1.50	2.00	2.50

ESTIMATING CHART 11-11 Labor Man-hours for Cast Steel Orifice Plates

Connection Type	Orifice Size, in											
	1	1¼	1½	2	2½	3	4	5	6	8	10	12
Screwed	2.95	3.20	3.50	5.00	6.00	6.50						
Slip-on	4.00	4.50	5.00	7.10	8.10	8.50	9.75	10.75	13.00	15.00	20.00	25.50
Welded neck	4.10	4.60	5.10	7.20	8.20	8.70	9.95	10.95	13.20	15.50	20.50	26.00

ESTIMATING CHART 11-12 Labor Man-hours for In-line Site Flow Indicators

Indicator Type	Indicator Size, in										
	¾	1	1¼	1½	2	2½	3	4	6	8	10	12
Screwed	1.00	1.15	1.30	1.40	1.70							
Flanged						1.55	1.65	2.60	3.30	4.75	6.80	7.00

ESTIMATING CHART 11-13 Labor Man-hours for Controllers

Controller Type	Man-hours*
Room thermostat and humidistat	1.10
Insertion and immersion thermostats	1.50
Remote-bulb thermostat	2.00
Multistage thermostat	1.75
Modulating pressure controller	2.00
Firestat	1.25

*Labor units include handling and setting in place only. If a pneumatic system is used, see Chap. 8 for estimating air piping.

temperature-control system. Since various types of control systems are commercially available, and these systems are usually composed of a variety of control components, this text does not provide detailed cost data for control instruments.

Labor Man-hours

Labor man-hours for the controlled devices which are installed by the mechanical contractor may be obtained from Estimating Charts 11-1 through 11-13. For automatic dampers, see Estimating Chart 7-16. The labor figures include handling, erecting, and making up the required joints for the controlled devices, based on normal working conditions. Correction factors may be applied to adjust these labor figures up or down according to actual job conditions.

12

TESTING AND BALANCING

GENERAL

The process of testing and balancing is necessary to assure that all building environmental systems are operating in accordance with design specifications. The process includes the following procedures:

1. Balancing of the air and water distribution
2. Adjustment of all systems to provide design quantities
3. Verification of the performance of all equipment and automatic controls
4. Measurement of electricity, sound, and vibration
5. Recording and reporting the results

TESTING

The process of testing includes the performance of both leakage tests and performance tests.

Leakage Tests

1. All piping systems must be leak-tested prior to the application of thermal insulation and/or concealment in shafts, hung ceilings, or other places where leakage cannot be readily observed. Any leaks which develop during the testing must be repaired, and the tests must then be repeated.

2. All piping systems must be subjected to a hydrostatic test of 8 h duration. Temporary test connections, air compressors, and the required testing equipment are provided by the mechanical contractor. The recommended test pressures for the various piping systems are as indicated in Table 12-1.

3. All concealed ductwork must be air-pressure tested prior to concealment. The test pressure is the design static pressure of the system plus

TABLE 12-1 Recommended Test Pressures for Piping Systems

System Classification	Pressure, psig
High-pressure steam and return	150
Low-pressure steam and return	100
Hot-water system	100
Chilled-water system	100
Condenser-water system	150
Fuel-oil system	200

25%, and the test duration is 1 h. The pressure loss must not exceed 1% of the test pressure.

4. The contractor must submit in writing, to the engineer who designed the job, assurance that the specified tests have been performed and that the systems are free of leaks. After the leakage tests are completed, no equipment, piping system, or duct system may be operated until all construction dirt and debris have been removed from equipment, piping, and ducts. A cleaning process is therefore required. It includes the cleaning of:

   a. *Piping systems.* All piping systems must be drained of test water, and clean water with a chemical detergent must be introduced into the various systems to remove all welding slag, cuttings, and debris. The cleaning process must have a duration of 48 h. All required temporary pipe connections, chemicals, etc., are to be provided by the mechanical contractor, as per the water-treatment manufacturer's recommendation. After the cleaning solution is drained from the systems, all strainer baskets must be removed and thoroughly cleaned.

   b. *Air systems.* All ductwork must be inspected for cleanliness, and debris such as mortar and bricks must be removed before the fans are operated. All coils must be thoroughly cleaned outside with a vacuum cleaner, so that construction dust and dirt are removed from the heat-transfer surface. All duct plenums and fan scrolls must be inspected, and all construction debris must be removed prior to fan operation. No fan may be operated without its filters in place. Dirty filters must be replaced before the system is turned over to the owner.

   c. *The cooling-tower basin and sumps.* These must be drained and inspected prior to turnover to the owner. All sludge and debris must be removed. Sump strainers must be removed and cleaned thoroughly. The distribution basin must be inspected, and any clogged orifices must be unplugged.

Performance Tests

After the various systems are placed in operation, the mechanical contractor makes the following performance tests:

1. All fans are tested for speed (rpm), static-pressure differential, and amperage draw of the fan motor.

2. All pumps are tested for speed, pressure differential, and amperage draw of the pump motor. Pump tests, including initial testing for the correct rotation direction (when making the electrical connections), must only be made when the pump is fully primed, with strainer in place.

3. Refrigeration machines are tested for temperature range and pressure differential of condenser and chilled water, and either steam pressure for absorption machines or compressor-motor amperage draw for electric chillers.

4. Cooling towers are tested for condenser-water temperature range, fan static pressure, and fan-motor amperage draw.

5. Filter banks are tested, when clean, for static-pressure differential.

6. Water coils are tested for both water- and air-pressure differential, water flow, and airflow.

7. All control thermostats and humidistats are tested for thermometer calibration. They are then tested to ensure that the component being controlled is performing its proper function.

All performance tests are conducted jointly by the equipment manufacturers and the mechanical contractor. The mechanical contractor submits the results of the performance tests to the engineer who designed the job, for approval.

BALANCING

All equipment, water-piping systems, and air-duct systems must be balanced so that design flow and temperature are achieved. The process of balancing can be divided as follows:

Water-Piping Systems

1. The water-circulating pumps for the various systems must be balanced, by use of their balancing cocks, to establish the design pressure differential.

2. In the chilled-water system, design flow can be established in the refrigeration machines by use of their balancing cocks. If the cooling loads are insufficient for temperature-range balance, the balance can be established by pressure differential across the equipment, and tempera-

ture-range checks can be made when the thermal loads reach design values. All cooling coils can be balanced by use of their balancing cocks. If the thermal loads are not established as yet, the pressure differential can be used for balancing, with the temperature range being checked at a later date.

3. In the condenser-water system, design flow can be established at the condensers and cooling-tower cells by use of balancing cocks, orifices, and meters provided in the design. Proper flow can be established in each distribution header of the cooling tower and across the condenser-water bypass valve.

4. In the hot-water system, design flow can be established in the heat exchangers by use of their balancing cocks. If the heating loads are insufficient for temperature-range balance, then balance can be established by pressure differential across the equipment, and temperature-range checks can be made when the thermal load reaches the design value. Hot-water coils can be balanced by use of their balancing cocks. If the thermal loads are not in the design range as yet, the pressure differential can be used for balancing, with the temperature range being checked at a later date.

Air-Duct Systems

1. Each fan must be balanced to its design static-pressure differential, speed, and amperage draw.

2. All filter banks, central coils, and fan-discharge ducts can be traversed with velometers and pressure-drop instruments to check fan air-volume delivery.

3. All automatic and manual volume dampers at central air-handling units must be adjusted and set so that design flow is established.

4. Main-trunk duct dampers must be adjusted and set, and trunk velometer and pressure readings must be taken to assure design flow in trunk ducts.

5. The distribution ductwork can be balanced, using the volume dampers, by first establishing the design flow in the various trunk ducts, and then by establishing the proper flow at the individual outlets, starting with the outlets nearest the duct shafts.

6. After all outlets, units, trunk duct, and main dampers have been set, the fans must be rechecked and adjusted, if necessary, to assure design air flow.

7. All dampers, when finally set, must be struck and marked to indicate the balanced setting.

Equipment Major equipment, such as refrigeration machines, refrigeration circuits, and condenser-water bypasses, can be balanced by the approved manufacturer's factory engineer, in the presence of the contractor's project engineer.

Recording and Reporting the Results

Balance reports, showing a log of readings, must be submitted for approval. They must show at least two sets of readings indicating design flow, and two sets of actual readings. The reports must also show test data for each system. A balancing plan, correlating layouts, and logs must also be submitted.

A basic concept of the balancing process is that there be one organization or person responsible for the complete balancing of all systems. The mechanical contractor can retain the services of an outside organization and/or consultant specializing in balancing work to perform the required balancing. All balancing must be done in the presence of the contractor's project engineer.

The organization which balances the system must provide all the instruments and equipment required to accomplish the balancing. These instruments may include velometer, pitot tube, anemometer, differential pressure gauges, manometer, thermometer, psychrometer, volt-ammeter, and revolution counter.

ESTIMATING THE COST OF TESTING AND BALANCING

The cost of testing and balancing varies with the type of building and the amount of work required by the drawings and specifications. However, for an average job including both heating and cooling systems, the cost of performance testing and balancing may be estimated on the basis of man-hours per ton of cooling load, as indicated in Estimating Chart 12-1.

ESTIMATING CHART 12-1   Labor Man-hours for Testing and Balancing Systems per Ton of Cooling Load

Building Type	Up to 100 tons	101–500 tons	Over 500 tons
Commercial, apartments	1.50	1.25	1.10
Educational, hospitals, hotels	1.75	1.50	1.25
Industrial	1.20	1.00	0.90
Laboratories	1.85	1.60	1.35

(Man-Hours for Testing and Balancing, per Ton)

ESTIMATING CHART 12-2 Labor Man-hours for Testing and Balancing Various Types of Devices and Equipment

Item	Man-hours	Item	Man-hours
Air-conditioning units	12.00	Heating-ventilating units	5.00
Air inlets (exhaust and return)	0.50	Induction units	1.25
Air outlets (supply)	0.80	Mixing boxes	1.00
Boilers	8.00	Pumps	5.00
Chillers	6.00	Radiators	0.90
Coils	1.25	Terminal reheat coils	1.25
Converters	6.00	Terminal air units	1.00
Cooling towers	8.00	Thermostats and humidistats	0.80
Expansion tanks	0.80	Valves (pressure-control and regulating)	1.00
Fans	2.00		
Fan-coil units	1.50	Water solenoid valves	0.30
Fin-tube radiation	0.80	Unit heaters	1.50

The information in the chart should only be used for rapid estimating and for checking. The specifications must be considered carefully when these labor figures are used. The contractor's markup must be added to the figures shown. If a testing-and-balancing organization performs the work, the mechanical contractor will still add a markup (usually 7.5%) to the subcontractor's cost.

For system leakage tests and cleanup, the estimator should allow 5% of the total man-hours for each piping system. For ductwork systems, 2.5% of the total field man-hours should be allowed for each system.

For an accurate estimate of the performance-testing and balancing cost, a quantity takeoff is required. All equipment and air devices which are to be tested and balanced should be listed by transposing from the takeoffs for the various systems. The estimator should then apply a man-hour figure to each item included in the testing-and-balancing list. (The labor figures given in Estimating Chart 12-2 can be used for this purpose. Each figure includes the testing and balancing of the air and/or water side, depending on the unit being tested.) The estimator must then transpose the total man-hours for each system to the job estimating sheets (refer to Fig. 1-4), calculate the labor cost, and apply the overhead cost and profit.

13

MOTORS AND MOTOR STARTERS

MOTORS

Motors for all equipment are normally furnished by the equipment manufacturers. The motors must be built in accordance with the latest standards of the National Electrical Manufacturers Association (NEMA). They must be of sufficient capacity to operate associated driven devices under the conditions of operation without overload, and must be of at least the horsepower indicated on the drawings and/or specifications. The motors must be provided with adequate starting and protective equipment.

Single-phase motors may be of the capacitor-start, split-phase, or shaded-pole type, as specified for individual applications. Three-phase motors may be of the squirrel-cage or wound-rotor induction type, according to starting torque and current characteristics as specified for individual applications. The NEMA classification of motor enclosures applies when motor types are specified as open, dripproof, splashproof, totally enclosed, or the like. Motors can be rated for continuous duty and under full load; the maximum rise in temperature may not exceed 40°C for the open type, 50°C for the dripproof and splashproof types, and 55°C for the explosionproof and totally enclosed types.

Motors may be equipped with ball or roller bearings with pressure grease lubrication and provisions for relief of grease. Sleeve bearings may also be used with an approved method of oil lubrication. Motors must operate without objectionable vibration or noise. Belt-connected motors must be equipped with adjustable bases and setscrews to maintain belt tension.

Unless otherwise specified or indicated, motors ½ hp and larger may operate on three-phase, 60-Hz, 480- or 460-V alternating current (ac). Motors smaller than ½ hp may operate on single-phase, 60-Hz, 120-volt ac. Motor type, horsepower, speed, and current characteristics are usually specified in the detailed specifications for individual equipment items or indicated on the schedules.

Manufacturers' prices for individual pieces of equipment usually include the motor price; however, the prices of various types of motors are indicated in Chap. 4, in the pump estimating section (refer to Estimating Charts 4-6 and 4-7). The various equipment labor figures indicated in this manual include the labor required for handling and erecting motors. Unless otherwise specified, final connections to all motors are the responsibility of the electrical contractor.

MOTOR STARTERS

Motor starters for equipment may comply with the following:

1. Starters for single-phase, 120-V motors may be of the toggle-switch type with integral thermal-overload protection, manual reset, and pilot light, unless otherwise specified.

2. Starters for three-phase motors of 100 hp and less may be of the combination switch-and-fuse magnetic across-the-line type, with integral or remote push-button station, pilot lights, etc. Magnetic starters are provided with a built-in overload-protection device for each of the three phases and a low-voltage protection release.

3. Starters for three-phase motors above 100 hp may be of the reduced-voltage, primary-resistance type with integral or remote push-button station, pilot lights, etc. Starters are provided with a built-in overload-protection device for each of the three phases and a low-voltage protection release. All starters must, however, be built in accordance with the latest standards of NEMA and other applicable electrical codes.

4. All starters must be furnished complete with a minimum of two sets of auxiliary contacts, unless otherwise specified.

The mechanical estimator should read the specifications for motor starters to determine whether the mechanical contractor or the electrical contractor must furnish the motor starters. If the mechanical contractor is to furnish the starters, the material cost of the various motor starters must be added to the job cost. A quantity takeoff is therefore required. All motor starters and related devices which come within the scope of the work should be listed, and a material price applied to each. Prices can be obtained from manufacturers' price lists. The installation and final connection of motor starters are the responsibility of the electrical contractors; however, the mechanical contractor may submit the necessary wiring diagrams to the electrical contractor to ensure that motors, starters, interlocks, etc., are properly installed.

14

FOUNDATIONS AND VIBRATION ISOLATION

GENERAL Supporting structures are used to support equipment, to safely withstand stresses to which they may be subjected, and to distribute the load and impact properly over the building structure. They should be designed to be stiff enough to support equipment without undue deflection. Vibration isolators should be placed between the equipment and the supporting structure to avoid problems due to the transmission of vibration and structure-borne noise. The various types of vibration isolators are steel springs, rubber isolators, glass-fiber loads, felt bands, and air springs. Equipment can be installed directly on the isolators, except where concrete or a structural base is required for component support and alignment. Equipment with large inherent unbalanced forces generally requires concrete inertia block (reinforced as required) to prohibit excessive movement. Equipment such as certain large fans, pumps, and compressors, as well as high-pressure fans, generally requires inertia block. All concrete inertia bases should be suitably reinforced in accordance with the requirements of the vibration-isolator manufacturer.

Floor-mounted equipment is generally erected on a concrete pad over the complete floor area of the equipment. The specifications should include the desired height of the concrete pad. Concrete pads should extend 6 in beyond the machine base in all directions, and steel dowel rods should be inserted into the floor to anchor the pads. Machine anchor bolts should be set in pipe sleeves at least two sizes larger than the anchor bolts. Ceiling-supported equipment may be hung by means of vibration-isolation hangers consisting of a steel housing or retainer and a steel spring.

ESTIMATING

Vibration Isolators

The material costs of the various types of vibration isolators may be obtained from vibration-isolator manufacturers. The total material cost can then be computed. All equipment labor figures given in this manual include provision for the installation of vibration isolators.

Fabrication of Structural-Steel Bases

Structural-steel bases are generally furnished by equipment manufacturers. If the equipment to be mounted is not furnished with integral structural bases, approved structural bases must be fabricated. If a fabricated steel base is required, a quantity takeoff must be made. The required quantities of the various structural elements must be determined, and the weight of these elements computed. Then material costs must be obtained from steel manufacturers. The labor man-hour figures in Estimating Chart 14-1 include the labor for handling, fabricating, and installing structural bases.

Concrete Foundations

For concrete foundations, quantity takeoff with the following information is required:

1. The physical dimensions of each individual foundation or pad
2. The sizes and quantities of the various anchor bolts
3. The required reinforcement bars
4. The types of concrete finishes
5. The volume of concrete, in cubic yards, computed from the physical dimensions
6. The computed areas of wood forms, in square feet
7. The computed surface areas of concrete topping finishes
8. The computed weights of steel reinforcement bars

ESTIMATING CHART 14-1 Labor Man-hours for Structural-Steel Bases

Weight of Structural-Steel Base, lb	Man-hours per 100 lb		
	Fabricate	Erect	Total
Up to 100	3.00	2.50	5.50
200	2.40	2.00	4.40
400	2.30	1.90	4.20
600	2.25	1.75	4.00
800	2.25	1.50	3.75
1000	2.25	1.50	3.75

ESTIMATING CHART 14-2 Labor Man-hours for Concrete Foundations

Classification	Man-hours
Setting forms	0.20/ft²
Reinforcing concrete	0.02/lb
Placing concrete	2.00/yd³
Top finishes, 1 in thick	0.07/ft²
Setting, bolting, and sleeves	0.15 each

Consider a concrete pad of dimensions $L \times W \times H$, where L is the length in feet, W is the width in feet, and H is the height in inches. The following formulas are of help in the computations listed above:

$$\text{Concrete volume} = \frac{L \times W \times H/12}{27} \quad \text{yd}^3 \tag{14-1}$$

$$\text{Wood-form area} = \frac{(L + W) \times H}{6} \quad \text{ft}^2 \tag{14-2}$$

$$\text{Topping finish area} = L \times W \quad \text{ft}^2 \tag{14-3}$$

Lumber prices are usually quoted in board feet. To find the number of board feet in a given piece of lumber whose thickness and width in inches and length in feet are known, one may use the formula

$$\text{Board feet} = \frac{\text{thickness (in)} \times \text{width (in)} \times \text{length (ft)}}{12} \tag{14-4}$$

To estimate material costs, the estimator must obtain the prices of the items on the takeoff sheet. Manufacturers quote prices on the following bases:

- Concrete mix: on a per-cubic-yard basis (1 ft³ of concrete weighs about 150 lb)
- Wood forms: on a per-square-foot basis
- Rebars: on a per-pound basis
- Anchor bolts: on a per-bolt basis
- Top finishes: on a per-square-foot-of-surface basis

The figures given in Estimating Chart 14-2 may be used to estimate labor man-hours.

15
PAINTING

GENERAL The surfaces of equipment, piping, ductwork, and miscellaneous ironwork which come within the scope of the painting work should be painted in accordance with design specifications. Painting generally requires two operations:

1. *Surface preparation.* All surfaces to be painted should be clean, smooth, dry, and free of dust, dirt, or any defects before paint is applied.

2. *Paint application.* All painting should be done in a neat and workmanlike manner. Brush painting is generally recommended for all painting at the job site. All surfaces should be given one or more coats of paint as specified. The paints in present use can be classified as follows:

   a. Prime-coat paints include zinc chromate for ironwork and zinc dust for galvanized materials.

   b. Semigloss enamel paint is used for finish coats.

The various piping systems are usually painted in accordance with the pipe-identification color code given in Table 15-1. All piping in accessible areas should be identified by a stenciled legend and flow arrow. In general, numbered tags should be provided on all special fittings, valves, and other operating devices, as well as equipment.

ESTIMATING Sometimes the mechanical estimator is assigned to estimate the painting work. If so, a quantity takeoff is required. All equipment, piping, ductwork, etc., which come within the scope of work should be listed. The estimating quantities for painting can be measured directly on the job drawings, or the measurements made during the takeoff procedures for the various systems may be used. In estimating the amount of paint required for various items, the actual area of the surface to be painted should be calculated, as follows:

TABLE 15-1 Piping-System Color Code

Piping-System Service	Color
Chilled water	Blue
Cold water	Green
Condensate drain	Tan
Condenser water	Gray
Gas	Purple
Fuel-oil supply	Black
Fuel-oil return	Silver
Heating water	Orange
Refrigeration	Gold
Steam and condensate	Red

1. For *ductwork*, the surface area can be calculated by using the information given in Chap. 9 for calculating the surface area of ductwork.

2. For *pipe*, the surface area can be calculated by using the following formula:

$$S_p = 0.26 \times d \times L \qquad (15\text{-}1)$$

where S_p = surface area of pipe, ft²
d = outside diameter of pipe, in
L = length of pipe, ft

3. For *radiators*, the specified area (square feet) of radiation can be used.

4. For *structural steel*, the surface area is calculated by finding the combined area of all structural shapes (refer to published structural-steel tables). However, the weight of structural steel can also be used in paint estimates.

5. For *enclosures*, the surface area is calculated by finding the combined area of all sides. Each side surface area is calculated by multiplying its length by its width.

6. For *equipment*, the surface area can be calculated by assuming that the unit is enclosed in a cubical container. The dimensions of this container are the physical dimensions of the unit to be painted. The surface area of the container is found as noted above for enclosures.

7. For *tanks*, the surface area is calculated by using the following formulas:

   a. For *cylindrical tanks*,

   $$\text{Surface area} = 1.57(2DL + D^2) \quad \text{ft}^2 \qquad (15\text{-}2)$$

ESTIMATING CHART 15-1 Labor Man-hours for Brush Painting

1. Miscellaneous Iron	Man-hours/ton
Prime coat	1.75
Finish Coat	2.25
Total	4.00
2. All Metal Surfaces	Man-hours/ft²
Prime coat	0.006
Finish coat	0.007
Total	0.013

  b. For *spherical tanks*,

  $$\text{Surface area} = 3.14 D^2 \qquad \text{ft}^2 \qquad (15\text{-}3)$$

  where L = length of cylindrical tank, ft
  D = tank diameter, ft

Painting costs include labor and materials. The material cost depends on the kind of paint specified. The estimator should obtain the cost of painting materials from manufacturers' price lists; labor man-hours can be obtained from Estimating Chart 15-1. The man-hour figures are based on average conditions and include the labor for surface preparation and paint application.

16 RIGGING

GENERAL

Rigging services are often required for loading, unloading, moving, and setting heavy and bulky equipment. Since rigging involves a risk to human life, local codes usually require that rigging work be conducted by a specialty contractor for reasons of safety. Riggers are often licensed by local government agencies. Safe lifting equipment and procedures are necessary to ensure that there will be no rigging hazards. Owing to such local codes, mechanical contractors sublet rigging work to rigging contractors. A detailed rigging estimate must be prepared by the rigging contractor. It requires detailed information such as the weight and size of the equipment that will come within the scope of the work, the various types of loading carriers, the required moving destination, the locations of the individual items of equipment in the building, as well as the schedule of delivery and erection of equipment. An estimating contingency value should be added to the rigging estimate to cover delays due to late delivery, accidents, lifting-equipment breakdowns, and bad weather.

For the purpose of rigging estimation, heating and cooling equipment may be classified according to the categories indicated in Table 16-1. The various types of lifting equipment which may be used in rigging are the winch truck, fork-lift truck, derrick crane, and gin pole.

TABLE 16-1 Equipment Classifications for Rigging

Category	Equipment Classification	Weight
1	Lightweight and compact	Up to 1500 lb
2	Lightweight and bulky	Up to 1500 lb
3	Heavyweight and compact	1–10 tons
4	Heavyweight and bulky	1–10 tons
5	Heavyweight and compact	10–50 tons
6	Heavyweight and bulky	10–50 tons

ESTIMATING CHART 16-1   Labor Man-hours for Various Rigging Operations

Rigging Operation	\multicolumn{6}{c}{Labor Man-hours per Ton}
	\multicolumn{6}{c}{Equipment Category}
	1	2	3	4	5	6
1. Unload from open trailer with power rigging equipment to:						
a. Ground adjacent to trailer	2.30	3.10	1.95	2.00	1.20	1.55
b. Supporting base located in lower level	2.70	3.45	2.10	2.15	1.40	1.90
c. Supporting base located in intermediate level	3.45	4.25	2.30	2.70	1.75	2.50
2. Unload from enclosed trailer to ground by:						
a. Using power rigging equipment to drag equipment from inside through trailer opening	2.90	3.85	3.10	3.45	2.30	3.10
b. Jacking, bulling, skidding on small roller, then dragging to trailer opening and removing by power equipment	3.70	5.25	4.00	4.50	3.25	4.05
c. Same as above except remove by hand or slide	3.85	7.00				
3. Manual moving operations:						
a. Jack up and insert skids or place rollers	0.60	0.75	0.50	0.58	0.39	0.50
b. Move on skids or rollers for 100 ft	1.05	1.15	0.46	0.66	0.39	0.58
c. Jack up or down and place or remove cribbing, per foot of height	0.60	0.75	0.50	0.58	0.39	0.46
d. Bull, move and/or turn up to 10 ft	1.35	1.55	0.95	1.05	0.77	0.90
4. Move by using rigging equipment:						
a. Lift and transport for 100 ft	0.40	0.45	0.36	0.42	0.32	0.34
b. Unload, set up on base	0.35	0.40	0.24	0.25	0.20	0.22
c. Build up cribbing, set equipment on top prior to lowering or horizontal positioning	0.70	0.85	0.58	0.66	0.46	0.54
5. Line up:						
a. Rough	0.80	0.80	1.25	1.25	1.50	1.50
b. Precise	1.50	1.50	2.50	2.50	3.50	3.50

ESTIMATING Rigging costs generally include the following items:

1. The cost of renting rigging equipment
2. The cost of an equipment operating crew
3. The cost of a rigging crew

Where rigging-equipment rental rates are required, the estimator should obtain these rates from equipment-rental agencies or from available published information; this manual does not include rigging-equipment rental rates.

The man-hour figures in Estimating Chart 16-1 include the labor required for loading, unloading, moving, and setting a piece of equipment. They are given as man-hours per ton and include an allowance for the equipment operating crew. They are average rates, subject to adjustment for the particular job conditions, type of rigging equipment available, and type of equipment which comes within the scope of the work. The equipment categories are defined in Table 16-1.

17

EXCAVATION AND BACKFILL

GENERAL Where the installation of underground tanks, piping, and/or conduits is required, the estimator must refer to the excavation and backfill section of the specifications to determine the kind of soil and the scope of earthwork. Excavation can be done with hand tools or power shovels or by blasting, depending on the kind of soil to be excavated. Soils may be classified as in Table 17-1. Excavation work may include the following items:

- Sheeting, bracing, and shoring to protect the sides of the excavation
- Pumping to keep excavated areas free of water
- Removal, piling, and transporting of excavated materials

However, some of these items may not be required by certain jobs. It is necessary to know which items may be encountered.

TABLE 17-1 Soils and Excavation Methods

Soil Category	Method of Excavation
Light	Earth easily shoveled and requiring no loosening; sandy soil is an example.
Medium	Earth easily loosened by hand tools. If power shovel is used, no preliminary loosening is required. Ordinary soil is an example.
Heavy	Soil can be excavated by power shovel, with no preliminary loosening required. Stiff clay soil is an example.
Hard	Soil requires light blasting and power machinery.
Rock	Requires blasting before removal.

TABLE 17-2 Typical Trench Configurations

Trench Configuration	Soil Condition	Cubic Yards per Foot
1. Rectangle with recommended width and depth, in feet.	Extremely stable, perhaps because shoring is used	$\dfrac{DW}{27}$
2. Trapezoid a. The depth and the bottom width are the same as in a rectangular trench, and the top width equals the bottom width plus the depth, in feet.	Reasonably stable	$\dfrac{D(2W+D)}{54}$
b. Same as *(a)* except that the top width equals the bottom width plus twice the depth, in feet.	Unstable, such as sand	$\dfrac{D(W+D)}{27}$

The upper portions of trenches and the entire excavations around piping, conduits, tanks, and maholes must be backfilled. Particular care must be exercised to backfill simultaneously on both sides of tanks and piping. Backfill should be free of rocks, debris, and corrosive material. Backfill may be either borrowed material (gravel) or excavated material. Tanks and piping must be inspected before backfilling.

ESTIMATING The unit used to compute quantities of excavation and backfilling work is the cubic yard: however, the foot may also be used for single-pipe trenches. In estimating trenches, the depth, width, and length of the trench are measured directly from the plans. The configuration of the trench should

TABLE 17-3 Recommended Trench Dimensions for Various Pipe Sizes

Pipe Size, in	Bottom Width, ft	Depth, ft
Up to 12	2.0	6.0
14–18	3.0	6.0
20–24	4.0	6.0
30–36	5.0	6.0

ESTIMATING CHART 17-1   Labor Man-hours for Excavation and Backfill

1. Labor Man-hours for Hand Excavation	
Soil Category	Man-hours per Cubic Yard (average for three lifts)
Light	1.75
Medium	2.50
Heavy	3.75
Hard	5.25
2. Labor Man-hours for Power-Shovel Excavation	
Soil Category	Man-hours per Cubic Yard (includes operating crew)
Light	0.05
Medium	0.08
Heavy	0.10
Hard	0.12
Rock	0.14
3. Labor Man-hours for Backfilling (average for all soil categories)	
Operation	Man-hours per Cubic Yard
Hand place	0.50
Hand spread	0.38
Machine backfill (bulldozer)	0.04
4. Labor Man-hours for Compacting	
Operation	Man-hours per Cubic Yard
Hand compacting	0.55
Pneumatic compacting	0.25

be determined as well. Trenches may be classified according to soil stability as shown in Table 17-2. Table 17-3 gives the recommended trench width and depth for various pipe sizes.

The volume (cubic yards) of excavation minus the volume of the buried pipe or tank gives the total volume of backfill required. The volume of the pipe or tank, in cubic yards, equals its diameter (in feet) squared, multiplied by the length of the pipe or tank (in feet) and multiplied by 0.03.

In estimating excavation work, the quantity of each type of soil should be listed separately. The necessity for shoring, pumping, and/or removal of excavating material must be determined.

Estimating Chart 17-1 gives the labor required for excavation and backfill, based on average conditions. The man-hour figures do not include blasting, hauling, or unloading. They include the operations of excavating and dumping, placing on side lines, or loading into trucks for hauling. If power equipment is used, the renting cost should be added to the earthwork costs.

18

CONCEPTUAL ESTIMATES

GENERAL A conceptual estimate is used for cost management during planning and early design stages. It is normally comprised of two types of estimates—cost model and schematic—both of which were discussed in Chap. 1. The conceptal estimate gives the project cost based upon parameters expressed in dollars per gross square foot of building area or other appropriate units. The cost per square foot can be obtained from actual data for previous jobs. This chapter gives general estimating and design data for the purpose of developing an accurate conceptual estimate.

GENERAL ESTIMATING DATA

Estimating Chart 18-1 indicates the unit costs of HVAC systems per gross square foot of building area for various types of buildings located in major cities. Estimating Chart 18-1 also indicates the HVAC systems most frequently used in each type of building. The information is based on the mid-1976 construction costs; it is given for reference only and should not be used for bidding purposes.

GENERAL DESIGN DATA

General It is frequently desirable to have design estimating data at hand, especially for use in working up schematic cost models and design-development cost estimates. The following design data will enable the estimator to obtain a broad, general picture of the size of the equipment and materials required for a project. The information was obtained from analyses of actual load estimates and is sufficient for determining general figures. For accurate load calculation, the reader should refer to the Appendix.

ESTIMATING CHART 18-1 HVAC Systems Total Costs per Square Foot of Gross Area (as of 1976) for the Purpose of Conceptual Estimates

Building Type	HVAC Systems	Atlanta, Ga.	Baltimore, Md.	Boston, Mass.	Chicago, Ill.	Cleveland, Ohio	Dallas, Tex.	Denver, Colo.	Detroit, Mich.	Minneapolis, Minn.	New Orleans, La.	New York, N.Y.	St. Louis, Mo.	San Francisco, Calif.	Seattle, Wash.	Washington, D.C.
\multicolumn{17}{c}{**1. Apartments**}
Low rise (5 story)	L.P. single duct, fan coil units—four-pipe	5.35	5.50	6.00	5.80	5.70	5.25	5.65	5.95	5.50	5.25	6.25	5.95	6.35	5.85	5.80
High rise (10 story)	L.P. single duct, fan coil units—four-pipe	5.15	5.25	5.75	5.60	5.45	5.00	5.45	5.70	5.30	5.00	6.00	5.70	6.05	5.65	5.60
\multicolumn{17}{c}{**2. Banks**}
Main	Combination of H.P. and L.P. single ducts with terminal reheat, fan coil units—four-pipe	8.60	8.75	9.55	9.30	9.10	8.40	9.05	9.50	8.80	8.40	10.00	9.50	10.10	9.35	9.30
Branch	L.P. single duct with reheat coils, fan coil units—four-pipe	6.00	6.15	6.70	6.50	6.35	5.90	6.30	6.65	6.15	5.90	7.00	6.65	7.10	6.55	6.50
\multicolumn{17}{c}{**3. College**}
Classroom	L.P. single duct with reheat coils, unit ventilators—four-pipe	7.95	8.10	8.85	8.60	8.40	7.75	8.40	8.80	8.15	7.75	9.25	8.80	9.35	8.65	8.60
Dining hall	L.P. single duct with reheat coils, fin-tube radiation	6.45	6.55	7.15	7.00	6.80	6.30	6.80	7.10	6.60	6.30	7.50	7.10	7.60	7.05	7.00
Laboratory	Combination of H.P. and L.P. single duct with terminal reheat, fin-tube radiation	10.30	10.50	11.45	11.15	10.90	10.10	10.85	11.40	10.55	10.10	12.00	11.40	12.15	11.20	11.15
Library	L.P. single duct with reheat coils, unit ventilators, and fan coil units—4-pipe	7.75	7.90	8.60	8.35	8.20	7.55	8.15	8.55	7.95	7.55	9.00	8.55	9.10	8.40	8.35
Student union	L.P. single duct with reheat coils, fan coil units—four-pipe	8.60	8.75	9.55	9.30	9.10	8.40	9.05	9.50	8.80	8.40	10.00	9.50	10.10	9.35	9.30
Auditorium	L.P. single duct with reheat coils, fin-tube radiation and cabinet heaters	10.30	10.50	11.45	11.15	10.90	10.10	10.85	11.40	10.55	10.10	12.00	11.40	12.15	11.20	11.15
Gymnasium	L.P. single duct with some reheat coils, and some radiation	7.75	7.90	8.60	8.35	8.20	7.55	8.15	8.55	7.95	7.55	9.00	8.55	9.10	8.40	8.35
\multicolumn{17}{c}{**4. Commercial**}
Department store	Variable-volume single duct with VAV boxes	6.45	6.55	7.15	7.00	6.80	6.30	6.80	7.10	6.60	6.30	7.50	7.10	7.60	7.05	7.00
Shopping center	Packaged roof—top A.C unit with L.P. single duct	4.55	4.60	5.05	4.90	4.80	4.40	4.75	5.00	4.65	4.40	5.30	5.00	5.35	4.95	4.90

ESTIMATING CHART 18-1   HVAC Systems Total Costs per Square Foot of Gross Area (as of 1976) for the Purpose of Conceptual Estimates *(Continued)*

Building Type	HVAC Systems	Atlanta, Ga.	Baltimore, Md.	Boston, Mass.	Chicago, Ill.	Cleveland, Ohio	Dallas, Tex.	Denver, Colo.	Detroit, Mich.	Minneapolis, Minn.	New Orleans, La.	New York, N.Y.	St. Louis, Mo.	San Francisco, Calif.	Seattle, Wash.	Washington, D.C.
5. Dormitories																	
Dormitories	L.P. single duct, fan coil units—four-pipe, and radiation	6.00	6.15	6.70	6.50	6.35	5.90	6.30	6.65	6.15	5.90	7.00	6.65	7.10	6.55	6.50
6. Hospital																	
Generally small	L.P. single duct with reheat coils, fan coil units—four-pipe	8.40	8.55	9.30	9.05	8.90	8.20	8.80	9.25	8.60	8.20	9.75	9.25	9.85	9.10	9.05
Generally large	Combination of H.P. and L.P. single duct with terminal reheat, radiation, induction units—four-pipe	8.60	8.75	9.55	9.30	9.10	8.40	9.05	9.50	8.80	8.40	10.00	9.50	10.10	9.35	9.30
Medical center	Combination of H.P. and L.P. single duct with terminal reheat, dual duct, induction units—four-pipe	10.30	10.50	11.45	11.15	10.90	10.10	10.85	11.40	10.55	10.10	12.00	11.40	12.15	11.20	11.15
Nursing home	Incremental units, L.P. single duct with reheat coils	6.45	6.55	7.15	7.00	6.80	6.30	6.80	7.10	6.60	6.30	7.50	7.10	7.60	7.05	7.00
7. Laboratory																	
Research	Combination of H.P. and L.P. single duct with terminal reheat, radiation, fume exhaust, and steam supply	10.75	10.95	11.95	11.60	11.40	10.50	11.30	11.90	11.00	10.50	12.50	11.90	12.60	11.70	11.60
8. Manufacturing																	
Heavy	Heating and ventilation	4.00	4.10	4.45	4.30	4.25	3.90	4.20	4.40	4.10	3.90	4.65	4.40	4.70	4.35	4.30
Light	Heating and ventilation	3.75	3.80	4.15	4.05	3.95	3.65	3.95	4.15	3.85	3.65	4.35	4.15	4.40	4.10	4.05
Process	Packaged roof-top units, L.P. single duct with reheat coils, heating, ventilation, process steam, exhaust	4.70	4.80	5.25	5.10	5.00	4.60	4.95	5.20	4.85	4.60	5.50	5.20	5.55	5.15	5.10
9. Motel/Hotel																	
Low rise	L.P. single duct with reheat coils, fan coil units—two-pipe & radiation	6.45	6.55	7.15	7.00	6.80	6.30	6.80	7.10	6.60	6.30	7.50	7.10	7.60	7.05	7.00
High rise	L.P. single duct with reheat coils, fan coil units—two-pipe, radiation	6.00	6.15	6.70	6.50	6.35	5.90	6.30	6.65	6.15	5.90	7.00	6.65	7.10	6.55	6.50

ESTIMATING CHART 18-1 HVAC Systems Total Costs per Square Foot of Gross Area (as of 1976) for the Purpose of Conceptual Estimates *(Continued)*

Building Type	HVAC Systems	Atlanta, Ga.	Baltimore, Md.	Boston, Mass.	Chicago, Ill.	Cleveland, Ohio	Dallas, Tex.	Denver, Colo.	Detroit, Mich.	Minneapolis, Minn.	New Orleans, La.	New York, N.Y.	St. Louis, Mo.	San Francisco, Calif.	Seattle, Wash.	Washington, D.C.
	**10. Museum**																
Museum	L.P. single duct with reheat coils and multizone ducts, forced radiation	9.45	9.60	10.50	10.20	10.00	9.25	9.95	10.45	9.70	9.25	11.00	10.45	11.10	10.30	10.20
	**11. Office (public)**																
Low rise	H.P. single duct with terminal reheat, fin-tube radiation, and cabinet heaters	5.90	6.00	6.60	6.40	6.30	5.80	6.25	6.55	6.05	5.80	6.85	6.55	6.95	6.45	6.40
High rise	H.P. single duct with terminal reheat, fin-tube radiation, and cabinet heaters	7.75	7.90	8.60	8.35	8.20	7.55	8.15	8.55	7.95	7.55	9.00	8.55	9.10	8.40	8.35
	**12. Office (private)**																
Low rise	L.P. single duct with reheat coils, fin-tube radiation, and convectors	5.15	5.25	5.75	5.60	5.45	5.00	5.45	5.70	5.30	5.00	6.00	5.70	6.05	5.65	5.60
High rise	H.P. single duct with terminal reheat, fin-tube radiation, and cabinet heaters	6.90	7.00	7.65	7.45	7.30	6.70	7.25	7.60	7.05	6.70	8.00	7.60	8.10	7.50	7.45
	**13. Parking garage**																
Enclosed	Heating and ventilation	1.42	1.45	1.58	1.54	1.51	1.39	1.50	1.58	1.46	1.39	1.65	1.58	1.68	1.55	1.54
	**14. School**																
Primary	L.P. single duct with reheat coils, unit ventilators—four-pipe and heating-ventilation	5.60	5.70	6.20	6.05	5.90	5.45	5.90	6.15	5.70	5.45	6.50	6.15	6.55	6.10	6.05
Secondary	H.P. single duct with terminal reheat, radiation, heating-ventilation for gymnasium	6.90	7.00	7.65	7.45	7.30	6.70	7.25	7.60	7.05	6.70	8.00	7.60	8.10	7.50	7.45
Vocational	L.P. single duct with reheat coils, fan coil units—four-pipe and radiation	7.30	7.45	8.10	7.90	7.75	7.15	7.70	8.10	7.50	7.15	8.50	8.10	8.60	7.95	7.90
	**15. Theater**																
Theater	L.P. single duct with some reheat coils, radiation	10.30	10.50	11.45	11.15	10.90	10.10	10.85	11.40	10.55	10.10	12.00	11.40	12.15	11.20	11.15
	**16. Warehouse**																
Warehouse	Heating and ventilating	3.75	3.80	4.15	4.05	3.95	3.65	3.95	4.15	3.85	3.65	4.35	4.15	4.40	4.10	4.05

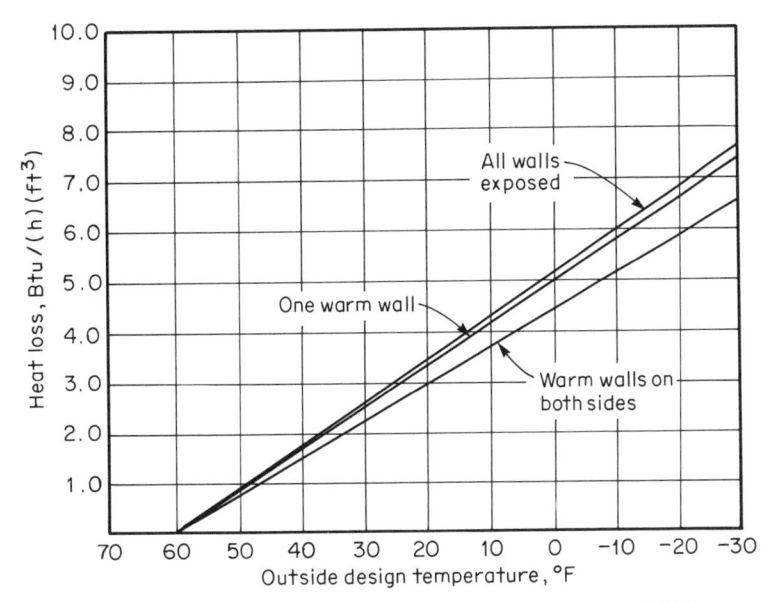

FIGURE 18-1 Chart for estimating heat loss at 60°F room design temperature for one-story building with flat roof. (For skylight in roof add 7½ percent.)

Heating-Load Estimating Data[1]

Figures 18-1 through 18-7 provide a method of estimating heating loads. The charts can be used to obtain the heat loss in Btu's per hour per cubic foot of building volume. Their accuracy is believed to be ±5 to 10%.

Data Commonly Used for Steam Calculations

- Square feet of equivalent direct radiation (EDR)
  EDR steam: 240 Btu/(h)/(ft^2) of emitting surface
  EDR water: 150 Btu/(h)/(ft^2) of emitting surface

- Boiler horsepower
  33,500 Btu/h
  33.5 MBH
  140 ft^2 EDR steam
  223 ft^2 EDR water
  34.5 lb/h steam

- Pounds per hour of steam
  970 Btu/h
  4.04 ft^2 EDR steam
  6.47 ft^2 EDR water

[1]Clifford Strock and Richard L. Koral, *Handbook of Air Conditioning, Heating and Ventilating*, 2d ed., Industrial Press, New York, 1965, has been used as a reference with permission.

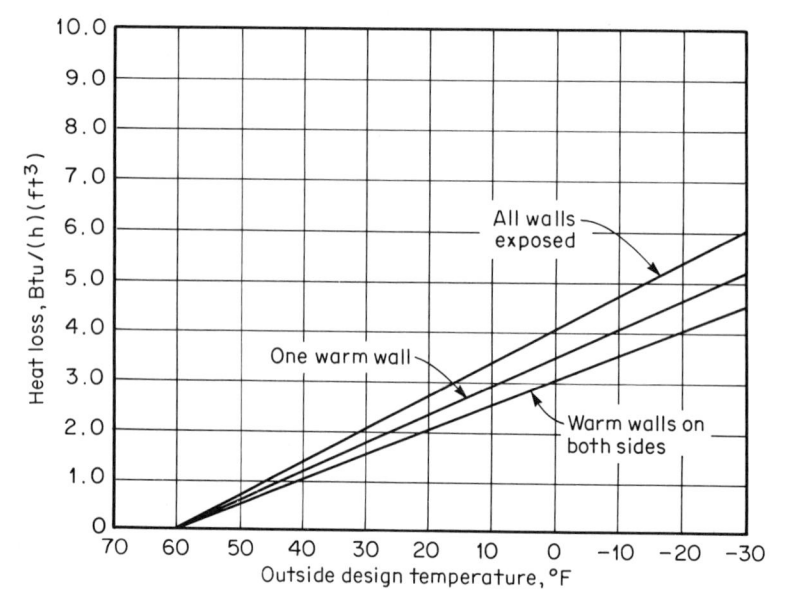

FIGURE 18-2 Chart for estimating heat loss at 60°F room design temperature for one-story building with heated space above.

Estimating the Amount of Ventilation Air

The amount of ventilation air can be computed on the basis of:

1. Number of air changes per hour
2. CFM per occupant
3. CFM per square foot of floor space

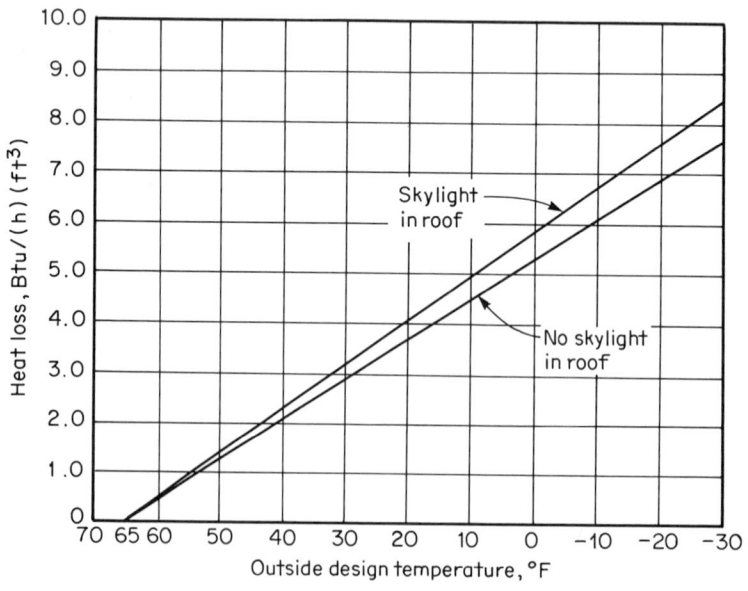

FIGURE 18-3 Chart for estimating heat loss at 65°F room design temperature for one-story building.

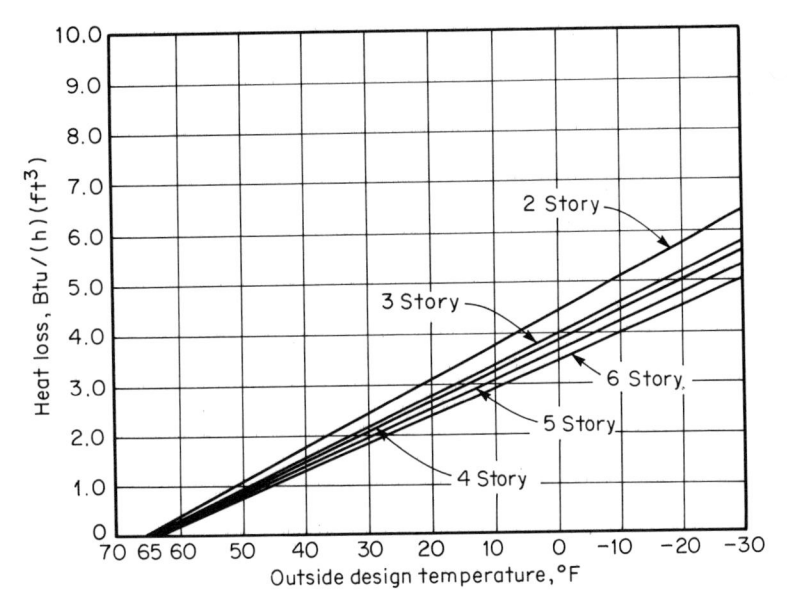

FIGURE 18-4 Chart for estimating heat loss at 65°F room design temperature for multistory building.

The volume of ventilation air for various buildings can be obtained from Table 18-1.

The ranges indicated in Table 18-1 are quite wide. Careful thought is required to select a particular value within each range to fit each case.

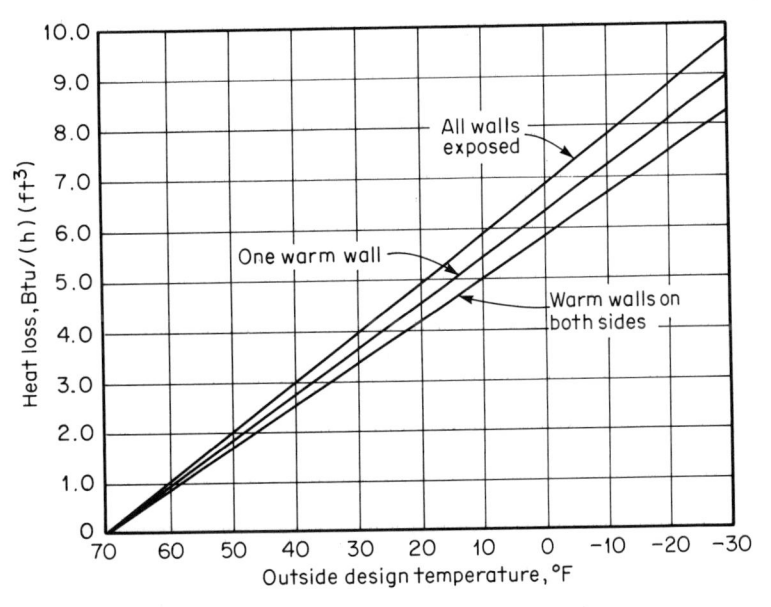

FIGURE 18-5 Chart for estimating heat loss at 70°F room design temperature for one-story building with flat roof.

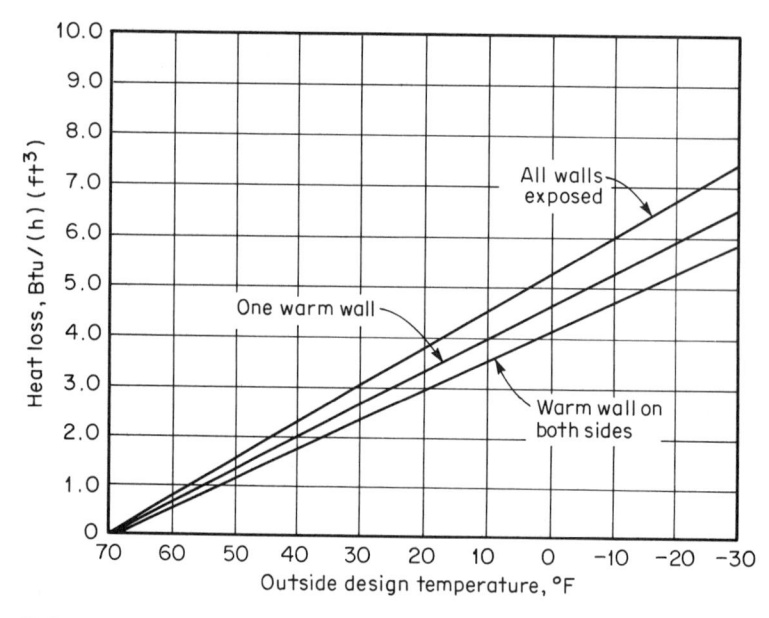

FIGURE 18-6 Chart for estimating heat loss at 70°F room design temperature for one-story heated space above.

Cooling-Load Estimating Data[2]

Tables 18-2 through 18-5 provide information for estimating cooling loads. The cooling-load data were obtained from analyses of actual load estimates. In most cases, the design criteria were:

[2]From *The abc's of Air Conditioning*, Carrier Corporation, 1972. Reproduced by permission of Carrier Corporation.

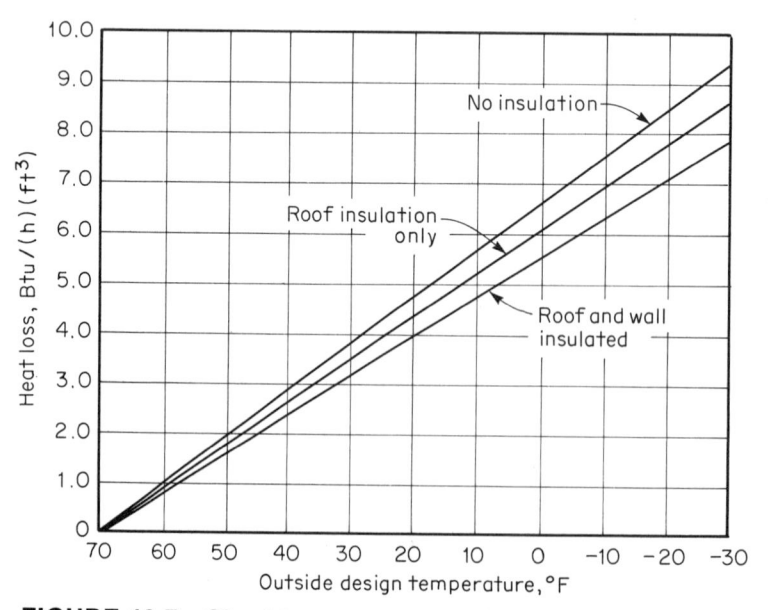

FIGURE 18-7 Chart for estimating heat loss at 70°F room design temperature for entire building (multistory). (Add 10% for bad north and west exposures.)

TABLE 18-1 Volume of Ventilation Air for Various Buildings

Functional Areas	Ventilation Air	Unit
Auditoriums, churches, dance halls	4–30 10–65 1.5–2	air changes* per hour cfm/occupant cfm/ft²
Barbershops and cafes	7.5	air changes per hour
Bedrooms	1	air changes per hour
Billiards and bowling	6–20	air changes per hour
Classrooms		
Colleges	25–40	cfm/occupant
Schools	30–40	cfm/occupant
Classrooms (schools)	2	cfm/ft²
Corridors	4 ½	air changes per hour cfm/ft²
Dining rooms	4–40 1.5	air changes per hour cfm/ft²
Garages	6–12	air changes per hour
Guest rooms	3–5	air changes per hour
Gymnasiums	12 1.5	air changes per hour cfm/ft²
Halls (residence)	1–3	air changes per hour
Kitchens	4–60 2–4	air changes per hour cfm/ft²
Laboratories	6–20	air changes per hour
Living rooms (residence)	1–2	air changes per hour
Lobbies	3–4	air changes per hour
Locker rooms	2–10 2	air changes per hour cfm/ft²
Lounges	6	air changes per hour
Mechanical rooms	3–12	air changes per hour
Operating rooms	50	cfm/occupant
Projection booths	30 1.5	air changes per hour cfm/ft²
Reading rooms	3–5	air changes per hour
Stores (retail)	6–12 4	air changes per hour cfm/ft²
Toilets		
Private	1–5	air changes per hour
Public	10–30 2	air changes per hour cfm/ft²
Waiting rooms	4–6	air changes per hour

Air changes is a method of calculating the amount of infiltration or ventilation air in terms of the number of room volumes per hour.

TABLE 18-2 Refrigeration for Central Heating and Cooling Plants

Classifications	Refrigeration, ft²/ton		
	Low	Average	High
Urban districts	475	380	285
College campuses	400	320	240
Commercial centers	330	265	200
Residential centers	625	500	375

Outside design conditions: 95°F dry bulb and 75 or 78°F wet bulb

Inside design conditions: 76 to 80°F dry bulb and 50% relative humidity

Systems: all air

Water Rates

The following formula may be used to obtain the water rate in gallons per minute:

$$\text{gpm} = \frac{\text{total load (Btu/h)}}{500 \times \text{TD}} \tag{18-1}$$

where TD is the temperature difference between the entering and leaving water temperatures. Table 18-6 gives the approximate water rates commonly used in water systems.

Air-Conditioning Systems and Applications

Air-conditioning systems can be categorized according to the fluids used, as follows:

1. All-air systems
2. Air-water systems
3. All-water systems
4. Refrigerant systems

Each type has certain functional and economic advantages when used in specific applications.

All-Air Systems

All-air distribution systems may be low-velocity (low-pressure) or high-velocity (high-pressure) systems or a combination of the two. They may be (see Fig. 18-8):

TABLE 18-3 Cooling-Load Low Check Figures

Building Classification	Occupancy, ft²/person	Lights, W/ft²	Refrigeration, ft²/ton	Air quantities, cfm/ft²
Apartment (high-rise)	325	1.0	450	0.7
Auditoriums, churches, theaters	15	1.0	400	1.0
Educational facilities (schools, colleges, universities)	30	2.0	240	0.9
Manufacturing				
Light	200	9.0*	200	1.6
Heavy	300	15.0*	100	2.5
Hospitals				
Patient rooms	75	1.0	275	0.4†
Public areas	100	1.0	175	1.0
Hotels, motels, dormitories	200	1.0	350	1.0
Libraries and museums	80	1.0	340	1.0
Office buildings				
General	130	4.0	360	0.25† exterior / 0.80 interior
Private	150	2.0		0.25†
Residential				
Large	600	1.0	600	0.7
Medium	600	0.7	700	0.6
Restaurants				
Large	17	1.5	135	1.4
Medium			150	1.5
Shopping centers				
Department stores				
Basement	30	2.0	340	0.7
Main floor	45	3.5	350	0.9
Upper floors	75	2.0	400	0.8
Malls	100	1.0	365	1.1

*Includes other loads.
†Based on induction system.

- Single-duct, low-velocity systems with terminal-reheat units
- Single-duct, high velocity systems with pressure-reducing units
- Single-duct, variable-volume systems
- Dual-duct systems with mixing boxes
- Multizone systems

TABLE 18-4 Cooling-Load Average Check Figures

Building Classification	Occupancy, ft²/person	Lights, W/ft²	Refrigeration, ft²/ton	Air quantities, cfm/ft²
Apartment (high-rise)	175	2.0	400	1.0
Auditoriums, churches, theaters	11	2.0	250	2.0
Educational facilities (schools, colleges, universities)	25	4.0	185	1.4
Manufacturing				
Light	150	10.0*	150	2.5
Heavy	250	45.0*	80	4.0
Hospitals				
Patient rooms	50	1.5	220	0.5†
Public areas	80	1.5	140	1.1
Hotels, motels, dormitories	150	2.0	300	1.3
Libraries and museums	60	1.5	280	1.2
Office buildings				
General	110	6.0*	280	0.5† exterior / 1.1 interior
Private	125	5.8		0.5†
Residential				
Large	400	2.0	500	1.0
Medium	360	1.5	550	0.9
Restaurants				
Large	15	1.7	100	1.75
Medium			120	1.50
Shopping centers				
Dept. stores				
Basement	25	3.0	285	1.0
Main floor	25	6.0*	245	1.4
Upper floors	55	2.5	340	1.0
Malls	75	1.5	230	1.8

*Includes other loads.
†Based on induction system.

Air-Water Systems

Only a small part of the cooling and heating of the conditioned area is effected by air brought from the central units. The major part of the room thermal load is balanced by hot or chilled water circulated by a coil. The different air-water types are (see Fig. 18-9):

TABLE 18-5 Cooling-Load High Check Figures

Building Classification	Occupancy, ft²/person	Lights, W/ft²	Refrigeration, ft²/ton	Air quantities, cfm/ft²
Apartment (high-rise)	100	4.0	350	1.5
Auditoriums, churches, theaters	6	3.0	90	3.0
Educational facilities (schools, colleges, universities)	20	6.0	150	2.0
Manufacturing				
Light	100	12.0*	100	3.8
Heavy	200	60.0*	60	6.5
Hospitals				
Patient rooms	25	2.0	165	0.7†
Public areas	50	2.0	110	1.3
Hotels, motels, dormitories	100	3.0	220	1.5
Libraries and museums	40	3.0	200	1.4
Office buildings				
General	80	9.0*	190	0.8† exterior / 1.8 interior
	100	8.0		0.8†
Residential				
Large	200	4.0	380	1.5
Medium	200	3.0	400	1.3
Restaurants				
Large	13	2.0	80	2.4
Medium			100	2.0
Shopping centers				
Dept. stores				
Basement	20	4.0	225	1.2
Main floor	16	9.0*	150	2.0
Upper floors	40	3.5*	280	1.2
Malls	50	2.0	160	2.5

*Includes other loads.
†Based on induction system.

- Induction units, either bypass-air or water-control
- Fan-coil units with supplementary air
- Radiant ceiling panels with supplementary air

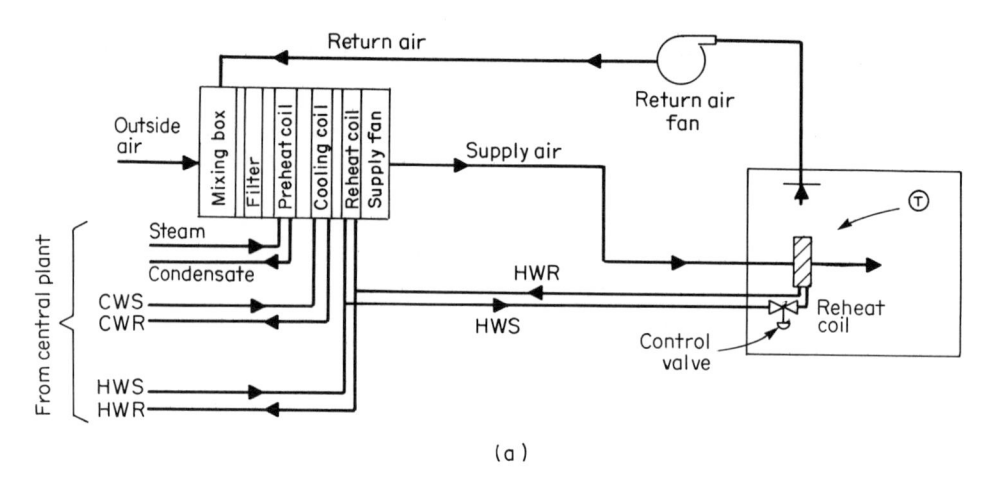

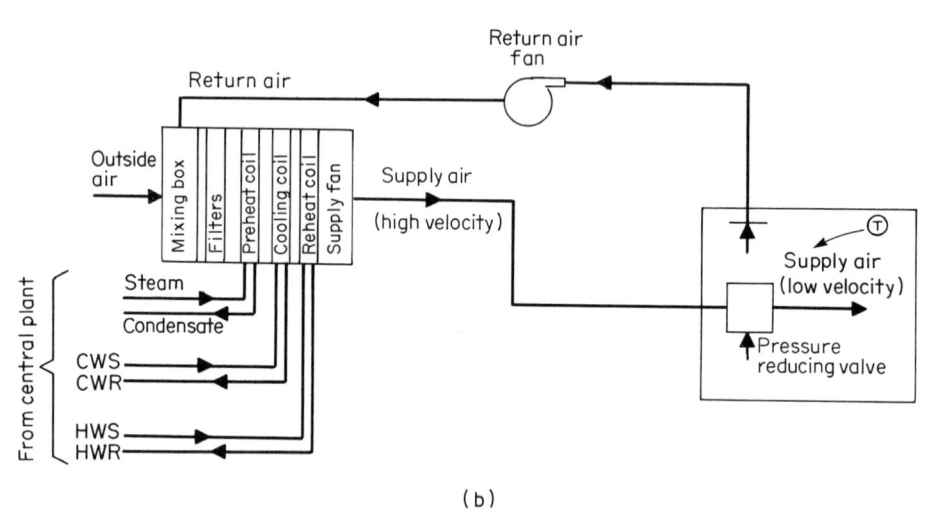

FIGURE 18-8 Typical all-air systems. (a) Single-duct system with reheat; (b) single-duct high-velocity system.

All-Water Systems

All-water systems are those with fan-coil units or unit-ventilator units which may be connected to one of the following circuits (see Fig. 18-10):

- Two-pipe system
- Three-pipe system
- Four-pipe system

Refrigerant Systems

Refrigerant systems are those which utilize self-contained units, such as:

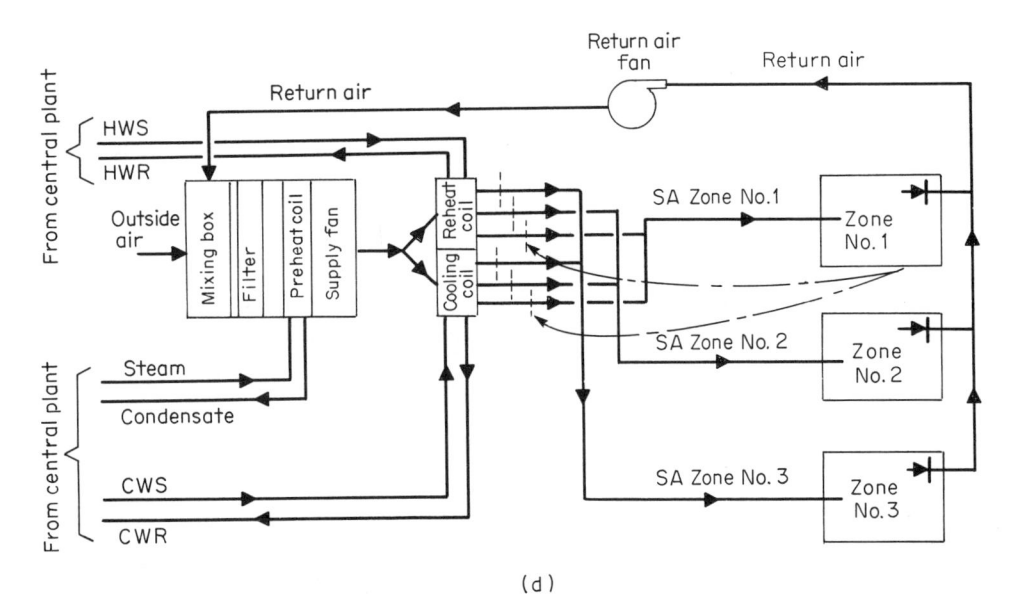

FIGURE 18-8 Typical all-air systems. *(c)* Single-duct variable-volume system; *(d)* multizone system.

- Window units
- Through-the-wall units
- Rooftop units
- Split air-conditioning units (Fig. 18-11)
- Heat-pump units

Functional Comparison of Systems

It is possible to set up a few rules which can be used as guideposts in selecting the proper system. The factors that affect comfort are, in order of

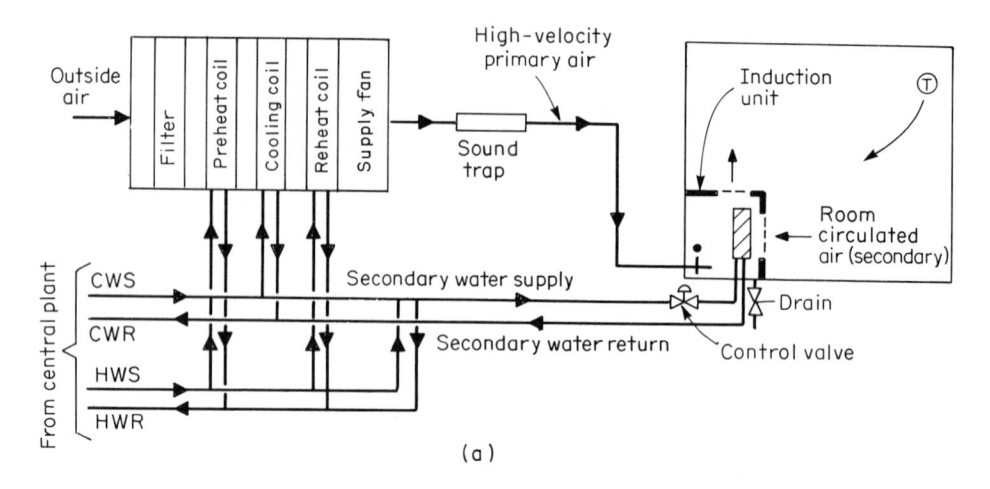

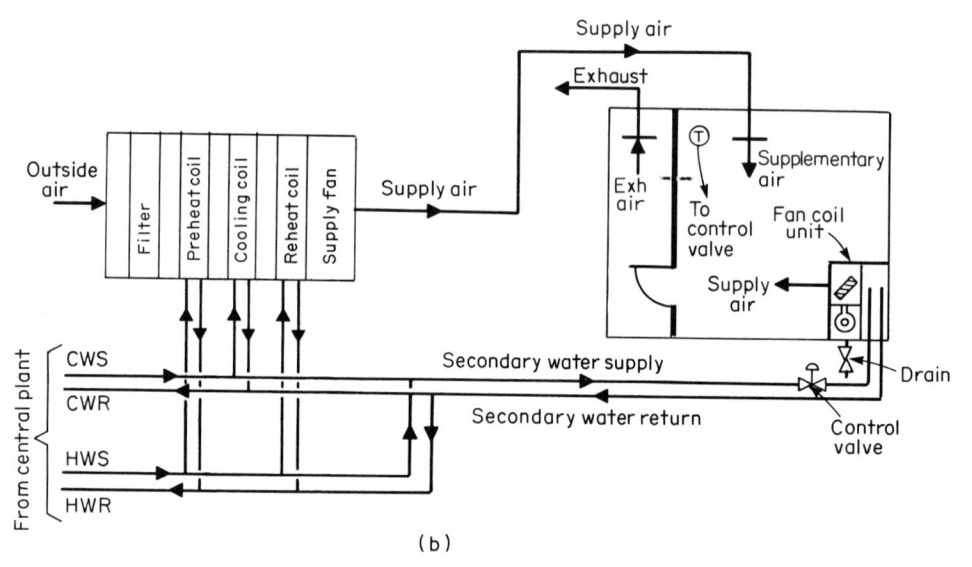

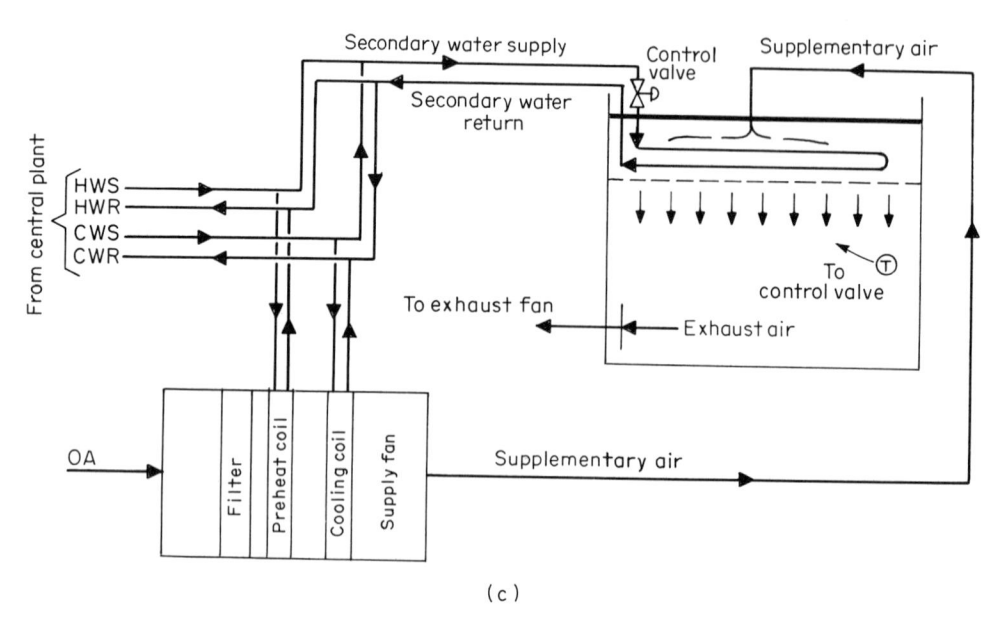

FIGURE 18-9 Typical air-water systems. *(a)* **Induction-unit system;** *(b)* **fan-coil-unit system with supplementary air;** *(c)* **radiant-panel system with supplementary air.**

Conceptual Estimates | 261

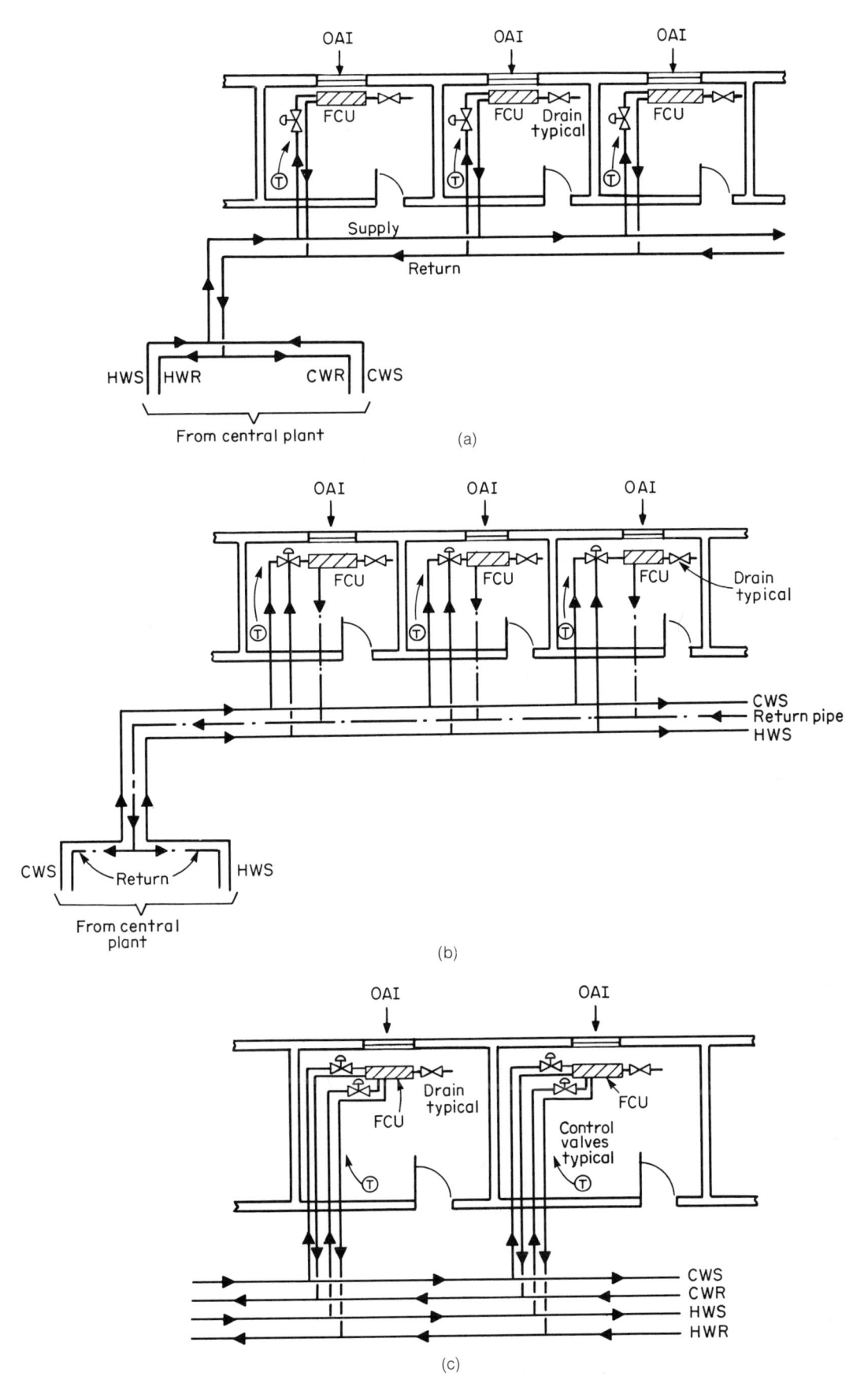

FIGURE 18-10 Typical all-water systems. Fan-coil units with (a) two-pipe system, (b) three-pipe system, (c) four-pipe system.

TABLE 18-6 Water Rates for HVAC Water Systems

System	Temperature Difference, °F	Water Rate
Chilled water	10	2.4 gpm/ton
Condenser water	10	3.0 gpm/ton
Hot water	20	3.35 gpm/bhp

importance, temperature, air quality, noise level, air movement, humidity, and radiant effects. With these factors as a yardstick (and disregarding economics), the systems may be rated functionally as in Table 18-7. All-air systems are more efficient for buildings having low cooling loads, whereas modern buildings with large glass areas and high lighting loads can more economically be served by some type of air-water system.

Applications The following is a guide to commonly used air-conditioning systems, based on building functions. Economics and the design objectives dictate the choice of system and modifications to be included. Systems other than those listed may be used, but often at higher cost.

1. *Individual room or zone units*

   a. *D-X self-contained, up to 2 tons.* Application: residential, specialty stores

   b. *D-X self-contained, over 2 tons.* Application: residential, specialty shops and stores, restaurants, television studios, computer rooms, country clubs, shopping centers, factories, motels

   c. *Fan-coil units (all water).* Application: residential, apartment buildings, hospitals, hotels, motels, dormitories, schools and colleges

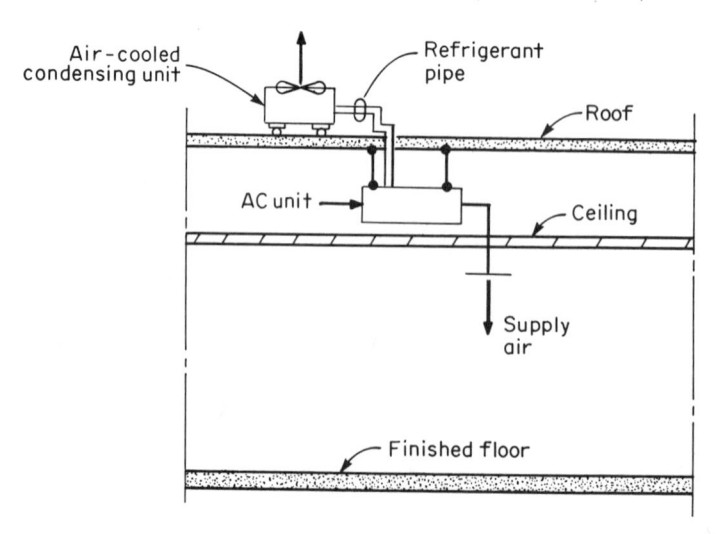

FIGURE 18-11 Split air-conditioning system (refrigerant).

TABLE 18-7 Functional Ratings of Systems

System	Category
1. Single duct with reheat	All air
2. Dual duct or multizone	All air
3. High velocity with reducing boxes	All air
4. Induction	Air-water
5. Radiant panel	Air-water
6. Fan coil with primary air	Air-water
7. Fan coil and outside air	All water
8. Fan coil and supplementary air from interior	All water
9. Package unit	Refrigerant
10. Single duct	All air
11. Self-contained	Refrigerant

2. *Central-station systems (all-air)*

   a. *Variable volume.* Application: office buildings, schools, colleges, laboratories

   b. *Bypass single airstream.* Application: restaurants, radio and television studios, clubs, theaters, auditoriums, factories, colleges and schools, libraries, department stores, shopping centers

TABLE 18-8 Space Requirements for HVAC Systems

Component	Space Requirement
Central cooling and heating plant	0.2–1.0% of gross area
Air handling	
Mechanical rooms	2.0–3.5% of served area
For induction units	0.50–1.5% of served area
Duct and piping risers*	Included in mechanical-room area
For induction units (two-pipe)	0.25–0.35% of gross cubage (volume)
For induction units (four-pipe)	0.3–0.4% of gross cubage
Ductwork distribution	0.8–1.3% of gross cubage
Outlets and terminal for fan-coil units	1.0–1.5% of served area
Outlets and terminal for induction units	2.0–3.0% of served area
Piping system	0.03–0.04% of gross cubage
For fan-coil units (two-pipe)	0.1–0.2% of served area
For fan-coil units (four-pipe)	0.25–0.3% of served area

*A *riser* is a pipe or duct extending one floor or more.

      c. *Reheat at terminal.* Application: restaurants, schools and colleges, laboratories

      d. *Reheat zone in duct.* Application: restaurants, radio and television studios, country clubs, factories, hospitals, schools and colleges, museums, libraries, laboratories

3. *Central-station systems (air-water)*

      a. *Multizone single duct.* Application: residential, restaurants, radio and television studios, country clubs, churches, office buildings, museums, libraries, shopping centers, laboratories

      b. *Induction units.* Application: office buildings, hotels, dormitories, apartment buildings, hospitals, laboratories

      c. *Fan-coil units with O.A.* Application: hotels, dormitories, apartment buildings

Space Requirements

Table 18-8 gives the space requirements for HVAC systems.

19

CHANGE ORDERS AND ALTERATION AND RENOVATION WORK

CHANGE ORDERS

A change order is considered to be any change in the design or specification bid documents issued by the architect or the engineer after bids have been awarded. The change may require that the contractor add to or omit some of the work agreed upon in the contract. The change may be required to correct an error, add, omit, or change a portion of the work, coordinate field installation, or satisfy an owner requirement. If a change order involves changes in design, the architect or engineer may either indicate the change on new drawing(s) or revise the bid drawing(s), according to the volume of work involved. In revised drawings, the change is usually indicated by a heavy outline with the number of the revision enclosed in a triangle. The number in the triangle is dated and keyed to a revision list in the title block of the drawing.

Change orders are taken off by the same procedure used for the original estimate. If a change order involves completely new work, the estimator can proceed in estimating the change order. The price of the change order is then compared with the original estimate in order to determine the net increase or credit. If the change order is partial, it is only necessary to repeat the takeoff of the items included in the affected area(s) in the original drawings. Next, the estimator does a takeoff of the items included in the revised area(s) in the change-order drawings. It is important to prevent takeoff duplication. This can be done by taking off only those items that are affected by the change in both sets of drawings. Once the takeoff is completed, the estimator calculates the material and labor costs for both takeoffs and compares the direct cost of the change order with that of the original bid documents. The estimator then adds the change-order overhead costs (engineering, estimating, etc.) to the net difference as computed above, and applies to this sum an appropriate markup margin, in order to obtain the total amount of the change order. Depending on whether the change-order amount is greater than or less than the original amount, a sum is added to or deducted from the total cost of the original contract. Where only deductions are involved, the amount of the change order should not include the contractor's markup.

ALTERATIONS AND RENOVATIONS

Alterations and renovations in an existing building may include the following items:

1. Additions to existing systems
2. Removal of existing items
3. Removal and relocation of existing items
4. Removal and replacement of existing items
5. Extensions to existing systems
6. Alterations and/or renovations to existing items
7. Cutting and/or patching

The architect or engineer may direct the contractor to perform alteration and renovation work in an occupied building in more than one stage. In this case, the estimator must produce an estimate for each stage, with material prices and labor rates based on the midpoint of each stage.

In alteration and renovation work, the estimator must inspect the existing building to become thoroughly familiar with the job conditions. The estimate must include all the work indicated in the bid documents. In addition, an allowance must be included for items which may not be indicated in the documents but which were noted during inspection of the existing building. The estimate may also include an allowance for protection of adjacent work, removal of rubbish and/or salvage, daily cleanup, restrictions during normal working hours, productivity losses, etc.

Overtime may be required in alteration and renovation work. If so, the estimator must use an appropriate factor to reduce labor productivity during the overtime period, because the output of the crew usually decreases during this period. The estimate must include overtime premiums in accordance with local wage-rate agreements. The estimator may apply a higher markup margin to the direct cost of alteration and renovation work than is applied to new work.

20

VALUE ENGINEERING FOR HEATING, VENTILATING, AND AIR CONDITIONING

GENERAL

Value engineering (VE) is becoming more widely accepted in the construction industry. Today less than 1% of private commercial construction is done with VE, but before long most federal government building contracts will require it. Buildings such as schools and sewage-treatment plants, which are frequently funded in large part by government grants, will soon require value engineering. Value engineering offers some real hope for cutting the cost of new construction without reducing performance and quality, or without increasing life-cycle costs (initial construction costs, and costs of maintenance, operation, and replacement over a specific number of years). Construction savings in the neighborhood of 10% are very possible with VE. The General Services Administration rewards the contractor with 50% of the savings resulting from contractor-originated improvements after the design has been approved.

The concepts of value engineering are most effective if applied during the design process, since the results of the analysis lead the design team to the most economical design for the project.

Definition

Value engineering. The systematic review of the value effectiveness of every dollar spent, as well as the elimination or simplification of anything that adds unnecessary cost to an item (above the cost of its function). This is a total, or life-cycle, cost approach which involves consideration of all the costs relating to initial construction as well as the costs of maintenance, operation, and replacement.

Alternative solutions may be developed for specific functions through the use of the most recent technical information regarding materials and methods. In contrast to item-oriented cost-cutting (using smaller quantities or fewer or cheaper materials), value engineering analyzes the function by asking:

- What is it?
- What does it do?
- What does it cost?
- What must it do?
- What other material or method could be used to do what it must do?
- What would the alternative material or method cost?

VALUE-ENGINEERING JOB-PLAN PHASES

The following four phases are the most commonly used in an organized approach to value-engineering studies and reviews.

Collection of Relevant Information

The information required typically includes the following major items:

1. Owner program requirements and criteria previously furnished to the engineer
2. Plans and specifications
3. Available cost estimates
4. Design data and applicable standards and criteria

This information is used to accomplish the following results:

1. Division of the job information into logical cost-model breakdowns (see Fig. 20-1) for areas or systems, such as heating, ventilating, and air conditioning, plumbing, structural, architectural, and electrical systems
2. Development of a list of potential high-cost areas such as those with high total cost and/or high ratio of cost to worth
3. Development of a list of functions to serve as a base for the remainder of the job plan work
4. Start of an "idea listing" (see Fig. 20-2a and b) of item-oriented ways of reducing costs
5. Selection of the highest-cost areas for in-depth analysis

Creativity

1. Use of brainstorming and related creativity techniques to develop alternative ways of accomplishing the real needs, i.e., the basic functions of the system or item being analyzed

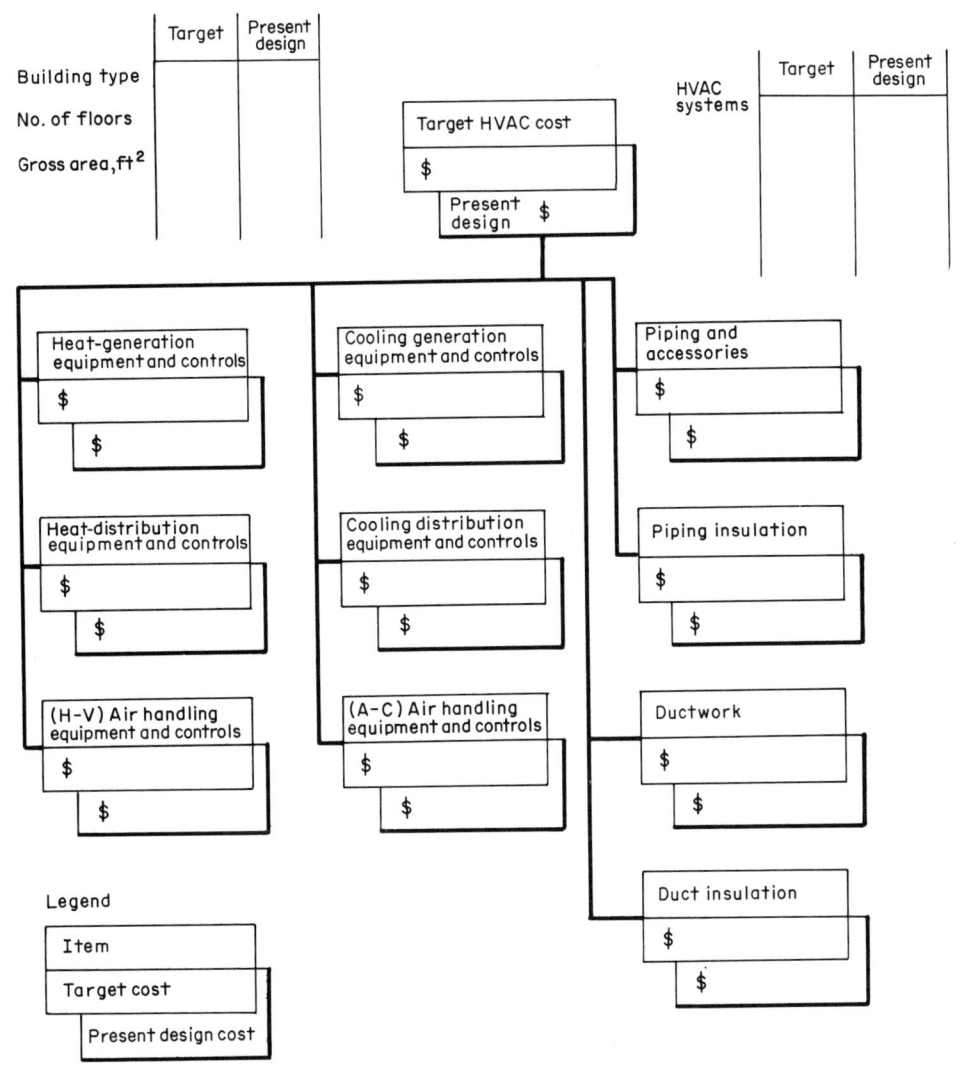

FIGURE 20-1 HVAC cost model. (For HVAC parameters, see Figures 1-3 and 1-4.)

2. Addition of item-oriented cost-reduction ideas, as they occur, to the idea listing started earlier

Analysis and Evaluation

1. In-depth analysis and screening of ideas generated in the creativity phase, using as guides the general feasibility, advantages and disadvantages, ability to meet the owner-user functional and aesthetic criteria, initial cost-saving potential, and maintenance-operation-replacement cost-saving potential

2. Fine tuning of the relatively few ideas which pass the analysis and screening steps described above

Proposal and Implementation

1. Preparation and submission to the client of proposals based on the best ideas

2. Implementation, follow-up, and control

TYPICAL HIGH-COST AREAS

The value-study review team usually examines high-cost areas through the following approach:

1. Rank the heating, ventilating, and air-conditioning systems in terms of their cost; higher-cost items are often worth an analysis.

2. List the function of each heating, ventilating, and air-conditioning system component and look for unnecessary functions, high-cost functions, and functions with a high ratio of cost to worth.

3. Look at a cost model containing actual cost estimates and target or idealistic costs for the basic heating, ventilating, and air-conditioning systems, in order to find the items which have high cost in relation to base cost (see cost model, Fig. 20-1).

4. Look for items with large life-cycle-cost implications, i.e., maintenance, operation, and replacement costs.

5. Review previous studies of similar facilities.

6. Note repetitive items, for which small changes can be multiplied.

7. Note more complex items, for which simplication can frequently save money.

VALUE ENGINEERING AND THE ESTIMATOR

Normally, value engineering is most effective if conducted by a team consisting of value-engineering specialist, architect, structural engineer, mechanical engineer, and cost expert. However, an estimator may also do value engineering on an informal, bits-and-pieces basis, as part of the day-to-day work, because:

1. The estimator, and particularly the senior estimator, has broad experience in estimating, which leads to the ability to analyze the functions of heating, ventilating, and air-conditioning systems in terms of their costs.

2. The estimator, through past experience with similar systems or applications, may be well suited to recommend the most economical system or system component without sacrificing basic function or required supporting functions.

Project		Item		Date
Description (Include reference to applicable spec. or drawing)	Estimated Potential Saving		Remarks	
	Initial	Owning and Operating		

FIGURE 20-2a Idea listing sheet.

Project OPERATIONS BUILDING		Item HVAC SYSTEM		Date Feb. 6, 1976
Description (Include reference to applicable spec. or drawing)	Estimated Potential Saving		Remarks	
	Initial	Owning and Operating		
H-1 Distribute air in the control room by means of the floating floor. Use the ceiling as the return plenum.	$3,000	N/A	Present design-air distribution is in ceiling of control room; floating floor is already in.	
H-2 Delete extra run of air return duct in areas which use the ceiling as a return air plenum and which also extend the return duct up to individual spaces.	$3,000	N/A		
H-3 Use a 78°F room temperature. Design dry bulb temperature for outside air should be 92°F. Make 25% reduction in the capacity of the cooling equipment-i.e.,chiller, cooling coils, tower and pump.	$10,000	Yes	Present design-design room temperature is 75°F, and design dry bulb temperature for outside air is 95°F. Advantages-Conserve energy; reduce cooling load by 25%. Comments-This is in accordance with GSA recommendations for energy conservation.	

FIGURE 20-2b Idea listing sheet taken from actual value engineering workshop.

3. The estimator, in developing a job estimate, is in a good position to spot areas which may require value analysis.

Sources of Additional Information

This chapter contains the highlights of value engineering and its application to heating, ventilating, and air-conditioning systems. For more detailed information about this subject, the *Value Engineering Handbook* by the General Services Administration (GSA), and the books by Dell'Isola and Miles (listed in the Bibliography) are recommended.

21
SYSTEMS OF WEIGHTS AND MEASURES

GENERAL The various estimating data given in this text are based on the U.S. Customary System of units (pound, foot, etc.). These units are identical with the corresponding English units with the exception that the British imperial gallon is equivalent to 1.2 U.S. gallons, and the British imperial bushel is equivalent to 1.03 U.S. bushels. Since all the major countries of the world, except the United States and Canada, either were using or had decided to use the modernized metric system, the National Bureau of Standards (NBS) recommended in July 1971, that the United States should gradually become a metricated country, with a changeover period of 10 years. With this in mind, users of this manual should know how to convert from the U.S. Customary System (USCS) to the modernized metric system, which is known as the *International System* of units (SI). This chapter presents the various factors for conversion from USCS to SI units.

TABLE 21-1a Basic SI Units

Classification	Unit of Measure	Symbol
Length	Meter	m
Mass	Kilogram	kg
Time	Second	s
Electric current	Ampere	A
Temperature (thermodynamic)	Kelvin	K
Amount of a substance	Mole	mol
Luminous intensity	Candela	cd

TABLE 21-1b Supplementary SI Units

Classification	Unit of Measure	Symbol
Plane angle	Radian	rad
Solid angle	Steradian	sr

TABLE 21-1c   Derived SI Units

Classification	Unit of Measure	Symbol
Area	Square meter	m²
Volume	Cubic meter	m³
Density	Kilogram per cubic meter	kg/m³
Velocity	Meter per second	m/s
Volumetric flow rate	Cubic meter per second	m³/s
Force	Newton (kg·m/s²)	N
Pressure	Newton per square meter	N/m²
Energy	Joule (N·m)	J
Power	Watt (J/s)	W
Heat flux density	Watt per square meter	W/m²
Heat-transfer coefficient	Watt per square meter-kelvin	W/(m²·K)
Heat capacity (specific)	Joule per kelvin	J/K
Capacity rate	Watt per kelvin	W/K
Thermal conductivity	Watt per meter-kelvin	W/(m·K)
Quantity of electricity	Ampere-second	A·s
	Coulomb	C
Electromotive force	Volt (W/A)	V
Electric resistance	Ohm (V/A)	Ω
Luminance	Candela per square meter	Cd/m²

THE INTERNATIONAL SYSTEM OF UNITS

In 1960, the General Conference on Weights and Measures adopted SI units for international use, to replace the old metric systems, i.e., the centimeter-gram-second, meter-kilogram-second, and meter-kilogram-second-ampere systems. The International System includes basic units, sup-

TABLE 21-2   Prefixes for SI Units

Prefix*	Multiple or Submultiple	Symbol
Mega	1,000,000	M
Kilo	1,000	k
Hecto	100	h
Deka	10	da
Deci	0.1	d
Centi	0.01	c
Milli	0.001	m

*Other prefixes are also used. However, the prefixes listed above are the most important in HVAC work.

TABLE 21-3 Length Equivalents

One	Is Equal to	One	Is Equal to
Foot	12 inches	Inch	0.0254 meters
Yard	3 feet	Foot	0.3048 meters
Mile	5280 feet	Yard	0.9144 meters
Kilometer	1000 meters	Mile	1.6090 kilometers

TABLE 21-4 Area Equivalents

One	Is Equal to	One	Is Equal to
Square foot	144 square inches	Square inch	6.541×10^{-4} square meters
Square yard	9 square feet	Square foot	0.0929 square meters
Acre	43,560 square feet	Square yard	0.8361 square meters
Hectare	2.471 acres	Acre	4,047 square meters
		Hectare	10,000 square meters

TABLE 21-5 Volume Equivalents

One	Is Equal to	One	Is Equal to
Cubic foot	1728 cubic inches	Cubic foot	0.02832 cubic meter
Cubic yard	27 cubic feet	Cubic yard	0.7646 cubic meter

TABLE 21-6 Capacity Equivalents

One	Is Equal to	One	Is Equal to
U.S. gallon	0.1337 cubic foot	U.S. gallon	3.785 liters
Liter	0.03531 cubic foot	Liter	0.001 cubic meter

TABLE 21-7 Flow Equivalents

One	Is Equal to	One	Is Equal to
Cubic foot per minute	4.72×10^{-4} cubic meter per second	Gallon per minute	6.309×10^{-5} cubic meter per second

TABLE 21-8 Mass Equivalents*

One	Is Equal to	One	Is Equal to
Ounce	16 drams	Ounce	0.02835 kilogram
Pound	16 ounces	Pound	0.4536 kilogram
Pound	7000 grains	Short ton	907.2 kilograms
Short ton	2000 pounds	Long ton	1016 kilograms
Long ton	2240 pounds	Metric ton	1000 kilograms
Metric ton	2205 pounds	Pound per square foot	4.8824 kilograms per square meter

*USCS units are based on avoirdupois weight.

TABLE 21-9 Velocity Equivalents

One	Is Equal to	One	Is Equal to
Foot per second	60 feet per minute	Foot per minute	0.00508 meter per second
Mile per hour	88 feet per minute	Mile per hour	1.609 kilometers per hour

TABLE 21-10 Pressure Equivalents

One	Is Equal to	One	Is Equal to
Inch of water @ 60°F	0.03609 psi*	Inch of water @ 60°F	248.84 newtons per square meter
Foot of water @ 39.2°F	0.43310 psi*	Foot of water @ 39.2°F	2,988.98 newtons per square meter
Inch of mercury @ 32°F	0.4912 psi*	Inch of mercury @ 32°F	3,386.39 newtons per square meter
Atmosphere	14.70 psi*	psi*	6,894.76 newtons per square meter
Atmosphere	1.0333 kgf† per square centimeter	kgf† per square centimeter	98,066.50 newtons per square meter

*psi = pound-force per square inch.
†kgf = kilogram-force.

TABLE 21-11 Temperature Equivalents

To Convert from	To	Use	To Convert from	To	Use
Fahrenheit	Rankine	R = °F + 459.67	Fahrenheit	Celsius	°C = 5/9 (°F − 32)
Fahrenheit	Celsius	°C = 5/9 (°F − 32)	Celsius	Kelvin	K = °C + 273.15
			Rankine	Kelvin	K = 5/9 R

TABLE 21-12 Energy Equivalents

One	Is Equal to	One	Is Equal to
Btu (mean)	778.104 foot-pounds	Foot-pound	1.3557 joules
U.S. horsepower-hour	2544.65 Btu	Btu (mean)	1054.90 joules
Kilowatthour	3412.66 Btu	Watthour	3600.00 joules
Btu (mean)	252.0 calories	Calorie	4.186 joules

TABLE 21-13 Power Equivalents

One	Is Equal to	One	Is Equal to
U.S. horsepower	550 foot-pounds per second	U.S. horsepower	745.70 watts
U.S. horsepower	0.70685 Btu per second	Metric horsepower	735.50 watts
U.S. horsepower	76.04 kilogram-meter per second	Boiler horsepower	9809.50 watts
Boiler horsepower	33,475 Btu per hour	Btu per hour	0.2929 watt
Ton of refrigeration	12,000 Btu per hour	Ton of refrigeration	3516.8 watts
Metric horsepower	75 kilogram-meters per second	Calorie per second	4.184 watts

TABLE 21-14 Examples

Item	U.S. Customary Units	SI Units
1. For a packaged cast-iron, hot-water, gas-fired residential boiler with 500 net MBH (146.45 kW), the unit of material cost is:	$4.04/MBH	$4.04/0.2929 kW = $13.79/kW
2. For a packaged centrifugal chiller of 500 tons (1758.40 kW) capacity, the unit of material cost is:	$110/ton	$110/3.5168 kW = $31.28/kW
3. For a low-pressure galvanized steel duct, the labor unit for fabrication and installation is:	0.12 man-hour/lb	0.12/0.4536 kg = 0.265 man-hour/kg
4. For rigid fiberglass duct insulation (cold service), the labor unit for installing 1-in-thick insulation is:	0.08 man-hour/ft^2	0.08/0.0929 m^2 = 0.861 man-hour/m^2

plementary units, and derived units (see Table 21-1a through c). Only those units which concern the HVAC estimator are included here.

Prefixes for SI Units (Multiples and Submultiples)

The prefixes denoting decimal multiples and submultiples of SI units are indicated in Table 21-2.

Conversion Factors

Refer to Tables 21-3 through 21-13. Also see Table 21-14 for examples of the use of conversion factors.

22

ABBREVIATIONS AND SYMBOLS

The abbreviations and symbols used in HVAC drawings and specifications are given in this chapter.

ABBREVIATIONS

Organizations and Agencies

The names of technical societies, trade organizations, and governmental agencies are abbreviated as follows in HVAC documents:

AABC	Associated Air Balance Association
ABMA	American Boiler Manufacturers Association
AGA	American Gas Association
ADC	Air Diffusion Council
AMCA	Air Moving and Conditioning Association
ANSI	American National Standards Institute
ASHRAE	American Society of Heating, Refrigerating and Air Conditioning Engineers
ASME	American Society of Mechanical Engineers
ASTM	American Society for Testing Materials
FIA	Factory Insurance Association
FM	Factory Mutual Insurance Company
IBR	Institute of Boiler and Radiator Manufacturers
MCAA	Mechanical Contractors Association of America
NFPA	National Fire Protection Association
NBS	National Bureau of Standards
NEC	National Electrical Code NFPA pamphlet No. 70
NEMA	National Electrical Manufacturers Association
SBI	Steel Boiler Institute
SMACNA	Sheet Metal and Air Conditioning Contractor's National Association
UL	Underwriters' Laboratories Incorporated
AWWA	American Water Works Association

General Abbreviations

HVAC terms and units are abbreviated as follows. SI abbreviations are shown in parenthesis when they differ from the common HVAC abbreviations.

ac	Alternating current
ACU	Air-conditioning unit
AHU	Air-handling unit
amp (A)	Ampere
ATC	Automatic temperature control
atm	Atmospheric
auto	Automatic
avg	Average
AWG	American wire gauge
BE	Beveled end
BF	Board foot
bhp	Brake horsepower or boiler horsepower
bldg	Building
bsmt	Basement
Btu	British thermal unit
Btuh (Btu/h)	British thermal unit per hour
°C	Degree centigrade (celsius)
CC	Cooling coil
C-C	Center to center
cc (cm³)	Cubic centimeter
cf (ft³)	Cubic foot
cfm (ft³/min)	Cubic feet per minute
CH	Chiller
CI	Cast iron
cm	Centimeter
const	Construction
CS	Commercial standard
CT	Cooling tower
cu ft (ft³)	Cubic foot
cu in (in³)	Cubic inch
cu yd (yd³)	Cubic yard
CV	Constant volume
CY (yd³)	Cubic yard
db (dB)	decibel
DBT	Dry-bulb temperature
dc	Direct current
DD	Dual duct
deg (°C or °F)	Degree
dia	Diameter
dia-in	Diameter-inch

disch	Discharge
dist	Distribution
dn	Down
DPT	Dew-point temperature
DWV	Drainage, waste, vent
D-X	Direct-expansion
EDR	Equivalent direct radiation
eff	Efficiency
exh	Exhaust
exp	Expansion
°F	Degree Fahrenheit
F&T	Float and thermostatic
FBM	Foot board measure
FCU	Fan-coil unit
FtoF	Face to face
FIG	Figure
flex	Flexible
flr	Floor
FOR	Fuel oil return
FOS	Fuel oil supply
fpm (ft/min)	Feet per minute
FS	Fuel supply
ft	Foot
ga	Gauge or gage
gal	Gallon
gph (gal/h)	Gallons per hour
gpm (gal/min)	Gallons per minute
hex	Hexagonal
horz	Horizontal
H.P.	High pressure
hp	Horsepower
hr (h)	Hour
ht	Height
HV	Heating and ventilating
HVAC	Heating, ventilating, and air conditioning
IB	Iron body
ID	Inside diameter
in	Inches
IPS	Iron-pipe size
IU	Induction unit
jnt	Joint
K	Degree Kelvin
kg	Kilogram
kv (kV)	Kilovolt
kw (kW)	Kilowatt
lb	Pound

ld	Load
LF	Linear feet
LH	Latent heat
L.P.	Low pressure
liq	Liquid
m	Meter
M (k)	Thousand
max	Maximum
MB	Mixing box
MBH	Thousand British thermal units per hour
M/C	Machine
mech	Mechanical
M&F	Male and female
MIN	Minimum
min	Minute
mm	Millimeter
mph (mi/h)	Miles per hour
MPS (m/s)	Meters per second
mtd	Mounted
MZ	Multizone
No.	Number
NP	Nominal pipe size
OAI	Outside-air intake
OD	Outside diameter
O&P	Overhead and profit
OSA	Outdoor supply air
OS&Y	Outside screw and yoke
oz	Ounce
P	Pump
pct	Percentage
PE	Plain end
PG	Pressure gage
pneu	Pneumatic
pr	Pair
PRV	Pressure-reducing valve
PSF (lb/ft^2)	Pounds per square foot
psi (lb/in^2)	Pounds per square inch
PVC	Polyvinyl chloride
R	Degree Rankine
RA	Return air
Rej.	Reject
ret	Return
RF	Refrigeration machine
RH	Relative humidity
rm	Room
rnd	Round

rpm (r/min)	Revolutions per minute
SA	Supply air
sch	Schedule
scr	Screw
sec (s)	Second
SH	Sensible heat
spec	Specification
sp gr	Specific gravity
sq	Square
SR	Steam return
sq in (in^2)	Square inch
sq ft (ft^2)	Square foot
sq yd (yd^2)	Square yard
SP	Static pressure
SS	Stainless steel
std	Standard
stm	Steam
supp	Supply
SZ	Single-zone
T (ton)	Ton
T&C	Thread and coupling
temp	Temperature
thermo	Thermostat or thermometer
TR	Ton of refrigeration
USG	United States gauge
USS	United States standard
V	Valve
V	Volt
VAV	Variable air volume
vel	Velocity
vert	Vertical
vol	Volume
VV	Variable volume
WBT	Wet-bulb temperature
wt	Weight
xhvy	Extra heavy
yd	Yard
yr	Year
%	Percentage
#	Number or pound

MECHANICAL SYMBOLS

Figure 22-1 indicates the standard symbols for mechanical systems.

Symbol	Description	Symbol	Description
————	Sanitary drain	—LPS—	Steam (low pressure)
--SS--	Sub soil drain	—MPS—	Steam (med. pressure)
--S--	Storm drain	—HPS—	Steam (high pressure)
—AW—	Acid waste	—LPR—	Condensate (low pressure)
--AV--	Acid vent	—MPR—	Condensate (med. pressure)
------	Vent	—HPR—	Condensate (high pressure)
—CW—	Cold water	—HWS—	Heating water supply
—DS—	Distilled water	—HWR—	Heating water return
—DW—	Drinking water	—EGS—	Ethylene glycol supply
—DWC—	Drinking water circulating	—EGR—	Ethylene glycol return
—HW—	Hot water	—CWS—	Chilled water supply
—HWC—	Hot water circulating	—CWR—	Chilled water return
—HW 180°—	Hot water 180°	—CPD—	Condensate or vacuum pump discharge
—HWC 180°—	Hot water 180° circulating	—CD—	Condensate drain
—TW—	Tempered water	—FOS—	Fuel oil supply
—TWC—	Tempered water circulating	—FOR—	Fuel oil return
—F—	Fire protection water service	--FOV--	Fuel oil vent
—SM—	Sprinkler main	—RL—	Refrigerant liquid line
—SP—	Stand pipe	—RS—	Refrigerant suction line
—G—	Natural gas	—RD—	Refrigerant hot gas discharge line
—A—	Compressed air	—CS—	Condenser water supply
—PN—	Pneumatic tube	—CR—	Condenser water return
—V—	Vacuum line	—BBO—	Boiler blow off
—OX—	Oxygen line	—ES—	Exhaust steam
—LPG—	L.P. gas	—HCWS—	Heating & cooling water supply
▬▬▬▬	Sanitary drain underground	—HCWR—	Heating & cooling water return
◄──	Direction of flow	—⦿—	Balancing valve
⌇⌇⌇	Flexible connector	→⊂	Running trap
─⊽─	Strainer	CO or CO	Clean out
─∣∣─	Union	─◡─	Vacuum breaker
─▭─	Expansion joint	─▷◁─	Gate valve
─⊘─	Automatic control valve	─⊙─	Globe valve
─⊠─	Pressure regulating valve	─⌐─	Check valve
─Ω─	Safety relief valve		
─⊬─	Blow off valve	─Ⓡ─	Pressure relief valve
─⊠#/h─	F & T trap (cap #/hr)	─⋈─	Stop & waste valve
─⊗─	Thermostatic trap	─Ȼ─	Stop & waste valve (in riser)
─⊙─	Radiator valve	─✳─	Automatic 3-way valve
─⌀─	Flow control valve	─●─	Gas cock

FIGURE 22-1 Standard mechanical symbols.

Abbreviations and Symbols | 285

Symbol	Description	Symbol	Description
	Air bleeder valve (radiant panel)		Fire alarm valve (sprinkler)
	Air vent		Cold weather valve (sprinkler)
	Gauge cock		Alarm gong (sprinkler)
	Solenoid valve	HB	Hose bibs
	Thermostatic expansion valve (Refrigerant)	WH	Wall hydrant
	Back pressure valve (refrigerant)	FH	Fire hydrant
	Sight glass	MH	Manhole
	Pipe or round duct riser	SW	Street washer
GCO	Grade cleanout		
WC	Water closet (tank type)	BT	Bath tub
WC	Water closet	SHD	Shower drain
UR	Urinal	FD	Floor drain
LAV	Lavatory	RD	Roof drain
KS	Kitchen sink	DS	Down spout
SS	Service sink	UH-1	Unit heater-propeller type
CSS	Clinic service sink	CUH	Cabinet unit heater
CRS	Classroom sink	9'0"-1-MBH	Fin tube (Figures = elements length, mark & MBH)
DF	Drinking fountain	C-1	Convector
EWC	Electric water cooler	UV-1	Unit ventilator
SM	Shower head	ID	Indirect waste drain
CS	Counter sink		
12/6	Rectangular duct–first figure is side show	12 X 6 G-1 / 250 / 6'-0"	Return or exh. grill (reg. similar) Figures = size, mark, CFM & distance above floor
12"	Round duct (figure = size)		Round ceiling diffuser
	Canvas connection		Square ceiling diffuser (supply)
	Volume damper		Round ceiling diffuser (supply & return)
	Turning vanes		Square ceiling diffuser (supply & return)
	Extractor		Square ceiling diffuser 3 way throw
	Supply duct		Square ceiling diffuser 2 way throw
	Return or exhaust duct		Motorized damper
ST	Sound trap		Gravity damper
12 X 6 R-1 / 250 / 6'-0"	Supply register (grill simular) figures = size, mark, cfm & distance above floor	FD-1	Fire damper
	Fire hose cabinet	1	Heating riser No.
F-1	Special fixture	1	Exhaust fan riser No.
	Column number	1	Detail number
11	Plumbing riser No.	M-2	Drawing number
101	Room number		

FIGURE 22-1 Standard mechanical symbols. *(Continued)*

APPENDIX

LOAD CALCULATIONS

CLIMATE CONTROL AND COMFORT

Comfort air conditioning, or climate control, means *the process of treating air so as to control simultaneously its temperature, humidity, cleanliness, and distribution* within well-defined limits and by systems which do not contribute objectionable noise. The human body generates surplus heat energy which must be dissipated by conduction, convection, and/or radiation. The heat dissipation and generation rates must be equal or we are uncomfortable.

AMBIENT TEMPERATURE AND RELATIVE HUMIDITY

The ambient or *dry-bulb* temperature affects the rate of heat dissipation by conduction. The ambient relative humidity affects the amount of heat being dissipated by evaporation from the body. The ambient air motion affects both the moisture (latent-heat) dissipation rate and the amount of dry heat dissipated by convection. The surface temperature of the structural surroundings affects radiant-heat losses from the body. When the outside wall and glass are at lower temperature than interior surfaces, a contrast in radiant loss is detectable and a feeling of discomfort is noticed.

HEATING-LOAD CALCULATIONS

The heating system for a building should be capable of supplying enough heat to replace structural losses and to heat outside air introduced for ventilation as well as infiltrated air. The heat generator should be capable of supplying the demand of the heat-emitting devices and other loads, of allowing for heat losses from piping and ducts, and of warming up from a cold start.

Heat-loss calculations are based upon a condition of steady flow and the difference between the inside and outside temperatures. The heat-transfer rate (per hour) through any structural element is given by the formula

$$Q = UA(t_i - t_o)$$

where Q = heat transferred to structural element, Btu/h
U = overall heat-transmission coefficient or rate of heat flow, Btu/(h)(ft^2)(°F)
A = surface area of structural element, ft^2
t_i = inside temperature of surface, °F
t_o = outside temperature of surface, °F

The coefficient U may be applied to composite as well as homogeneous structural material. U is, consequently, the reciprocal of the sum of all the individual resistances of the component materials, air spaces, and inside and outside surfaces; thus,

$$U = \frac{1}{1/f_i + x/k + x_n/k_n + 1/c + 1/c_n + 1/a + 1/a_n + 1/f_o}$$

or

$$U = \frac{1}{R_1 + R_2 + R_3 + R_4 + R_5 + R_6} = \frac{1}{R_T}$$

where f = surface or film conductance of air adjacent to the surface, Btu/(h)(ft^2)(°F)
$f_i = f$ for an inside surface
$f_o = f$ for an outside surface
k = thermal conductivity, quantity of heat in Btu's transmitted by conduction per hour through one square foot of homogeneous material for each Fahrenheit degree per inch of thickness
c = thermal conductance; same as k except for a stated thickness and not per inch of thickness
a = thermal conductivity of an air space, quantity of heat in Btu's transferred per hour across an air space of one square foot for each degree Fahrenheit of temperature difference between the surfaces bounding the air space
x = thickness of material whose conductivity is k, in
R = resistance to flow of heat; reciprocal of conductance or conductivity
R_T = sum of resistances of individual components

Tables of conductance and conductivity for many types of structural materials and components appear in the *ASHRAE Handbook of Fundamentals*. For winter heat loss, U values are usually calculated on the basis of 15-mi/h wind velocity outside and still air inside. For those conditions,

$f_o = 6.00 \qquad f_i = 1.46$

$\dfrac{1}{f_o} = 0.17 \qquad \dfrac{1}{f_i} = 0.68$

If the U value for 15 mi/h is known, the following formula can be used to find the U value for the same structural element at any other wind velocity:

$$U_x = \frac{1}{1/f_x + 1/U_{15} - 0.17}$$

Values of f_x for other wind velocities are given in Table A-1.

Infiltration Losses

These result from the leakage of outside air, mainly through cracks and openings around window and doors, with consequent displacement of an equal volume of room air. The volume of air that must be heated to room temperature can be estimated from the table of infiltration through windows and doors which appears in the *ASHRAE Handbook of Fundamentals*. This volume must be accounted for in calculating the total heat load.

It is usually assumed that air can enter only two sides of a rectangular building, since heated air must leave from the opposite sides. Consequently, the combination of two adjacent sides which gives the greatest heat load is commonly used as the basis for estimating infiltration. The following formula expresses the infiltration loss by air change:

$$H = 0.24 d(t_i - t_o)nv$$

where H = infiltration loss, Btu/h
n = number of air changes per hour (may be obtained from HVAC texts)
v = volume of room, ft³
d = density of air, usually 0.075 lb/ft³
t_i = inside design air temperature, °F
t_o = outside design air temperature, °F

Substituting $d = 0.075$ gives

$$H = 0.018 C_v(t_i - t_o)$$

where C_v is the rate of air infiltration in cubic feet per hour.

TABLE A-1 Surface Conductance of Adjacent Air for Several Wind Velocities

Wind Velocity, mi/h	f_x, Btu/(h)(ft²)(°F)	Wind Velocity, mi/h	F_x, Btu(h)(ft²)(°F)
0	1.46	15	6.00
5	3.20	20	7.30
7.5	4.00	25	8.60
10	4.60	30	10.00

There are additional causes of infiltration. One is the chimney effect in multistory buildings, which is an influx of cold air at the lower levels to replace heated air leaving the upper levels. Another is the mechanical withdrawal of air through the exhaust system, which causes a reduction of air pressure in the building and increases the rate of flow of cold air into the building at doors and windows.

Ventilation Air Losses

These are introduced when outside air is drawn into the building to carry away or reduce the concentration of heat, moisture, odors, or dust. The amount of outside air drawn in should be kept to a minimum, since the use of outdoor air for ventilation is an expensive process, in winter and in summer.

The amount of outside ventilation air introduced into specific functional areas can be obtained from various texts, handbooks, and codes. The following formula is used to calculate the quantity of heat lost through ventilation airflow:

$$Q = 1.08R(t_i - t_o)$$

where Q = heat for ventilation air, Btu/h
R = rate of ventilation airflow, cfm
t_i = temperature of indoor air, °F
t_o = temperature of outside air, °F

Pitched-Roof Losses

The heat quantity involved in a pitched roof with ceiling is calculated with the formula

$$Q = U_c A(t_c - t_o)$$

where Q = heat transferred, Btu/h
A = ceiling area, ft²
t_c = temperature of air at ceiling, °F
t_o = temperature of outside air, °F
U_c = combined ceiling and roof coefficient

U_c is calculated with the formula

$$U_c = 1/R_c + R_r/n$$

where R_c and R_r are the resistances of ceiling and roof, respectively, and n is the ratio of the roof area to the ceiling area.

Other Losses For *partitions and floors*, where a difference in temperature exists, still-air coefficients are used to determine the value of U for heat-loss calculations.

For *exterior walls*, the areas of the walls less the areas of windows and doors are the net wall areas used in calculating heat losses.

Basement floors of uninsulated concrete placed on the ground have a heat loss of approximately 0.10 Btu/(h)(ft²) for each Fahrenheit degree difference between the basement-air and ground temperatures (the latter can be assumed to be 50°F). Basement walls starting at ground level have a heat loss per unit area that is twice the floor loss. The floor loss of a small building with a concrete floor built on the ground can be computed from the exposed perimeter of the floor slab (see *ASHRAE Handbook of Fundamentals*).

Building insulation. Owing to high fuel costs and energy shortages, a careful consideration should be given to building insulation to conserve fuel. *Vapor barriers* are required wherever winter design temperatures are 20°F or lower, to prevent condensing and freezing in the structural elements. The *relative humidity* that can be tolerated during cold weather is dictated by the formation of condensation on single-pane window glass. To permit a higher relative humidity, storm sash or thermopane should be used.

In many installations, such as theaters, classrooms, auditoriums, and some industrial applications, it is unnecessary to supply heat during occupancy. However, heat must be supplied to maintain a comfortable temperature before occupancy. In some cases, the heat from lights and occupants may necessitate cooling during much of the heating season.

Example A-1 The following data were taken by a heating engineer during a survey of a client's office in New York City. The only heat loss is through one wall facing north. The exposed wall measures 20 × 10 ft. The wall has one single-pane, ⅛-in-thick window, 20 ft² in area. The wall is constructed of 4-in face brick, 1 in air space, ¾-in wood-fiber insulating board ($d = 15$ lb/ft³), ⅜-in gypsum lath on 2 × 4 in studs, and ½-in plaster. The outside design temperature is 0°F, and the inside design temperature is 70°F. Infiltration is assumed to be 10 cfm.

a. Calculate the room heat loss.

b. How many pounds per hour of steam at 2 psig are required to heat the room?

c. How much air at 100°F is required to heat the room?

Solution   From the *ASHRAE Handbook of Fundamentals*:

U(glass) $= 1.13$
k for face brick $= 9.20$ $x = 4$ in
k for common brick $= 5.00$ $x = 4$ in
k for fiber board $= 0.34$ $x = ¾$ in
k for plaster $= 8.00$ $x = ½$ in
k for gypsum lath $= 3.30$ $x = ⅜$ in
a for air space (any thickness) $= 1.10$
f_i for air film (inside) $= 1.46$
f_o for air film (outside) at 15 mi/h $= 6.00$

Since

$$U(\text{wall}) = \frac{1}{R_T}$$

we have

R for outside air film $= 1/6$ $= 0.167$
R for face brick $= 4/9.2$ $= 0.434$
R for common brick $= 4/5$ $= 0.800$
R for air space $= 1/1.1$ $= 0.909$
R for fiber board $= 0.75/0.34$ $= 2.200$
R for air space $= 1/1.1$ $= 0.909$
R for gypsum lath $= 0.38/3.3$ $= 0.114$
R for plaster $= 0.5/8$ $= 0.063$
R for inside air film $= 1/1.46$ $= 0.606$
R_T (total) $= 6.201$

Therefore,

U (wall) $= 1/6.201 = 0.161$
Net wall area = gross wall area − glass area
             $= 200 - 20 = 180$ ft²

a. The room heat loss is calculated as follows:

Heat transferred through wall $= U_w A_w (t_i - t_o)$
              $= 0.161 \times 180 \times (70 - 0) = 2028$ Btu/h
Heat transferred through glass $= U_g A_g (t_i - t_o)$
              $= 1.13 \times 20 \times (70 - 0) = 1582$ Btu/h
Total transferred heat load $= 2023 + 1582 = 3610$ Btu/h
Infiltration load $= R \times 1.08 \times (t_i - t_o)$
              $= 10 \times 1.08 \times (70 - 0) = 756$ Btu/h
Room heat loss $= 3610 + 756 = 4366$ Btu/h

b. From the steam tables: At a pressure of 2 psig or 2 + 14.7 = 16.7 psia,

Heat of vaporization = 966.1 Btu/lb
Pounds of steam per hour = room heat loss/966.1
= 4366/966.1 = 4.5 lb/h

c. The air supply at 100°F is calculated as

cfm = room heat loss / 1.08 $(t_a - t_r)$
= 4366 / 1.08 (100 − 70)
= 4366 / 1.08 × 30 = 135 cfm

DEGREE-DAYS AND FUEL CONSUMPTION[1]

Each degree of declination below 65°F in mean outdoor temperature, averaged over a 24-h period, is a degree-day. For example, a mean temperature of 45°F is 20 degree-days per day. Usually, heat is not required when the mean temperature for a 24-h period is 65°F. The main use of degree-days is in estimating fuel requirements and overall heating-system efficiency. Normal degree-days and design outside temperatures for different cities appear in the *ASHRAE Handbook of Fundamentals*.

The heat lost from the building during a whole heating season can be expressed by the formula

$$H = \frac{24\, hd\, (t_i - t_a)}{t_i - t_o}$$

where H = seasonal heat loss, Btu
h = hourly heat loss from building for design conditions, Btu
d = number of days in heating season
t_i = inside design temperature, °F
t_o = outside design temperature, °F
t_a = average outside temperature for heating season, °F
24 = hours per day

For an inside design temperature of 70°F, the formula becomes

$$H = \frac{24hd(70 - t_a)}{70 - t_o}$$

The average outside temperature for a heating season may be expressed as

$$t_a = 65 - \frac{D}{d}$$

---

[1] Clifford Strock and Richard L. Koral, *Handbook of Air Conditioning, Heating and Ventilating*, 2d ed., Industrial Press, New York, 1965, has been used as a reference, with permission.

where D is the number of degree-days in the heating season. Substituting in the seasonal heat-loss formula gives

$$H = \frac{24hd}{70 - t_o}\left[70 - \left(65 - \frac{D}{d}\right)\right]$$
$$= \frac{24h(5d + D)}{70 - t_o}$$

or

$$\frac{1000H}{D} = \frac{24(5d + D)1000\,h}{(70 - t_o)D}$$

The term $24(5d + D)1000h/(70 - t_o)D$ can be calculated for any given locality. It is called K and is the unit heat requirement for that locality; its units are Btu's per degree-day per 1000 Btu's per hour, and it is the heat loss at the design conditions. The values of K and D are given in tables of data on normal heating seasons in United States cities in HVAC texts and handbooks.

The working formula for the seasonal heat loss is then

$$H = KD\frac{h}{1000}$$

That is, for a system designed for stated outdoor and indoor temperatures, the fuel consumption per 1000 Btu's of heat loss is nearly proportional to the degree-days.

Example A-2 What would be the heat required for a residence in New York, where the design heat loss is 70,000 Btu/h, there are 241 days in the heating season, the average outside temperature during the heating season is 44°F, and the outside design temperature is 0°F?

Solution

$d = $ 241 days
$t_a = $ 44°F
$t_o = $ 0°F
$h = $ 70,000 Btu/h
$H = \dfrac{24hd(70 - t_a)}{70 - t_o}$
$= \dfrac{24 \times 70{,}000 \times 241(70 - 44)}{70 - 0}$
$= 150{,}384{,}000$ Btu

which is the heat loss for the whole season. Assuming 535 Btu/ft³ of gas at an efficiency of 80%, the gas consumption per season would be

$$\frac{150{,}384{,}000}{535 \times 0.80} = 351{,}364 \text{ ft}^3$$

The example can also be solved by obtaining the values of K and D from a textbook and using the formula

$$H = KD \frac{h}{1000}$$

For New York,

$K = 424.7 \quad D = 5050$

Therefore,

$H = 424.7 \times 5050 \times 70 = 150{,}131{,}450$ Btu/year

and the gas consumption per season is

$$\frac{150{,}131{,}450}{535 \times 0.80} = 350{,}774 \text{ ft}^3$$

COOLING AND DEHUMIDIFYING CALCULATIONS

As a first step, outdoor and indoor design conditions should be established. *Outdoor design conditions*, i.e., dry-bulb and wet-bulb temperatures for outside air, are in-town conditions that are commonly measured at local airports and averaged over a period of 5 years. Tables appearing in the *ASHRAE Handbook of Fundamentals* give summer outdoor design conditions. *Indoor design conditions* for comfort air conditioning, i.e., dry-bulb temperature and relative humidity, usually vary with the function and quality of the building. Recommended dry-bulb temperature and relative humidity are indicated in Table A-2.

Indoor design conditions by building application can be obtained from tables appearing in HVAC textbooks.

Load Calculation

Peak loads may not occur simultaneously in the various spaces in a building, owing to sun effects and differences in the type of load. Thus, for a central unit or plant serving these spaces, the load is not necessarily the sum of the peak loads of the various spaces.

TABLE A-2 Recommended Indoor Design Conditions

Building Quality	Dry-Bulb Temperature, °F	Relative Humidity, %
Low	80	55–60
Normal	78	50–55
High	75	45–50

The sun time of the peak loads for spaces with different exposures should be determined. Then, calculation of the space load involves the addition of the various loads that are expected to occur at the sun time of the peak load (not the peak load in each category). The load categories are:

1. Transmission, including the effect of the sun on opaque construction
2. Radiant solar energy through glass
3. Room loads, sensible and latent
4. Excess outside air load, sensible and latent
5. Heat gain into ductwork and piping
6. Heat introduced into the system by the air-conditioning equipment itself

Transmission Calculation

The calculation of the heat transmission through building envelope is similar to the calculation of heat loss, except:

- The outside film coefficient is based on a 7½-mi/h wind velocity.
- Equivalent total temperature differences should be used to determine the heat gains through the building envelope. These can be found from the effect of the sun on opaque construction, using the *sol-air temperature* and the *time lag* in periodic heat flow.

The sol-air temperature is defined as the outside-air temperature that would produce the same total transmission through the wall or roof as is actually the case with the sun heating the outside surface of the wall or roof. The time lag is defined as the time required for heat to travel through a wall, from the exterior to the interior face. Massive walls with considerable insulation have a long time lag; light walls have a short time lag, glass and sheet metal have practically none. Heat gains through walls and the roof can be represented by the formula

$$H = AU(TD)$$

where H = heat gain, Btu/h
TD = equivalent temperature difference
A = surface area, ft^2

Radiant Solar-Energy Loads through Glass

These are usually significant percentages of the total loads; in air-conditioned spaces, the glass should be shaded to reduce heat gains through the glass. To determine the solar heat entering through the building windows, walls, and roof, the estimator should consult the tables appearing in the *ASHRAE Handbook of Fundamentals*.

Room Loads

These include the heat produced by occupants, lights, and equipment, and are divided into two categories:

1. Loads that affect occupants; their sensible components are used for determining the quantity of circulated air required by the room.

2. Loads that do not affect occupants but are imposed upon the equipment, such as the heat convected by light fixtures which are used as return inlets.

Room loads are divided into sensible heat and latent heat. Sensible heat can be felt, and it changes the temperature of substances absorbing it. Latent heat signifies the rate of release of water vapor into the space by occupants, cooking equipment without hoods, certain industrial processes, and outside air that does not pass through the dehumidifier.

Excess Outside Air

Outside air is required for ventilation and pressurization of the building and is usually a specific percentage of the circulated air or the number of air changes per hour, with a minimum cfm/occupant. Excess outside air loads are divided into sensible heat and latent heat. These should be calculated from outside and inside design conditions.

Room Sensible-Heat Factor

The amount of circulated air required by a room depends on the difference between the room design dry-bulb temperature and the temperature of the conditioned air entering the room. It can also be governed by the room sensible-heat factor, which is the ratio between the room sensible heat and the room total heat; i.e.,

$$RSHF = \frac{RSH}{RSH + RLH}$$

where RSHF = room sensible-heat factor
RSH = room sensible heat, Btu/hr
RLH = room latent heat, Btu/hr

Safety factors should be applied to calculated loads in accordance with engineering practices; moderate values should be used. The following formulas are frequently of use:

$$\text{Total air supply (cfm)} = \frac{\text{TSH}}{1.08(t_r - t_s)}$$

where TSH = total sensible-heat, Btu/h
t_r = room dry-bulb temperature, °F
t_s = dry-bulb temperature for supply air, °F

$$\text{Tonnage equivalent to grand total} = \frac{\text{grand total load (Btu/h)}}{\text{12,000 Btu/h/ton of refrigeration}}$$

Latent heat = cfm × 4.5 × change in latent heat of air supply (Btu/lb)
Total heat = cfm × 4.5 × change in enthalpy of air supply (Btu/lb)

The usual cooling- and heating-load calculation form is designed to result in a room load, a required room circulated-air quantity, and the grand total load. A typical calculation form is shown in Fig. A-1.

Psychrometric Chart

The psychrometric chart is one of the most serviceable tools available to air-conditioning engineers. It is a graph of the properties of moist air at various conditions of temperature and humidity. Figure A-2 is a skeleton psychro-

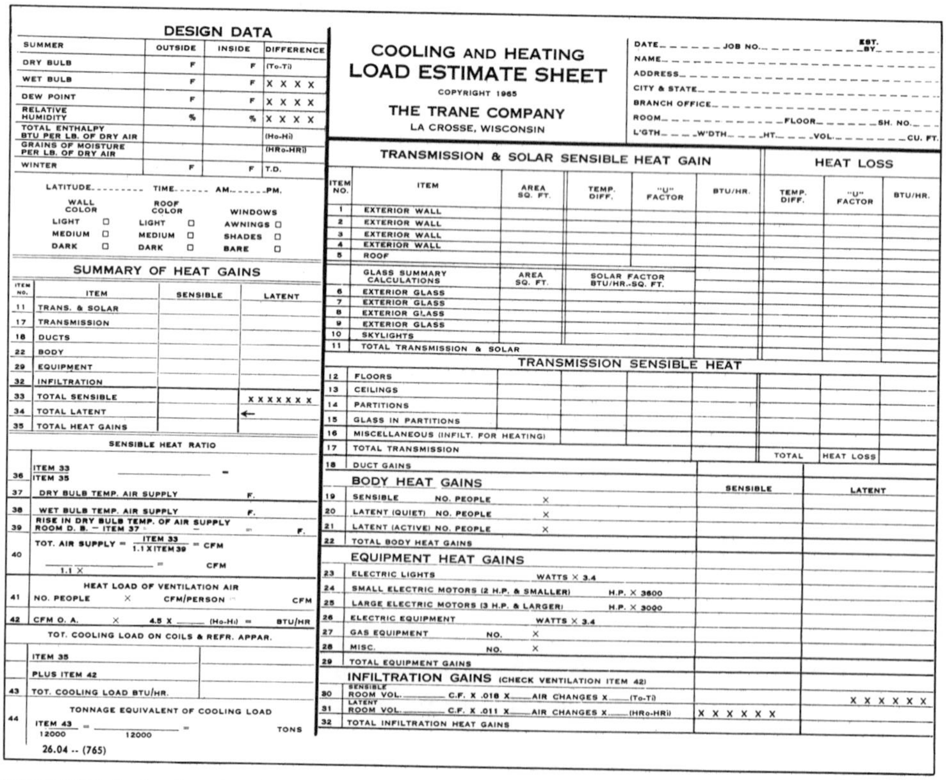

DESIGN DATA

SUMMER	OUTSIDE	INSIDE	DIFFERENCE
DRY BULB	F	F	(To-Ti)
WET BULB	F	F	X X X X
DEW POINT	F	F	X X X X
RELATIVE HUMIDITY	%	%	X X X X
TOTAL ENTHALPY BTU PER LB. OF DRY AIR			(Ho-Hi)
GRAINS OF MOISTURE PER LB. OF DRY AIR			(HRo-HRi)
WINTER	F	F	T.D.

LATITUDE .... TIME ...... AM ...... PM.

WALL COLOR: LIGHT ☐ MEDIUM ☐ DARK ☐
ROOF COLOR: LIGHT ☐ MEDIUM ☐ DARK ☐
WINDOWS: AWNINGS ☐ SHADES ☐ BARE ☐

SUMMARY OF HEAT GAINS

ITEM NO.	ITEM	SENSIBLE	LATENT
11	TRANS. & SOLAR		
17	TRANSMISSION		
18	DUCTS		
22	BODY		
29	EQUIPMENT		
32	INFILTRATION		
33	TOTAL SENSIBLE		X X X X X X X
34	TOTAL LATENT		←
35	TOTAL HEAT GAINS		

SENSIBLE HEAT RATIO

36 ITEM 33 / ITEM 35 ______ =
37 DRY BULB TEMP. AIR SUPPLY F.
38 WET BULB TEMP. AIR SUPPLY F.
39 RISE IN DRY BULB TEMP. OF AIR SUPPLY ROOM D. B. − ITEM 37 − = F.
40 TOT. AIR SUPPLY = ITEM 33 / 1.1 X ITEM 39 = CFM
______ / 1.1 X ______ = CFM

HEAT LOAD OF VENTILATION AIR
41 NO. PEOPLE × CFM/PERSON = CFM
42 CFM O. A. × 4.5 X ____ (Ho-Hi) = BTU/HR

TOT. COOLING LOAD ON COILS & REFR. APPAR.
ITEM 35
PLUS ITEM 42
43 TOT. COOLING LOAD BTU/HR.

TONNAGE EQUIVALENT OF COOLING LOAD
44 ITEM 43 / 12000 = ______ / 12000 = TONS

26.04 -- (765)

COOLING AND HEATING LOAD ESTIMATE SHEET
COPYRIGHT 1965
THE TRANE COMPANY
LA CROSSE, WISCONSIN

DATE ______ JOB NO. ______ EST. BY ______
NAME ______
ADDRESS ______
CITY & STATE ______
BRANCH OFFICE ______
ROOM ______ FLOOR ______ SH. NO. ______
L'GTH ____ W'DTH ____ HT. ____ VOL. ______ CU. FT.

TRANSMISSION & SOLAR SENSIBLE HEAT GAIN / HEAT LOSS

ITEM NO.	ITEM	AREA SQ. FT.	TEMP. DIFF.	"U" FACTOR	BTU/HR.	TEMP. DIFF. (HEAT LOSS)	"U" FACTOR (HEAT LOSS)	BTU/HR. (HEAT LOSS)
1	EXTERIOR WALL							
2	EXTERIOR WALL							
3	EXTERIOR WALL							
4	EXTERIOR WALL							
5	ROOF							
	GLASS SUMMARY CALCULATIONS	AREA SQ. FT.	SOLAR FACTOR BTU/HR.-SQ. FT.					
6	EXTERIOR GLASS							
7	EXTERIOR GLASS							
8	EXTERIOR GLASS							
9	EXTERIOR GLASS							
10	SKYLIGHTS							
11	TOTAL TRANSMISSION & SOLAR							
	TRANSMISSION SENSIBLE HEAT							
12	FLOORS							
13	CEILINGS							
14	PARTITIONS							
15	GLASS IN PARTITIONS							
16	MISCELLANEOUS (INFILT. FOR HEATING)							
17	TOTAL TRANSMISSION					TOTAL	HEAT LOSS	
18	DUCT GAINS							

ITEM NO.	ITEM	SENSIBLE	LATENT
	BODY HEAT GAINS		
19	SENSIBLE NO. PEOPLE ×		
20	LATENT (QUIET) NO. PEOPLE ×		
21	LATENT (ACTIVE) NO. PEOPLE ×		
22	TOTAL BODY HEAT GAINS		
	EQUIPMENT HEAT GAINS		
23	ELECTRIC LIGHTS WATTS × 3.4		
24	SMALL ELECTRIC MOTORS (2 H.P. & SMALLER) H.P. × 3600		
25	LARGE ELECTRIC MOTORS (3 H.P. & LARGER) H.P. × 3000		
26	ELECTRIC EQUIPMENT WATTS × 3.4		
27	GAS EQUIPMENT NO. ×		
28	MISC. NO. ×		
29	TOTAL EQUIPMENT GAINS		
	INFILTRATION GAINS (CHECK VENTILATION ITEM 42)		
30	SENSIBLE ROOM VOL. ____ C.F. X .018 X ____ AIR CHANGES X ____ (To-Ti)		X X X X X X
31	LATENT ROOM VOL. ____ C.F. X .011 X ____ AIR CHANGES X ____ (HRo-HRi)	X X X X X X	
32	TOTAL INFILTRATION HEAT GAINS		

FIGURE A-1 Typical cooling and heating load estimate sheet. (*Copyright ©The Trane Company,* 1965. *Used by permission.*)

1— Dry bulb temperature line
2— Humidity ratio line
3— Humidity ratio scale
4— Wet bulb temperature line
5— Specific volume line
6— Enthalpy scales
7— Dewpoint temperature scale
8— Relative humidity line
9— Vapor pressure scale
10— Sensible heat ratio scale

A— Index point for sensible heat ratio scale

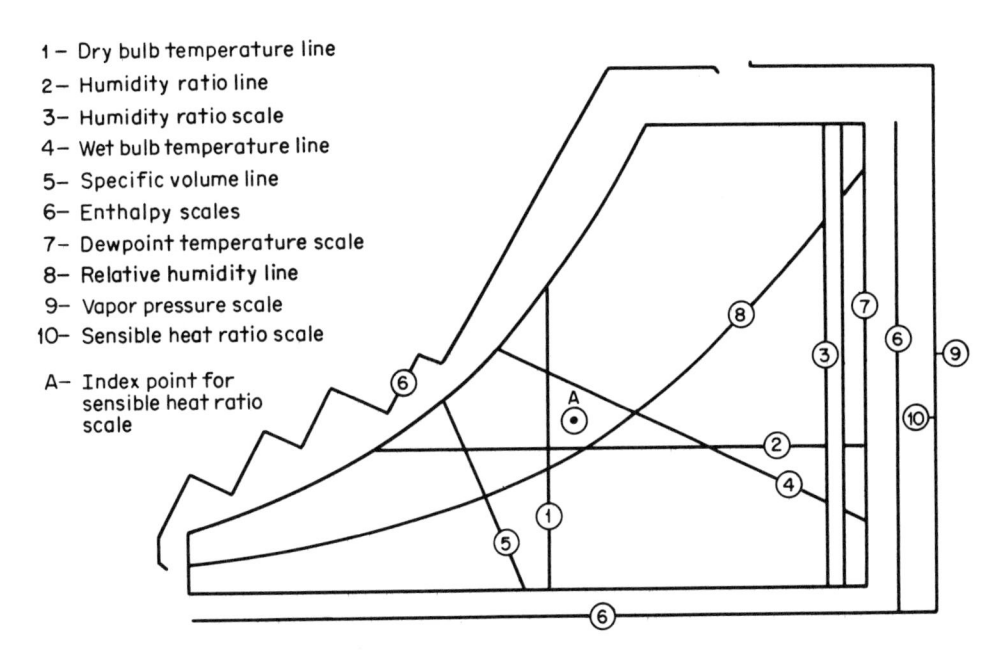

FIGURE A-2   Lines and scales on psychrometric chart. (*Copyright © The Trane Company, 1965, The Trane Air Conditioning Manual. Used by permission.*)

metric chart in which the lines and scales are shown. These lines and scales may be defined as follows:

(1) Dry-bulb-temperature (Fahrenheit) line.

(2) Humidity-ratio line, giving the weight of vapor in each pound of dry air. This also is known as the specific humidity, the number of grains of moisture per pound of dry air.

(3) Humidity-ratio scale. The humidity ratio at any point on the chart is read on the scale in grains per pound of dry air.

(4) Wet-bulb-temperature (Fahrenheit) line above 32°F.

(5) Specific volume, the number of cubic feet of mixture per pound of dry air.

(6) Enthalpy, representing the latent heat in 1 lb of dry air and the W grains of moisture associated with it, in Btu's per pound of dry air.

(7) Dew-point temperature (Fahrenheit).

(8) Relative humidity, the vapor pressure of the air divided by the saturated vapor pressure at the same dry-bulb temperature, in percent.

(9) Vapor pressure, the pressure exerted by water vapor in the air, in inches of mercury absolute.

(10) Sensible-heat ratio, the ratio of sensible heat to total heat in the process.

Point A is the index point for the sensible-heat-ratio scale. A typical psychrometric chart is shown in Fig. A-3.

Example A-3   An auditorium with a capacity of 500 persons is to be maintained at 80°F dry-bulb temperature and 50% relative humidity by supplying air to the auditorium at a dry-bulb temperature of 65°F. There is no direct solar load, but the following estimates have been made of heat gains other than those due to human occupancy:

Infiltration air: 500 cfm at 95°F dry-bulb temperature and 60% relative humidity

Transmission gain: 50,000 Btu/h sensible heat and 20,000 Btu/h latent heat

Determine:

a. The tonnage equivalent to the grand total heat
b. The supply state of the conditioned air, and the capacity of the supply fans in cubic feet per minute

Solution   The *ASHRAE Handbook of Fundamentals* gives the following heat gains for humans seated and at rest as in a theater or auditorium, adjusted for the normal distribution of males, females, and children:

Sensible heat gain per person = 200 Btu/h
Latent heat gain per person   = 130 Btu/h

The heat gains from occupancy are:

Sensible heat gain = 500 × 200 = 100,000 Btu/h
Latent heat gain   = 500 × 130 = 65,000 Btu/h

The heat gains due to infiltration air at 95°F dry-bulb temperature and 60% relative humidity are:

Sensible heat gain = cfm × 1.08 × $(t_o - t_i)$
                   = 500 × 1.08 × (95 − 80) = 8100 Btu/h
Latent heat gain   = cfm × 0.69 × $(W_o - W_i)$

where W_o = outdoor humidity ratio, grains/lb dry air
      W_i = indoor humidity ratio, grains/lb dry air
      0.69 = 4.5 lb/h × 1076 Btu/h/7000 grains/lb
      1076 = average heat removal required to condense 1 lb of water vapor from room

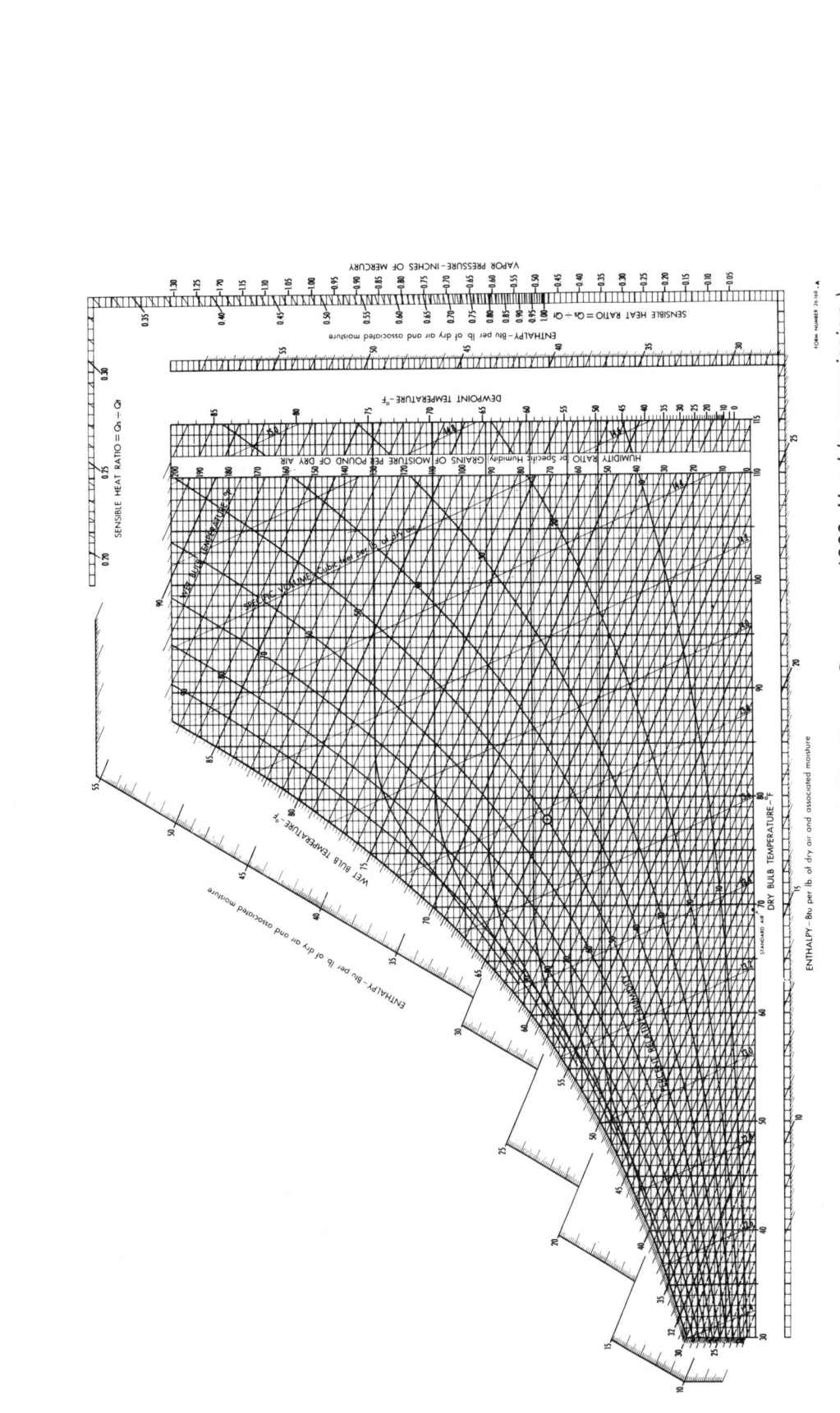

FIGURE A-3 Typical psychrometric chart. *(Copyright © The Trane Company 1960. Used by permission.)*

From the psychrometric chart,

$W_o = 153 \quad W_i = 77$

Latent heat gain = $500 \times 0.69 \times (153 - 77)$ = 26,100 Btu/h

The heat gains due to transmission are:

Sensible heat gain = 50,000 Btu/h
Latent heat gain   = 20,000 Btu/h

a. The total heat gains are

Total sensible heat gain = 100,000 + 8,100 + 50,000 = 158,100
Total latent heat gain   = 65,000 + 26,100 + 20,000 = 111,100
Total heat gain   = 269,200 Btu/h

Tonnage = $\dfrac{269,200}{12,000}$ = 22.43 tons

b. The required air quantity is calculated as

$$\text{cfm} = \frac{\text{room sensible heat}}{1.08(t_i - t_s)}$$

where t_s = dry-bulb temperature of conditioned-air supply = 65°F. Therefore,

$$\text{cfm} = \frac{158,100}{1.08(80 - 65)} = 9760$$

The supply state of the conditioned air is determined as

Room latent heat = cfm $\times$ 0.69 $\times (W_i - W_s)$

where W_s = humidity ratio at 65°F. Thus,

$111,100 = 9760 \times 0.69 \times (77 - W_s)$

Therefore,

W_s = 60.50 grains/lb dry air

From the psychrometric chart at 65°F dry-bulb temperature and a humidity ratio of 60.50, the supply air from the fan is found to be at

65°F dry bulb   58°F wet bulb   65% relative humidity

BIBLIOGRAPHY

Air Conditioning Manual, The Trane Co., La Crosse, Wisconsin, 1965.

ASHRAE Handbook of Fundamentals, 1972 ed., American Society of Heating, Refrigerating and Air Conditioning Engineers, New York.

ASHRAE Handbook—1975 Equipment Volume, American Society of Heating, Refrigerating and Air Conditioning Engineers, New York.

ASHRAE Handbook—1976 Systems Volume, American Society of Heating, Refrigerating and Air Conditioning Engineers, New York.

Baumeister, T.: *Marks' Standard Handbook for Mechanical Engineers*, 7th ed., McGraw-Hill, New York, 1967.

Building Construction Data, 1976 ed., R. S. Means, Duxbury, Mass.

Carrier Air Conditioning Co.: *Handbook of Air Conditioning System Design*, McGraw-Hill, New York, 1966.

Construction Labor Reports, 1976–1977 *Wage Rate Guide*, Bureau of National Affairs, Washington, D.C.

Current Construction Costs, 1976 ed., Lee Saylor, Walnut Creek, Calif.

Dellon, Alfred L.: *Managing for Project Control and Profitability*, 4th International Cost Engineering Symposium and 20th Annual Meeting, American Association of Cost Engineers, Boston, 1976.

Dell'Isola, Alphonse J.: *Value Engineering in the Construction Industry*, 2d ed., Construction Publishing, New York, 1974. (Now published by Van Nostrand Reinhold Co., New York.)

Dodge Digest (City Indexes and Codes), McGraw-Hill Information Systems, McGraw-Hill, New York (periodical publication).

Dodge Manual for Building Construction Pricing and Scheduling, 1976 ed., McGraw-Hill Information Systems, McGraw-Hill, New York.

Energy Conservation Design Guidelines for Office Buildings, General Services Administration, Public Building Service, Washington, D.C., January, 1974.

Engineering Design Manual For Pressurization and Air Control in Closed Hydronic Systems, ITT Fluid Handling Division, Morton Grove, Ill., 1966.

Fibrous Glass Duct Construction Standards, 3d ed., Sheet Metal and Air Conditioning Contractor's National Association, Vienna, Va., 1972.

Gladstone, John: *Mechanical Estimating Guidebook*, 4th ed., McGraw-Hill, New York, 1970.

High Velocity Duct Construction Standards, 2d ed., Sheet Metal and Air Conditioning Contractor's National Association, Vienna, Va., 1969.

Labor Estimating Manual, Mechanical Contractor's Association of America, Washington, D.C., 1976.

Labor Rates for the Construction Industry, 1976 ed., R. S. Means, Duxbury, Mass.

Low Velocity Duct Construction Standards, 4th ed., Sheet Metal and Air Conditioning Contractor's National Association, Vienna, Va., 1969.

McKee-Berger-Mansueto, Inc.: *Building Cost File* (four regional editions), Construction Publishing Company, New York, 1976. (Now published by Van Nostrand Reinhold Co., New York.)

———: *Design Cost File*, 1976 ed., Construction Publishing Company, New York. (Now published by Van Nostrand Reinhold Co., New York.)

Miles, L. D.: *Techniques of Value Analysis and Engineering*, 2d ed., McGraw-Hill, New York, 1972.

Page, John S.: *Estimator's Manhour Manual on Heating, Air Conditioning, Ventilating and Plumbing*, Gulf, Publishing, Houston, 1961. (Now published by R. S. Means, Duxbury, Mass.)

Sheet Metal Industry Uniform Trade Practices, Sheet Metal Contractor's Association of New York City (recently known as Sheet Metal Industry Promotion Fund of New York City) and Nassau-Suffolk Sheet Metal Workers and Roofers Employers Association, New York, May, 1969.

Strock, Clifford, and Richard L. Koral: *Handbook of Air Conditioning, Heating and Ventilating*, 2d ed., Industrial Press, New York, 1965.

Underground Heat and Chilled Water Distribution Systems, NBS Building Science Series 66, U.S. Department of Commerce, National Bureau of Standards, Washington, D.C., May, 1975.

Value Engineering Handbook, PBS P8000.1, General Services Administration, Public Building Service, Washington, D.C., March, 1973.

INDEX

INDEX

Abbreviations:
  general, 280–283
  for organizations and agencies, 279
Absorption water chillers, 51–55
  estimating data for, 55
Acoustic lining, 173–174, 184–185
  estimating data for, 184–185
Adjusting (*see* Testing and balancing)
Administrative overhead costs, 15
Air:
  combustion, through boilers, 25
  infiltration losses, 289–290
  ventilation: loads, 290, 297
    requirements, 250–251, 253
Air boots, troffer, 104
Air changes, 253*n*.
Air-condensing units, 57–58
Air conditioners, unitary, 111–121
  heat pumps, 112–118
    air-to-air, 112–113, 116, 118
    water-to-air, 113, 115, 117–118
  indoor self-contained, 111, 114
  rooftop, 111, 114
  split system, 115, 119–121
  window, 111, 115
Air-conditioning systems, 254–264
  air-water, 256–257, 260
  all-air, 254–255, 258–259
  all-water, 258, 261
  applications of, 262–264
  functional ratings of, 259, 263
  refrigerant, 258, 262
  space requirements for, 263–264
Air-cooled condensers, 56–58
Air-diffusing units, 98–106
  exhaust and return-air inlets, 105–106
  supply-air outlets, 102–105
  terminal control units, 98–101

Air-duct construction (*see* Ductwork)
Air-handling equipment, 107–134
  air-handling units, 107–123
  air-treatment equipment, 128–132
  dampers, 133–134
  fans, 119–129
  heat-recovery equipment, 117–119, 124
Air-handling units, central, 107–108
Air quantities, 253, 255–257
  for cooling, 255–257
  for ventilation, 253
Air separators, 77–78
Air-to-air heat-recovery wheels, 117–119, 124
Air-treatment equipment, 128–132
  humidifiers, 128–131
  sound attenuators, 130, 131
Air-velocity classifications, 110–111, 168·
  for ductwork, 168
  for fans, 110–111
Air vents, piping, 152, 165
All-air high-pressure units, 98–101
  constant-volume boxes, 101
  dual-duct mixing boxes, 100–101
  estimating data for, 99
  pressure-reducing valves, 99–100
  single-duct boxes, 100
  single-duct reheat boxes, 100
  variable-air-volume boxes, 100
Alteration and renovation work, 266
Apartments, budget costs for, 246
Axial fans, 126–128
  estimating data for, 127–128
  in-line centrifugal, 127
  propeller, 127
  tube, 126
  vane, 128
  (*See also* Fans)

Backfill (*see* Excavation and backfill)
Balancing (*see* Testing and balancing)
Banks, budget costs for, 246
Baseboard units, 81–84
  cast-iron radiant, 81
  finned-tube, 83–84
  electric, 83
  hot water or steam, 83–84
Bibliography, 303–304
Boiler breeching, 31–32
  (*See also* Ductwork)
Boiler efficiency, 25
Boiler feedwater units, 32–37
Boiler stack, 31–32, 34
Boilers, 23–34
  cast-iron, 26–29
  electric, 29–31, 33
  hot-water, 28–30
  ratings of, 24
  steam, 27
  steel, 27, 30–32
  types of, 23–24
Brazed joints, piping, 142–143
Budget costs for HVAC systems, 245–248
Burners, 25–26

Central heating-ventilating-air-conditioning units, 108–113
  component, 108–111
  estimating data for, 112–113
Centrifugal fans, 121–125, 127–129
  blade types, 121–125
  estimating data for, 125, 129
  with air-foil blades, 125
  power roof ventilators, 129
  (*See also* Fans)

307

Index | **308**

Centrifugal pumps, 65–76
　estimating data for, 66–70, 74–76
　　base-mounted, 68–70, 74–76
　　in-line, 66–67
　(See also Pumps)
Centrifugal water chillers, 50–51, 53
　estimating data for, 53
Change orders, 265–266
Chillers, water, 50–55
　absorption, 51–55
　mechanical compression, 50–53
　　centrifugal, 50–51, 53
　　reciprocating, 50–52
Coal, heat values for, 39
Coils, duct reheat, 88–89
　estimating data for, 89
　electric, 89
　hot-water or steam, 89
Collectors, solar, 45–48
Colleges, budget costs for, 246
Color code, piping-system, 233–234
Commercial buildings, budget costs for, 246
Compressors, types of mechanical water chillers by types of, 51
Conceptual estimates, 245–264
　cost data for, 245–248
　design data for, 245, 248–264
Condensate return pumps, 33–36
Condensers, 54–60
　air-cooled, 56–58
　evaporative, 57–60
　water-cooled, 56–57
Constant-volume boxes, 99, 101
Construction contracts, 20
　competitive bidding, 20
　negotiated contract, 20
Contractor, definition of, 2
Contractor's estimates, 2
Contractor's markup, 14–16, 21
　check figures for, 21
　general and administrative overhead, 15
　job overhead, 15
　profit, 16
Control systems, automatic, 199–220
　auxiliary equipment, 209–211
　central panels, 216
　controlled devices, 207–209
　controllers, 203–207
　　indicating and recording, 205
　　sensing elements, 203
　　transducers, 203–205
　　types of, 205–207
　cooling systems, control of, 212
　definitions of control terms, 200–202

Control systems, automatic (*Cont.*):
　estimating, 216–220
　feedback systems, 199–200
　heating systems, control of, 211–212
　labor man-hours for, 217–220
　sequence of operation, 213–216
　types of, 202
Convectors, 80–83
　estimating data for, 82–83
　electric, 83
　hot water or steam, 82
Conversion factors [see Metric units (SI)]
Converters (see Heat exchangers)
Cooling-distribution equipment, 93–106
　air-diffusing units, 98–106
　forced-convection units, 93–98
Cooling-generation equipment, 49–62
　absorption water chillers, 51–55
　condensers, 54–60
　cooling towers, 59–62
　mechanical compression water chillers, 50–53
Cooling-load calculations, 295–302
　indoor design conditions, 295
　load categories, 295–297
　　outside air, 297
　　radiant solar energy, 296
　　room, 297
　　transmission, 296
　psychrometric chart, 298–300
　sensible-heat factor, 297–298
Cooling-load check figures, 252, 255–257
Cooling towers, 59–62
　forced-draft, 62
　induced-draft, 60–62
　natural-draft, 59–60
Copper tube (see Piping, copper)
Cost(s):
　of estimating, 16, 20
　HVAC systems total, per square foot of gross area, 246–248
Cost model, 3, 269

Dampers, 133–134, 208–209
　automatic control, 208–209
　fire, 133–134
　manual splitter, 133
　volume, 133–134
Deaerators, 35–37
Degree-days and fuel consumption, 293–295
Design conditions, indoor, 295

Design-phase estimates, 2–3
　cost model, 3
　final (pre-bid), 3
　preliminary (design development), 3
　schematic, 3
Diffusers, 102–105
　linear slot, 104
Direct-fired hot-water generators, 28
Dormitories, budget costs for, 247
Drawings, HVAC, 8–9
　checklist for review of, 9
Dual-duct mixing boxes, 99–101
Ductwork (sheetmetal work), 167–185
　access doors, 174
　acoustically lined, 173–174, 184–185
　aluminum, 168–170, 184–185
　apparatus casings and plenums, 174
　belt guards, 174–175
　black-steel (iron), 167, 169, 171, 173, 184–185
　boiler breeching, 31–32
　configurations of, 169, 170
　construction of, 170–173
　copper, 167, 169, 184–185
　estimating, 177–185
　　computing weights of ducts, 180–183
　　labor man-hours for, 183–185
　　material costs, 183–184
　　takeoff and, forms for, 182–183
　　takeoff procedure, 177–180
　fibrous glass, 168, 173, 184–185
　field labor, 185
　flexible ducts, 168, 185
　galvanized steel, 167, 169–172, 184–185
　gauges, metal, 169–173
　　tables of weights, 169
　hangers, 175
　insulation, 191–197
　labor man-hours for, 183–185
　material prices, 183–184
　materials, 167–168
　measurement (takeoff) rules, 177, 180–181
　painting, 234–235
　shop labor, 185
　sizing, 175–176
　stainless steel, 167–169, 184–185
　surface areas, 175–179
　velocity and pressure classifications, 168
　weights, tables of, 169

Electric control systems, 202, 204, 208–211
Electronic control systems, 202, 204
Energy, solar, 45–48
Equipment insulation, 191–192, 194–197
Estimate development procedure, 4–16, 21
    calculating profit and overhead, 14–16, 21
    estimating labor costs, 13–14
    local market analysis, 11–12
    pricing of material, 12
    quantity takeoff, 9–11
    reviews: of drawings, 8–9
    of specifications, 4–8
Estimates, types of, 2–3
    conceptual estimates (*see* Conceptual estimates)
    contractor's estimate, 2
    design-phase estimate, 2–3
Estimating and estimators, 1–2
Estimating cost of estimating, 16, 20
Estimating forms, 16–20
Excavation and backfill, 241–243
    labor man-hours for, 242–243
        backfilling, 243
        compacting, 243
        excavating, 243
    methods of, 241–242
    trench configurations, 242
    trench dimensions, 242
Exhaust inlets, 105–106
Expansion joints, 151, 164
Expansion tanks, 74–77

Fan-coil units, 93–96
Fans, 119–129
    axial (*see* Axial fans)
    centrifugal (*see* Centrifugal fans)
    definitions of, 119–121
    power roof ventilators, 129
Filters, 109
Finned-tube units, 84–85
Fire dampers, 133–134
Forced-convection units, 84–89, 93–96
    cooling distribution, 93–96
        fan-coil, 93–95
        induction, 96–98
        unit ventilators, 94–96
    heat-distribution, 84–89
        duct reheat coils, 88–89
        unit heaters, 86–88
Forms:
    cooling and heating load, 298
    ductwork, 182–183

Forms (*Cont.*):
    estimating, 16–19
    piping and accessories, 154–157
    value engineering, 269, 271
Foundations and vibration isolation, 229–231
    concrete foundations, 230–232
    structural steel bases, 230–231
    vibration isolators, 229–230
Fuel consumption, 293–295
Fuel-oil storage tanks, 42–44
Fuel-oil transfer pumps, 42, 44
Fuel-oils, grade of, 39–40
Fuels, heat values for, 39–40
Furnaces, warm-air, 116–117, 123
    electric, 123
    gas-fired, 123
    oil-fired, 123

Garages, budget costs for, 248
Gases, fuel, heat values for, 39–40
Gauges and thermometers, 152–153, 165–166
General and administrative overhead costs, 15, 21
Grilles:
    exhaust, 105–106
    supply, 102–103, 105

Hangers and supports, 146–147, 161
Heat- and cooling-generation auxiliary equipment, 63–78
    hydronic specialties, 74–78
    pumps (*see* Pumps)
    water-treatment systems, 63–64
Heat-distribution equipment, 79–92
    forced-convection units, 84–89
    natural-convection units, 80–84
    radiant heating, 88, 90–92
Heat exchangers, 43–47
    steam-to-water, 45–47
    water-to-water, 43
Heat-generation equipment, 23–48
    boiler feedwater equipment, 32–37
    boilers, 23–34
    fuel storing and handling equipment, 39–44
    heat exchangers, 43–47
    pressure-reducing stations, 36, 38–39
    solar energy, 45–48
Heat pumps, 112–118
    (*See also* Air conditioners, unitary)

Heat-recovery equipment, 117–119, 124
    air-to-air heat-recovery wheels, 117–119, 124
    chiller with double-bundle condensers, 117
    reheat-recovery coils, 117
Heating-load calculations, 287–293
    infiltration losses, 289–290
    pitched-roof losses, 290
    ventilation air losses, 290
    walls, partitions, and floor losses, 291
Heating-load check figures, 249–252
Heating-ventilating units, 115–117, 122
Hospitals, budget costs for, 247
Hotels, budget costs for, 247
Humidifiers, 128–131
Humidistats, 206–207, 220
Hydronic specialties, 74–78
    air separators, 77–78
    expansion tanks, 74–77

Indirect hot-water heaters, 29
Induction units, 96–98
Infiltration losses, 289–290
Infrared heaters, 91–92
    electric, 91
    gas-fired, 92
    oil-fired, 92
Insulation, thermal, 187–197
    acoustic, 173–174, 184–185
    applications of, 189–193
    block, 194–196
    calcium silicate, 196
    duct, 191–197
    equipment, 191–192, 194–197
    estimating of, 193–197
    fiberglass, 194–195
    flexible (blanket), 194–195
    jackets, metal, 197
    labor man-hours for, 194–195
    material prices, 196
    materials for, 187–188
    physical and thermal properties of, 188
    piping, 189–191, 193–197
    requirement for HVAC system, 189–193
    takeoff procedure, 193–195
    thickness of, 188, 190–191
International System of units [*see* Metric units (SI)]

Job overhead costs, 15, 21

Index

Labor:
　how to estimate, 13–14
　prevailing rates for, in principal cities, 14
Labor man-hours (see specific product or operation)
Laboratories, budget costs for, 247
Leakage tests (see Testing and balancing)
Life-cycle costs, 267
Load calculations, 287–302
　cooling, 295–302
　heating, 287–295
　　degree-days, 293–295
Load check figures:
　cooling, 252–257
　heating, 249–252

Magnetic flowmeters, 219
Make-up air units, 115–117, 122–123
　heating-ventilating units, 115–117, 122
　warm-air furnaces, 115–117, 123
Manufacturing buildings, budget costs for, 247
Market analysis, 11–12
　escalation indexes, 11–12, 21
　market contingency, 12, 21
Markup, contractor's (see Contractor's markup)
Material, equipment and, how to price, 12
Mechanical symbols, 283–285
Metric units (SI), 273–278
　conversion factors, 275–278
　　area, 275
　　capacity volume, 275
　　energy, 277
　　examples of the use of, 277
　　flow, 275
　　length, 275
　　mass, 276
　　power, 277
　　pressure, 276
　　temperature, 276
　　U.S. customary units, 275–277
　　velocity, 276
　　volume, 275
Mixing plenums, 108
Motels and hotels, budget costs for, 247
Motor starters, 228
Motors, 65, 67, 69–71, 74–76, 227–228
　horsepowers: brake, 65
　　for pumps, 67, 69–71
　　standard, 74

Motors (Cont.):
　material costs, 75–76
　specifications, 227–228
Museums, budget costs for, 248

Natural (free) convection units, 80–84
　baseboard, 81–84
　convectors, 80–83
　finned-tube, 83–84
　radiators, small-tube, 80
Natural-draft cooling towers, 59–60

Office buildings, budget costs for, 248
Orifice flowmeters, 219
Orifice plates, 219
Overhead costs, 15–16, 21
Overtime, 266

Painting, 233–235
　brush, labor man-hours for, 235
　piping system color code, 233–234
　quantity takeoff, 233–235
Panel heating, 90
Panels, heating and cooling, 90
Parameters, HVAC system, 17–19, 21
Parking garages, budget costs for, 248
Performance tests (see Testing and balancing)
Physical properties of insulating material, 188
Piping, 135–166
　anchors, 161
　application of, 138
　black steel: threaded, 158
　　welded, 159
　brass, 137–139
　brazing, 142–143
　color code for, 233–234
　consumption of: brazing alloys, 143
　　solders, 144
　　solvent cement for PVC, 144
　copper, 137–138, 158–159
　estimating, 153–166
　expansion joints, 151, 164
　fittings, 140–141
　flanged joints, 145
　forms, takeoff and estimating, 154–157
　galvanized steel, 158

Piping (Cont.):
　hangers and supports, 146–149, 161
　insulation, 189–190, 193–197
　labor man-hours for, 158–166
　material prices, 166
　materials for, 136–138
　method of joining, 141–145
　painting, 233–235
　polyvinyl chloride (PVC), 138–143, 158–159
　sizing, 153–155
　sleeves, 161
　soldering, 142–143
　solvent cementing, 143–144
　specifications and standards for, 138–141
　steel, 136–137, 158, 160
　strainers, 151, 165
　system, distribution (see System, distribution piping)
　takeoff procedure, 153–157, 166
　thermometers and gauges, 152–153, 165–166
　traps, steam, 151–152, 165
　trenching (see Excavation and backfill)
　underground, 145–146, 160
　valves, 147–150, 161–163
　vents, air, 152–165
　victaulic joints, 145
　welding, 143–145, 160
Pneumatic control system, 202, 205–207, 209–210
Pressure-reducing stations, 36, 38–39
Profit, estimating, 16, 21
Psychrometric chart, 298–302
Pumps, 65–76
　applications of, 65
　capacities of: and motor horsepowers, 67, 69, 71
　　and pipe sizes, 66, 68
　centrifugal (see Centrifugal pumps)
　formulas for, 65
　fuel-oil transfer, 42, 44
　motor costs, 75–76

Quantity takeoff procedure for HVAC systems, 9–11

Radiant heating, 88, 90–92
　heating and cooling panels, 90
　infrared heaters, 91–92
　panel heating, 90
　snow melting, 90–91

Radiators (small-tube type), 80
Reciprocating water chillers, 50–52
Refrigerant systems, 258–259, 262
Refrigeration machines (see Chillers, water)
Registers:
  exhaust and return, 105–106
  supply, 102, 105
Reheat coils, duct, 88–89
Renovation work, 266
Rigging, 237–239
  weight categories, 237
Rigging operations, various, labor man-hours for, 238
  estimating costs of, 239
Roof ventilators, power, 127, 129
Rooftop units, unitary, 111, 114
Room air conditioners (window type), 111, 115

Schools, budget costs for, 248
Scope and design contingency, 11, 21
Self-contained air conditioners, indoor, 111, 114
Sheet-metal work (see Ductwork)
Site flow indicators, 220
Sleeves, pipe, 161
Slot diffusers, 104
Snow melting, 90–91
Soils, categories of, 241
Sol-air temperature, 296
Solar collectors, 45–48
Solar energy, 296
  applications of, 45–48
Soldering consumptions, 142, 144
Sound attenuators (traps), 130–131
Sound liner (see Acoustic lining)
Specifications for HVAC systems, 4–8
Split-system air-conditioning units, 115, 119–121
Stacks, boiler, 31–32, 34
Starters, motor, 228
Steam boilers, 27
Steam humidifiers, 128–131
Steam traps, 151–152, 165
Storage tanks, fuel-oil, 42–44
Strainers, 151, 165
Supply-air outlets (diffusers), 102–105
Symbols, mechanical, 283–285

System, distribution piping, 135–136
  direct return, 136
  four-pipe, 135
  one-pipe, 135
  reverse return, 136
  three-pipe, 135
  two-pipe, 135
Systems of weights and measures [see Metric units (SI)]

Takeoff sheets and tools, 10–11
Tanks, fuel-oil, 42–44
  estimating data, 44
  sizing, 42
  surface areas, 234–235
  volume, 243
Testing and balancing, 221–226
  air balance, 224
  estimating cost of, 225–226
  leakage test, 221–222
  performance test, 223
  recording and reporting the result of, 225
  water balance, 223–225
Theaters, budget costs for, 248
Thermometers and gauges, 152–153, 165–166
Thermostats, 205–206
  labor man-hours for, 220
Threading, pipe, 141–142
Traps, steam, 151–152, 165
Trenching (see Excavation and backfill)
Troffer air boots, 104
Tube-axial fans, 126

U-value, how to calculate, 288–289
Underground piping systems, 145–146, 160
Unit heaters, 86–88
Unit ventilators, 95–96
Unitary air conditioners (see Air conditioners, unitary)

Value engineering, 267–272
  approach to, studies and review, 268–270
  concepts of, 267
  definition of, 267–268
  and estimator, 270, 272

Value engineering (*Cont.*):
  forms for: HVAC cost model, 269
    idea listing sheet, 271
Valves:
  construction of, 147–150, 207–208
  functions of, 147, 207
  labor man-hours for, 161–163, 218–219
    angle, 161–162
    automatic flow control, 218–219
    ball, 161–162
    brass and bronze, 161
    butterfly, 162
    cast-iron, 162
    cast-steel, 162
    check, 161–162
    diaphragm, 162
    gate, 161–162
    globe, 161–162
    iron-body, 162
    plug, 161, 163
    pressure and temperature regulating, 218
    pressure regulating (PRV), 163
    relief, 163
    solenoid, 219
    temperature regulating, 218
Vane-axial fans, 126, 128
Vapor barrier, insulation with, 194
Variable-air-volume boxes, 99–100
Ventilation air:
  estimating amount of, 250–251
  load, 290, 297
  requirements of, 253
Venturi, cast-iron, labor man-hours for, 219
Vibration isolation, 229–230
Victaulic joints, 145

Wage rate in principal cities, 14
Warehouses, budget costs for, 248
Water chillers (see Chillers, water)
Water-cooled condensers, 56–57
Water-treatment systems, 63–64
Weights:
  concrete, 321
  ducting materials, 169
  ductwork, 180–182
Welding, pipe, 143–144, 160
  production rate of, 144, 160
Window air conditioners, 111, 115